THE

INTERNATIONAL SERIES

OF

MONOGRAPHS ON PHYSICS

GENERAL EDITORS

N F MOTT   SIR EDWARD BULLARD

THE INTERNATIONAL SERIES OF MONOGRAPHS ON PHYSICS

GENERAL EDITORS

N F MOTT
Henry Overton Wills Professor
of Theoretical Physics in the
University of Bristol

Sir EDWARD BULLARD
Director of the National
Physical Laboratory,
Teddington.

ELECTRICAL BREAKDOWN OF GASES

BY

J. M. MEEK

AND

J. D. CRAGGS

OXFORD

AT THE CLARENDON PRESS

1953

Oxford University Press, Amen House, London E C.4

GLASGOW NEW YORK TORONTO MELBOURNE WELLINGTON
BOMBAY CALCUTTA MADRAS KARACHI CAPE TOWN IBADAN

Geoffrey Cumberlege, Publisher to the University

PRINTED IN GREAT BRITAIN

PREFACE

THIS book has been written primarily for physicists and electrical engineers engaged in fundamental studies of the growth of electrical discharges in gases, but we hope that it will also be of use to those concerned with the development and application of gas-filled electronic tubes, with gaseous dielectrics, and with the many other technical problems in which electrical discharges are involved. We have included little original material and have confined ourselves largely to a review of some of the many fairly recent published papers, in particular those appearing subsequent to about 1930. Thus the pioneer work of Townsend and his school, on which the whole of gaseous discharge physics is so largely based, has received much less attention than would be the case in a more historical survey. Even with the above limitation the literature is so extensive that we have been able to give a detailed treatment of only a limited number of papers though we have tried to give comprehensive lists of references

In order to make the book as complete as possible we have included descriptions of experimental results obtained in conditions that are not so simple or well defined as would be desirable in studies of the fundamental processes involved. For example, we have referred to a number of experiments on the breakdown of gaps between a point and a plane, an electrode arrangement giving an extreme condition of an asymmetrical field but one in which the field distribution often cannot be calculated. Further, it may be considered preferable to study the mechanisms of breakdown in single pure gases rather than in impure or in mixed gases. The technical importance of air as a dielectric is such that we feel justified in giving it the full attention that it receives in this book, but some of the experimental results obtained with impure gases are of doubtful value, not only because the nature of the contaminants is often not specified but also because the effects of impurities vary so widely and cannot, in many circumstances, easily be estimated. Further investigations with pure gases at high pressures, i.e. of the order of 1 atmosphere or more, would clearly be valuable in many directions, but there are sometimes considerable technical difficulties in the production and manipulation of the large quantities of pure gases required for such studies. Modern work at low pressures, where high initial vacua are essential, invariably involves careful techniques of purification.

The mechanism of spark breakdown for high values of the product pd, pressure times gap length, is, we feel, still uncertain. For example, the recent experiments by Fisher and by Llewellyn Jones and their colleagues show that the onset of breakdown of uniform-field gaps in air, for $pd < 1,000$ mm. Hg $\times$ cm , is explicable in terms of a cathode-dependent secondary emission process Raether, in his cloud-chamber studies, finds that streamers do not appear unless $pd > 1,000$. Further experimental studies are required of the breakdown of gaps at higher values of pd before a decision can be made as to whether avalanche-streamer transitions occur. According to the approximate calculations of the space-charge field in an electron avalanche, transition to a streamer will not take place in uniform-field gaps in air unless pd is about 1,500 or more, though for gaps subjected to overvoltages streamers would be expected to occur at lower values of pd. In this book we have often used the term 'streamer' to describe the luminous filamentary discharges that are frequently observed to precede the growth of the spark channel, without any implications as to their mechanism of formation

The main part of the manuscript was written in the years 1948–51 and most of it was finally despatched to the publishers in the autumn of 1951 We have tried to include the more important later papers, up to about the end of 1952, in an appendix.

We have many acknowledgments to make It is clear that we have drawn greatly from the books by von Engel and Steenbeck, Thomson and Thomson, Loeb, and Cobine, and from the review article by Druyvesteyn and Penning We wish therefore to make special acknowledgment to these authors. We also thank many workers in the field of gaseous discharge physics for sending reprints and for helpful discussion, particularly Professor D. R. Bates, Professor K. G. Emeléus, Dr. A. von Engel, Dr. L Fisher, Dr. B. Ganger, Professor F. Llewellyn Jones, Professor L. B. Loeb, Professor H. S. W. Massey, Dr. H. Raether, and Professor J Sayers. Dr. H. Raether kindly supplied the photographs reproduced in Figs. 4.5, 4.6, and 4.8. Dr. Haynes sent us the original for Fig. 11.4 and Dr. A von Hippel lent us the blocks for Figs. 4.42, 4 43, 4.44, and 4.45. We are indebted to members of the Metropolitan-Vickers research laboratories, notably Dr. R. W. Sillars and Mr F. R. Perry, for permission to include some hitherto unpublished experiments in Chapters X and XII. Fig. 10.3 is a Crown Copyright and we are grateful to Her Majesty's Stationery Office for permission to reproduce it.

We are indebted to various colleagues for reading and commenting on chapters in manuscript and to Mr. W. Hopwood and Miss D. E. M. Garfitt for reading the proofs with us.

Finally, we thank the various officials of the Clarendon Press for their helpfulness and careful attention to detail in the production of this book.

J. M. M.
J. D. C.

DEPARTMENT OF ELECTRICAL ENGINEERING
LIVERPOOL UNIVERSITY
April 1953

CONTENTS

Plates 1–8 are to be found between pp. 176 and 177, 9–15 between pp. 432 and 433.

I

FUNDAMENTAL PROCESSES IN
ELECTRICAL DISCHARGES

INTRODUCTION

AN electrical discharge is usually built up mainly by electron collisions producing fresh ions and electrons in Townsend avalanches and proceeds as a transient discharge until a maintenance mechanism is established, as for instance by the cathode emission of electrons in an arc discharge. The discharge then becomes independent of the external sources of irradiation which supply the electrons necessary for the starting of the discharge. The efficiency of the electrons, as producers of fresh ions, depends on their energy and therefore on their mean free paths in the electric field. Since the electrons in a gaseous discharge are not monoenergetic their energy distributions must be deduced, directly or indirectly, if the discharge is to be quantitatively understood. Finally, in the stable state, an externally applied electric field is required to replace those electrons lost by recombination, diffusion and, sometimes, attachment These requirements must also be met during the transient state of the discharge if it is later to achieve self-maintenance.

Mott and Massey [1] have published the most extensive and authoritative work, to date, on collision processes. The book by Arnot [2] gives an excellent elementary account of experimental work on atomic collisions. Summarized accounts are also given in the various books on electrical discharges in gases, particularly those by Loeb [3], Seeliger [4], von Engel and Steenbeck [5], and Townsend [105].

While this chapter includes some reference to collision cross-sections and to general ionization and excitation mechanisms, it is mainly concerned with the processes listed below, which relate particularly to the gas discharges considered in the subsequent chapters.

(i) Photoionization, which is thought to be effective in avalanche and streamer processes (Chapters II, III, and VI).

(ii) Recombination, which is important in decaying plasmas (Chapter X), in all high ion density discharges, and in avalanche-streamer transitions [6] (Chapter VI).

(iii) Attachment and detachment, i.e. the mechanics of negative ions. This is important in negative point corona (Chapter III), and in discharges in electro-negative gases such as O_2, Cl_2, CCl_2F_2, etc.

(iv) Mobilities of charged particles, or the relation between the electric field in a discharge and the resulting drift velocities in the field direction of the particles

(v) Diffusion processes, which are related to the movements of particles between regions of varying particle concentrations, and are important in low-pressure discharges and in high-pressure transient discharges (Chapters VI and X).

(vi) Energy distributions, or the variations in energy amongst the members of a particle population in a discharge A knowledge of these distributions is fundamental in discharge physics since they enable the behaviour of swarms to be calculated in terms of cross-sections, etc., for individual processes.

(vii) Dynamic changes in distributions, or the partial establishment of equilibria in transient discharges.

(viii) The Townsend α-coefficient for the ionization of gas particles by electrons.

(ix) The γ-coefficient for the production of secondary electrons in the gas or at the cathode by various possible mechanisms.

Some of the many types of collision process may be summarized as in Table 1.1, where A, B are atoms, A^*, B^* are excited atoms, A^+, B^+ are positive ions, e is an electron, A^+, B^+, e are ions and an electron with appreciable kinetic energy, ν is the frequency, and h is Planck's constant

TABLE 1.1

$A+e \rightarrow A^*+e$	Excitation by electron impact.
$A+e \leftarrow A^*+e$	A collision of the second kind
$A+e \rightarrow A^++e+e$	Ionization by electron impact.
$A+h\nu \rightarrow A^*$	Photo excitation (absorption of light)
$A+h\nu \leftarrow A^*$	Emission of light
$A+h\nu \rightarrow A^++e$	Photo-ionization.
$A+h\nu \leftarrow A^++e$	Radiative recombination.
$A+B \rightarrow A^*+B$	Excitation by atom impact.
$A+B \rightarrow A^++e+B$	Ionization by atom impact
$A^*+B \rightarrow A+B^*$	Excitation by excited atoms.
$A^++B \rightarrow A+B^+$	Change of charge
$A^++B \rightarrow A^++B^++e$	Ionization by positive ion impact.

Collisions of the second kind (bracketing the last three rows)

Collision cross-sections

Free electrons, ions, photons, and excited atoms are all produced by collisions in a gaseous discharge. According to Arnot [2, p. 48] a collision can be defined as follows: if the relative distance between two particles is first decreased and then increased a collision has occurred if a physical change occurs in either of the particles during the process. If the mean distance between two collisions (of an unspecified, general kind) is λ and there are n particles passing through a gas per second per cm ^{2}, $n\,dx/\lambda$ collisions will be made in distance dx. Let us study, for example, the scattering of an electron beam on the assumption that each colliding electron is lost. Then the number lost over the distance dx per second per cm.^{2} is

$$dn = -n\,dx/\lambda.$$

If $n = n_0$ at $x = 0$ we have

$$n = n_0\,e^{-x/\lambda} = n_0\,e^{-\mu x} \tag{1.1}$$

and

$$\mu = \frac{1}{\lambda} = NQ, \tag{1.2}$$

where N is the number of atoms per cm.^{3} Q is the effective cross-section of the atom in cm ^{2} and is the area of cross-section which it would have if in the collision it behaved as a solid sphere on the kinetic theory. NQ is the sum of the effective cross-sections of all the atoms in 1 cm.^{3} of gas and is called the total effective cross-section. It is usually given for a gas at a pressure of 1 mm Hg and at $0°$ C., when the value of N is $3\,56\times10^{16}$. μ is called the absorption coefficient. Some kinetic theory cross-sections calculated from viscosity data [7] are given in Table 1.2.

TABLE 1 2

Gas			Kinetic-theory cross-section Q cm $^2\times10^{16}$
He	.	.	2 83
Ne	.	.	4 14
A	.	.	6 46
Kr	.	.	7 54
Xe	.		9 16
O$_2$	.	.	6 9
N$_2$		.	7 8
H$_2$			3 7
H$_2$O	.		5 8
Hg		.	10 2

In high-pressure discharges, particularly, most of the collisions are elastic since in normal conditions only those few electrons in the high energy tail of the distribution can cause excitation or ionization at a single impact. Elastic collisions are also especially important in decaying plasmas (see p. 415). At an elastic collision an electron loses, on the average, $2\cdot66m/M$ of its energy where M and m are the atomic or molecular mass and the electronic mass respectively.

Cravath's treatment [83] of the energy loss suffered by an electron in an elastic collision is derived on the assumption that the ions and molecules are smooth elastic spheres with Maxwellian velocity distributions. The final result for f, the average energy loss per collision expressed as a fraction of the average energy of the ions, is given by

$$f = \frac{8}{3} \frac{mM}{(m+M)^2}\left(1 - \frac{T_m}{T_i}\right), \qquad (1.3)$$

where m and M are ionic and molecular masses respectively and T_i and T_m are the ionic and molecular temperatures.

For $E/p \gg 0.1$, $T_m \ll T_i$ and the term T_m/T_i can be neglected [3]. Also, as $m \ll M$, the term $mM/(m+M)^2$ is essentially m/M. Hence equation (1.3) reduces to the approximate relation

$$f = 2.66 \frac{m}{M}.$$

The Cravath value for f is clearly of importance in considerations of the variation of electron temperature with time in a decaying plasma (see Chapter X) where the electron temperature is low enough to permit the neglect of inelastic collisions as a mode of energy loss. Such cases have been considered for T R. switches [84] and for certain specialized aspects of spark discharges [85].

Druyvesteyn and Penning [86] point out that for hard sphere collision where the gas atoms are at rest and the electrons have energy ϵ the mean energy loss per elastic collision is

$$\Delta\epsilon = -\frac{2m}{M}\epsilon.$$

For atoms having a Maxwellian distribution, with mean energy $\bar{\epsilon}_g = \frac{3}{2}kt$, and monoenergetic electrons as above, then

$$\Delta\epsilon = -\frac{2m}{M}\epsilon\left(1 - \frac{4\bar{\epsilon}_g}{3\epsilon}\right). \qquad (1 4)$$

If the electrons also have a Maxwellian distribution with mean energy $\bar{\epsilon}$, Cravath's result given in equation (1.3) is obtained.

The variation of the probability of collision at very low electron energies was first fully studied by Ramsauer and later by Ramsauer and Kollath [88, 90] and Brode [89] with remarkable results. Typical data are given in Figs. 1.1 to 1.3 [89] and in Figs. 1.4 to 1.8 [88]. Experimental values do not extend below about $\sqrt{V} = 0.25$, where V is in volts. The rise in the probability of collision P_c for the lowest energies is known as the Ramsauer effect. It is clear that the variations in P_c will seriously affect the low energy part of the electron distribution in a gas. For example, the sharp rise in P_c for high energy electrons in argon [89] causes a deficiency of electrons in that region in the distribution of energies.

The probability of collision P_c, as used by Brode [89], Ramsauer and Kollath [88, 90], and others, is defined as the number of collisions made by an electron in travelling 1 cm. through a gas at a pressure of 1 mm. Hg and a temperature of 0° C. It follows therefore that

$$P_c = \frac{1}{\lambda_e} = NQ,$$

where λ_e is the electron mean free path, N is the number of gas atoms

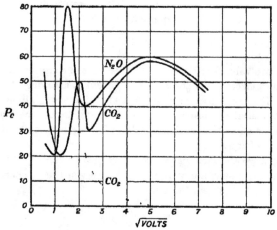

FIG 1.1. Probability of collision in N_2O and CO_2. The dotted curve is the probability of scattering in CO_2 at right angles to the electron beam.

per cm ^{3}, and Q is the effective cross-section of the atom in cm.^{2} At 1 mm. Hg and 0° C., $N = 3 \cdot 56 \times 10^{16}$, and therefore

$$Q = 0 \cdot 281 \times 10^{-16} P_c \text{ cm.}^2$$

In a gas at a pressure p mm. Hg the electron mean free path is λ_e/p. From equations (1.1) and (1.2) it follows that an electron beam passing through a gas is reduced in intensity from I_0 to I in a distance x where

$$I = I_0 e^{-pP_c x}. \tag{1.5}$$

If the electrons possess sufficient energy the gas atoms may become excited or ionized. The effective cross-sections Q_e and Q_i, for excitation and ionization respectively, are smaller than Q as many collisions occur in which the atom is neither excited nor ionized. The ratio of Q_e to Q is termed the probability of excitation at a collision. Similarly the ratio of Q_i to Q is termed the probability of ionization at a collision.

The numbers of excited atoms or ions formed by an electron in 1 cm. thickness of gas at 1 mm. Hg and 0° C. are respectively termed the efficiencies of excitation or ionization The probability of excitation or ionization at a collision is then found by dividing the efficiencies by the

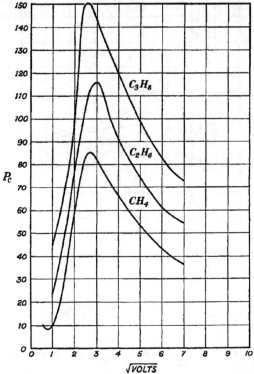

Fig. 1 2. Probability of collision in CH_4, C_2H_6, and C_3H_8

number of collisions made and is determinable in various ways. Darrow [7, p. 99] terms the quotient (cross-section for excitation/kinetic theory cross-section) as the efficiency of excitation although Arnot defines this quantity as the effective cross-section for excitation, as given above.

The probability of excitation by an electron is zero at the excitation potential [8] and rises to a maximum value lying generally between 0·001 and 0·1 at a small voltage above the excitation potential, thereafter falling gradually The effective cross-section is then $\sim 10^{-17}$ to 10^{-16} cm^2 Some critical potentials are listed in Table 1 3.

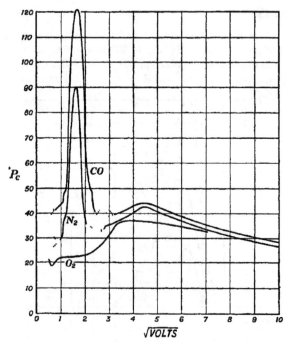

FIG 1.3 Probability of collision in CO, N$_2$, and O$_2$.

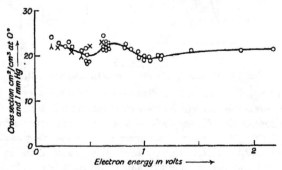

FIG 1.4. Elastic collision cross-sections in He.

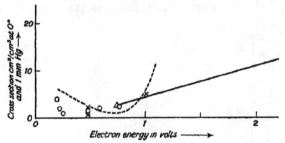

FIG. 1 5 Elastic collision cross-sections in A

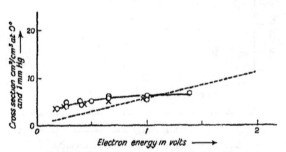

FIG. 1.6. Elastic collision cross-sections in Ne.

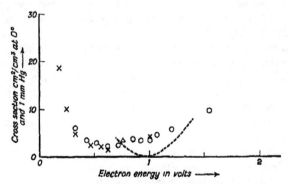

FIG 1.7. Elastic collision cross-sections in Kr.

Reference [88] gives details of the theoretical treatments (lines) and the different sets of experimental data (points) for Figs. 1.4 to 1.8,

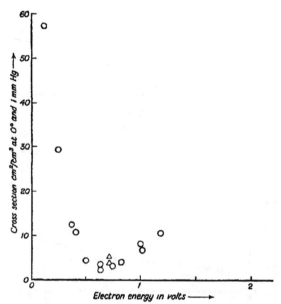

FIG. 1.8. Elastic collision cross-sections in Xe

Excitation by a photon is an unlikely process unless the photon energy is close in value to the excitation potential [2] It is important in the study of light emission from discharges in which self-absorption is important [9, 10] and is mentioned again in discussions of excitation temperatures (Chapter X).

Excitation by positive ions is generally not important in high pressure discharges since positive ions of energies $\sim$ 100 eV or more [7, p. 114] are required. In low pressure glow discharges, especially where Townsend's β coefficient (see Chapter II) is appreciable, the process may occur as the field strengths in such discharges are generally high.

Excitation by excited atoms, particularly if they are metastable, is sometimes effective and is known as a collision of the second kind. The extreme case of ionization of an atom, or molecule, A by a collision with a metastable atom B has been extensively studied by Penning [11] and others (see Chapter II). It is necessary then that the energy of excitation of the level in B should be greater than the ionization potential of A.

Ionization by electrons follows the same trends as the excitation process; the cross-section for ionization is zero at the ionization potential,

TABLE 1.3†

Atom or molecule	First excitation potential (eV)	Ionization potentials (eV)	
		I	II
A	11·56	15 8	27·5
	11 40 (metastable)		
	11 66 (metastable)		
Ag	3·1	7 6	21 4
Al	3 13	6 0	18·8
Ba	1 56	5 2	10 0
Br	.	11 8	19 2
C	.	11 3	24 8
Ca	1 9	6 1	11 9
Cd	3 78	9 0	16 9
	3·71 (metastable)		
	3 93 (metastable)		
Cl	..	12 9	23 7
Cu	1 4	7 7	20 2
F	..	17 4	34 9
Fe	.	7 8	16 5
H	10 2	13 6	
He	20 9		
	19 8 (metastable)	24 6	54 1
Hg	4 87	10 4	18 8
	4 64 (metastable)		
	5 44 (metastable)		
I	2 34	10 4	19·0
K	1 6	4 3	31 7
Kr	9 98	14 0	24 5
	9 86 (metastable)		
	10 51 (metastable)		
Li	1 8	5 4	75 7
Mg	2 7	7 6	15 0
Mo		7 1	..
N	6 3	14 5	29 6
Na	2 1	5 1	47 3
Ne	16 58	21 6	40 9
	16·53 (metastable)		
	16 62 (metastable)		
Ni	..	7 6	18 2
O	9 1	13 6	35 2
Pt	.	8 9	18 5
Rb	1 5	4 8	27 3
Sn	..	7 3	14 6
Sr	1 75	5·7	11·0
W	.	7 9	
Xe	8 39	12 1	21 2
	8 28 (metastable)		
	9 4 (metastable)		
Zn	4 01	9·4	18 0
	3 99 (metastable)		
	4 06 (metastable)		

† From Landolt–Bornstein Vol I (Part 1), 1950.

TABLE 1.3 (cont.)

Atom or molecule	First excitation potential (eV)	Ionization potentials (eV)	
		I	II
Br_2	.	13 3	
C_2		12	
CH_3Br		10 5	
CH_3Cl		11 2	
CH_3I		9 5	
CH_4		13 1	
CN		14	
CO	6 0	14·1	
CO_2	10 0	13·7	
CS_2	.	10 4	
Cl_2		13 2	
F_2	.	17 8 (calculated)	
H_2	11 2	15 4	
H_2O	7 6	12 6	
I_2	1 9	9 0	
N_2	6 1	15 5	
NO		9 5	
NO_2	.	11 0	
N_2O		12 9	
O_2	.	12 2	
SO_2		12 1	

The last two columns I and II refer to single and double ionization.

increases to a maximum, and then decreases slowly. The maximum value of the probability of ionization is generally greater than the corresponding quantity in excitation and lies between about 0·5 (argon) and 0·2 (helium) for some of the common gases, at about 100 eV The cross-sections, for singly ionized atoms, are therefore of the same order as the kinetic theory cross-sections. Ionization functions for various gases are shown in Figs. 1 9 and 1.10 [2]. The work of Bleakney and others on the formation of multiply ionized atoms in Hg, A, and Ne [12, 13] show that for most discharge problems these processes will be of negligible importance. This is not always the case, however, as the work of Bowen and Millikan [14], Edlén [15], and others on the formation of stripped atoms, etc , has shown. Ionization by high energy electrons, up to 30 kV, has been studied by various workers [7, 42].

Townsend's ionization coefficient α is a basic quantity in discharge physics and is defined as the number of new ions of either sign produced by an electron in moving through 1 cm in the direction of the applied field. The Townsend α cannot be directly linked with the probability of ionization, as defined above, since such data, deduced from work on collision processes, refer to monoenergetic electrons. In a gaseous discharge the existence of a finite spread in electron energies means that cross-section data can be used to determine α only if the form of the electron energy distribution is known. Studies of this kind were first made by Townsend and his school and will be referred to later in this chapter.

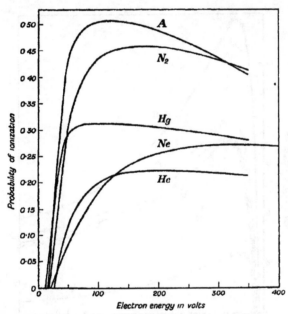

Fig 1 9. Curves showing probability of ionization at a collision as a function of the energy of the impacting electron in volts

Ionization by positive ions in a gas is an improbable process except at energies higher than those likely to be encountered in the forms of discharges considered in this book. The process is discussed further on page 285.

Photoionization

The process of photoionization is represented by

$$A + h\nu \rightarrow A^+ + e^-$$

and is the converse of radiative electron-ion recombination. The escaping electron has energy

$$\tfrac{1}{2}mv^2 = h\nu - h\nu_n$$

where ν_n is the series limit frequency. The process appears to be important in high pressure spark breakdown, as explained by the theoretical studies of Loeb, Meek, Raether, and others (Chapter VI), in certain other cases of corona discharges (Chapter III), and in ionosphere work [16, 17]. Theoretical treatments have been given by Kramers [35] (atomic hydro-

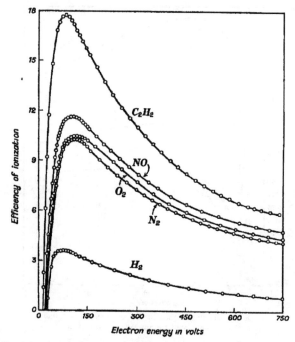

Fig 1.10 Curves showing the efficiency of ionization for some molecular gases.

gen), Trumpy (alkali metal vapours), [216] Vinti, Wheeler, and Su-Shu Huang (helium), and Bates (O, N, Ne, etc.) among others. Of this work, the results obtained by Su [18] for helium are probably the most accurate (Fig 1.11).

Bates's curves [16] for the absorption cross-sections per $2p$ electron in N, O, F, and Ne are given in Fig 1.12. The calculated absorption cross-sections at the spectral heads (Bates, private communication) are given in Table 1·4. Massey [19] gives a cross-section for H, at the spectral limit, of $0·6 \times 10^{-17}$ cm.^{2} (Fig. 1.13).

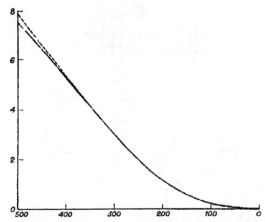

Fig 1.11 Atomic absorption coefficient for helium in units
of 10^{-18} cm ^{3} is plotted against the wave-length in angstroms
(abscissae) for the ground state The full line represents the
result derived from the matrix element of the dipole moment,
while the broken line represents that of the momentum.

TABLE 1 4

Atom	Cross-section in units of 10^{-17} cm.^{2}
N	2 2
O	0 45
F	0 61
Ne	0 58

Vinti [20] expresses his results as $\mu a = 8 \cdot 02 \times 10^{-18} \times K$ cm.$^{-1}$ per
atom, where $\mu a = \mu/N_0$ and $N_0 =$ number of atoms/cm.^{3} Vinti shows
that the work of Wheeler [21] is in good agreement with his own results.
Table 1.5 gives Vinti's values for K.

The experimental techniques for the determination of absorption
coefficients have been developed by Ditchburn and his school [22, 23]
and Lawrence et al. [24] for the alkali metals, using spectroscopic
methods. Despite the large volume of work on absorption and emission
spectra down to 1,000 Å, and the smaller volume of work down to 300 Å
by Lyman and later workers, there is little information on optical absorption coefficients at these wave-lengths with the exception of the work
of Schneider [25] on air. The techniques are difficult since, below
$\sim$ 1,250 Å (the fluorite limit), the source, absorption chamber, and

TABLE 1.5

Energy of electron in eV	Equivalent wavelength in angstroms	Constant K
0	504	1 05
0 5	395	0 59
1	325	0 39
1 6	267	0 26
2	239	0 21
2 5	212	0 16
3 0	190	0·13
4	157	0·087
5	134	0 062

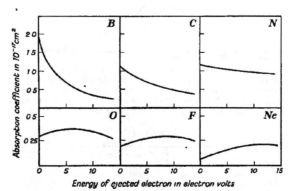

Fig. 1.12. Illustrating the continuous absorption coefficients (per $2p$ electron) of the normal atoms from boron to neon as functions of the energy of the ejected electron In terms of the wave number of the absorbed radiation the ranges covered are approximately as follows, in 10^4 cm.$^{-1}$

Boron	6 7 → 17 7	Oxygen	10·9 → 21 9
Carbon	9 1 → 20 1	Fluorine	13 6 → 24 6
Nitrogen	11 7 → 22 7	Neon	17 4 → 28 4

the evacuated spectrograph body must be in direct windowless communication. Table 1.6 gives the series limits for singly ionized atoms.

Electrical methods, all so far used at pressures ∼ 10 cm. Hg, are (a) the space charge neutralization technique [26, 27, 28], with which it is very difficult to obtain quantitative absorption data, (b) the direct ionization method in which the ions produced between two plates by a collimated beam of radiation passing through them are measured, and (c) Geballe's method [29]. The latter author determined the absorption coefficient in hydrogen of the photoionizing radiations produced, in the same gas chamber, by a Townsend discharge by measuring the

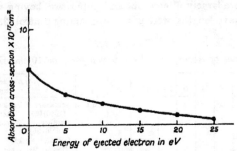

FIG. 1 13.   Photoionization cross-sections in atomic
hydrogen.

TABLE 1.6

Element	First ionization potential	Series limit wavelength in angstroms
He	24 5	504
Ne	21 5	570
A	15 7	790
H_2	15 6	790
H	13 5	915
N_2	15 6	790
N	14 5	850
O_2	12 5	990
O	13 6	910

photo-emission from two nominally identical brass plates placed at
different distances from the discharge.

The results of absorption experiments are expressed either in terms of
linear absorption coefficients μ or as cross-sections Q where, using the
usual symbols,

$$I = I_0 e^{-\mu x}. \tag{1.6}$$

If μ is measured at pressure p then $\mu = NQ$ where N = number of atoms
per cm.³ at pressure p.

The limited extent of the theoretical results, for simple monatomic
gases, is shown by Figs. 1.11 to 1.13. Apparently, owing to theoretical
difficulties, no calculated data on heavier atoms have been published but
this information would be of great value. The experimental data are
given in Table 1.7. The theoretical values (~ 40 cm.$^{-1}$) at $p = 10$
cm. Hg are much larger than the experimental values of Raether,
Greiner, Schwiecker, and Jaffe et al., probably because the methods used
involve such a large value of the product of the gas pressure times the
distance between the source and the detector that the absorbable

radiations were largely filtered out and impurities, having series limits at different wave-lengths, were possibly assuming control [33].

TABLE 1.7

Values of Absorption Coefficient μ for $p = 10$ cm. Hg

H₂	Air	O₂	H	N	O	Ho	No	A	p	Author	Experimental method and remarks
.	50	.							0 1	Schneider (1940) [25]	Vacuum spectro-meter
0 11	0 26	0 66							14 -50	Raother (1938) [30]	Wilson chamber
55								.	0 1	Geballo (1944) [29]	Photo cell (brass cathode)
0·18									10	Groiner (1933) [31]	Geiger-counter
3 6	7 0	.					.	.	1 5	Cristoph (1937) [32]	Geiger-counter
		.	22							Massey (1938) [19]	Calculated at series limit
.	.			80	16		21	.	.	Bates (1939) [16]	Calculated at series limit
.	.	.			31			.	.	Vinti (1933) [20]	Calculated at series limit
0·12	..	.	.			0 051		0 13	2 5-10	Jaffé et al (1949) [33]	Geiger-counter

p is the gas pressure in cm Hg used in the experiments

Milne showed in 1924 that, since the photoionization and radiative recombination processes were complementary the cross-sections could be related in a simple fashion. Thus if $F_n(v)$ is the probability of capture for an electron of velocity v, and α_ν is the absorption cross-section for photoionization by a quantum $h\nu$ we have

$$\frac{F_n(v)}{\alpha_\nu} = \frac{h^2\nu^2}{\pi m^2 v^2 c^2}. \tag{1.6a}$$

It is possible to derive α_ν in this way for hydrogen, from the known values of $F_n(v)$, provided that v is known from, for example, a knowledge of the electron temperature in the discharge conditions being studied. The Milne relation and Kramers's [35] calculated values of photoionization cross-sections in atomic hydrogen have been used by Cillié [34, 36] to compute the relative numbers of the electrons captured into the nth level of the hydrogen atoms, when the latter are in equilibrium with the free electrons at a given electron temperature. From these data, information on electron temperatures may, in principle, be obtained by studying the intensity decrements in the recombination continua such as that on the short wave-length side of the Balmer series limit in hydrogen, namely 3,647 Å [37, 76, 152].

The space charge neutralization method of Langmuir and Kingdon [40] has been used by Mohler [39] to study photoionization in hydrogen. With pure, dried hydrogen at a pressure of $\sim$ 0·02 mm. Hg the photo-ionization sensibly vanished Mohler concluded that there was no evidence that hydrogen emits radiation capable of ionizing the normal molecule and mentioned in support of this that no hydrogen lines had been identified beyond 885 Å [41]. Mohler pointed out, however, that whereas the ionization potential of H_2 is about 15·9 V [42], and whereas it would not be expected that H_2 or H would emit radiations of such quantum energies, there still remains the possibility of H_2^+ having excited states of energy greater than 15·9 eV. The production of such states would require a double excitation [39] which could be effected either by successive electron collisions at an energy $>$ 15·9 eV or by a single electron collision at an energy $>$ 31·8 eV. The possibility of a recombination process being important is neglected in this instance by Mohler and indeed would probably not be important at such a low pressure, since high ion concentrations could not be obtained, in contrast with the conditions obtaining in sparks at atmospheric pressure. Mohler [43] describes elsewhere also the detection of photoionization of a gas by a discharge in the same gas and reports positive results in argon and neon.

Ladenburg, Van Voorhis, and Boyce [38, 44] studied the absorption coefficient in oxygen of radiations having λ between 1,900 and 300 Å. From about 1,300 to 300 Å the measurements were only qualitative, but the data are interesting. There is a zone of strong absorption, with a maximum linear absorption coefficient $\mu = 490$ cm.$^{-1}$ (reduced to 76 cm. Hg and 0° C.) at $\lambda = 1,450$ Å. A second zone of absorption commences at 1,100 Å and is still strong at 300 Å. The window ($\lambda = 1,450$ to 1,100 Å) was transparent for a 4 m. path of O_2 at 0·25 mm. Hg. Preston [45] gives results for absorption in O_2, N_2, CO_2, and H_2O at $\lambda = 1,216$ Å. Dershem and Schein [46] have published absorption coefficients for the carbon $K\alpha$ line (44 6 Å) in He, N_2, O_2, Ne, A, and CO_2 (Table 1.8). ρ is the gas density.

TABLE 1.8

Absorber	μ/ρ
He	3,600
N_2	3,850
O_2	5,765
Ne	13,100
A	45,700
CO_2	4,780

Holweck [47] gives further information on the absorption of soft X-rays in gases.

Recombination

Where populations of positively and negatively charged particles coexist, processes of recombination may take place. A representative reaction is

$$A^+ + B^- \rightarrow AB + h\nu.$$

In this expression B^- may be an electron or a negative ion. Alternatively a third body C may be involved and the masses of A, B, and C may vary widely. In this case it is not necessary to introduce the radiation term $h\nu$.

Radiative recombination processes, especially the electron-ion reaction, are of considerable interest and it is thought that very short wave-length radiation may be so produced [6]. In cases where atoms are multiply ionized photons with sufficient energy to give photoionization may be produced. In the hypothetical case of a discharge in pure atomic hydrogen, however, such photoionizing radiations could still be produced by recombination. Recombination may also be important even where multiple ionization is possible, since cross-sections for the latter are small, and most discharges (except perhaps the 'hot sparks' used by Bowen et al. [14], Edlén [15] and others) have too low a value of E/p for many sufficiently energetic electrons to be produced. It is difficult in many cases to draw true comparisons, but the point is stressed here since some writers on discharge physics have occasionally dismissed photoionization mechanisms as being virtually impossible in pure gases. In mixed or impure gases, where the ionization potentials of the constituents are different, excited states in one gas are generally capable of producing photons of sufficient energy to ionize other constituents.

Loeb [3, p. 87] distinguishes several types of recombination process, viz. (a) initial recombination, (b) preferential recombination, (c) columnar recombination, and (d) volume recombination. Loeb also points out that possibly two types of electron recombination occur, i.e. preferential and volume, where an electron returns to its parent ion in the former case, and to another positive ion in the latter case. Volume recombination is obtained with positive and negative particles randomly distributed in space; it is the normal process, and includes electron and negative ion processes. Preferential recombination is that of the original members of an ion pair. The two remaining processes, columnar and initial, are not so clearly defined but refer respectively to the special conditions of highly ionized and restricted regions (α particle tracks) and to a case intermediate between (b) and (d) where the ions are not distributed uniformly but are yet too far apart to give preferential recombination.

The number of individual recombinations in time dt, for a region in which n_+ and n_- are uniform concentrations of particles per cm.^{3}, between times t and $t+dt$ is

$$dn = -\alpha n_+ n_- \, dt$$

where, by definition, α is the coefficient of recombination for the particular process involved. Use of the simple equation implies, of course, that n_+ and n_- are controlled only by the recombination process relevant to the value of α used [3, p. 89]. The dimensions of α are such that

$$\alpha = vQ$$

where, in the case of electron recombination, v is the electron velocity and Q the cross-section for the process.

If $n_+ = n_-$, and n_0 is the initial ion density at time $t = 0$,

$$n = \frac{n_0}{1 + n_0 \alpha t}, \tag{1.7}$$

provided again that α is constant for the ranges of variables considered If this is not so, then

$$\alpha_t = \frac{1}{t_2 - t_1}\left(\frac{1}{n_2} - \frac{1}{n_1}\right), \tag{1.8}$$

where α_t is the value of α for the time interval t_1 to t_2 during which n falls from n_1 to n_2

Few data exist on radiative electron-ion recombination, where the electron temperature is a controlling factor. Massey [19, p. 35] shows that the cross-section for electron capture into the ground state of H^+ is $\sim 3 \times 10^{-22}$ cm ^{2} at 5 eV For smaller electron energies this quantity tends to infinity as v^{-2}. The reaction for this case is

$$H^+ + e \rightarrow H + h\nu,$$

$$h\nu = \tfrac{1}{2}mv^2 + V_i,$$

and the spectrum is continuous from a long wave-length limit equivalent to V_i. Recombination into other levels is, of course, possible and the Balmer series limit continuum has been observed in hydrogen discharges by various workers [48, 49]. In gases where electron attachment is likely, the process becomes less important since ion-ion recombinations are, in general, more probable. The particular relevance of electron-ion recombination to certain special discharges will be discussed in Chapter II.

Kenty [51] and later Mohler [26, 27, 39, 50] and others [52] studied radiative electron/ion volume recombination in A, Cs, and Hg at low pressures, using spectroscopic [55] and probe techniques. For the range of gas pressures used Kenty [51] and Hayner [53] had earlier shown that the electron and ion concentrations are closely equal. In a particular case (Hg at 46–110 μ pressure) the mean electron or ion concentration was found by a probe method to be $1\cdot9\times10^{12}$/cm.^{3} and $\alpha = 3\cdot5\times10^{-10}$ cm.^{3}/ion-sec. in conditions where diffusion to the walls was unimportant. The electron temperature was about $1,000$–$3,000°$ in most of the cases considered. α was 10^{-10} cm.^{3}/ion-sec. for A, Cs, and Hg at 10–800μ pressure for the same range of electron temperatures.

Calculations for hydrogen give $\alpha = 1\cdot7\times10^{-12}$ cm.^{3}/ion-sec. at an electron energy of $0\cdot1$ eV and it would be expected that α for Cs would also be of this order. The discrepancy between the calculated and measured values is large although Mohler [50] points out that there is some pressure dependence of α even at 10μ in Cs so that the simple recombination process assumed is probably not the only reaction occurring. Massey [54] has referred more recently to this matter which is of importance in ionospheric studies.

Quantitative experimental work on radiative electron-ion recombination is rare, although such information is valuable in the study of discharges, especially at high pressures and current densities, when negative ion formation is improbable. Hence the experiments of Biondi and Brown [56] are of great interest. The method used [57] consists of measuring the time variation of the resonant frequency of a microwave cavity (about $3,000$ Mc./s) in which a discharge is initiated, maintained for a time, and then allowed to decay. The electron concentration, as a function of time after the termination of the discharge current, can then be determined

The results may be summarized as follows:

(a) In neon from $195°$ K. to $410°$ K. the value of α was

$$(2\cdot07\pm0\cdot05)\times10^{-7} \text{ cm.}^3/\text{ion-sec.,}$$

constant over the pressure range of 15–30 mm. Hg. At $77°$ K. the pressure dependence of α was appreciable, but the extrapolated zero pressure value was close to that given above.

(b) The corresponding (zero pressure) value for impure argon was 3×10^{-7} cm.^{3}/ion-sec. at $300°$ K. and for He a value of $1\cdot7\times10^{-8}$ is given.

(c) In H_2, at 2–18 mm. Hg, $\alpha = 24\cdot7\times10^{-7}$ cm.^{3}/ion-sec. and is independent of pressure. In N_2 at 2–18 mm. Hg $\alpha = 14\times10^{-7}$ cm.^{3}/ion-sec at low pressures and increases with increasing pressure. O_2 also showed pressure dependence of α over the pressure range 2–18 mm. Hg, the low pressure value of α being about $2\cdot7\times10^{-7}$ cm.^{3}/ion-sec. For H_2 there was also a temperature dependence of α. It is suggested by Biondi and Brown that the values of α for these diatomic gases are higher than for monatomic gases because, in the former case, excitation of vibrational and rotational states serves as an efficient means of absorbing the electron's initial kinetic energy.

In discussing these results, which should be compared with the lower experimental and theoretical values given by other workers [51, 54, 58 to 60], it is suggested [56, 57] that transitions of electrons into the higher excited states have previously been incorrectly underestimated. A method of calculation is given for He, Ne, and A which gives α values too high by a constant factor of about 11, i.e. the low value of He is consistent with this theory. Bates [62] suggests that molecular He ions are operative [68].

Holt *et al.* [71] have carried out work on neon at 10–30 mm Hg, using Brown's technique [57], at 3,000 Mc./s., and have supplemented the measurements with observations of the light emitted during the afterglow, i.e. after current zero. At 300° K., $\alpha = 1\cdot1\times10^{-7}$ cm.^{3}/ion-sec. for radiative electron/ion recombination. Spectroscopic observations of the afterglow in the visible and near ultra-violet showed that recombination continua were absent and the authors suggest therefore that transitions to the upper levels, which would give continua in the far infra-red and which were not sought, may be more important than is supposed [51]. The above authors [71] have also published work on He [128].

Stueckelberg and Morse [72] found that recombination into the 1s state of hydrogen gave the data given in Table 1.9 (taken from a curve).

TABLE 1.9

Electron energy eV	Cross-section cm.$^2 \times 10^{20}$
0 5	0 55
1 0	0 25
2	0·15
4	0·07
8	0·03
14	0·02

For $V = 0\cdot2$ volts, we have for the 2s, 2p, and 3d terms respectively, cross-sections of $1\cdot7\times10^{-21}$, $5\cdot5\times10^{-21}$, and $2\cdot6\times10^{-21}$ cm.^{2} Zanstra [60], analysing some experimental data of Rayleigh's [73] (see also

Craggs and Meek [74]) on long duration hydrogen afterglows, used the relation

$$\frac{dN}{dt} = -\alpha N^2 \quad \text{and} \quad \alpha = \sum_{n=1}^{\infty} \alpha_n. \tag{1.9}$$

The total number of recombination processes/cm.^{3}/sec. on any final level n is then

$$L = \alpha' N^2, \quad \text{where } \alpha' = \sum_{n=3}^{\infty} \alpha_n. \tag{1.10}$$

This represents also the number of quanta/cm.^{3}/sec. emitted as Balmer line radiation [75] where the values of α_n correspond to different levels and α is the total recombination coefficient as determined by electrical measurements of ion and electron concentrations. Cillié [34] and other astrophysicists have calculated α_n from

$$\alpha_n = 3 \cdot 208 \times 10^{-6} M(n, T),$$

where

$$M(n, T) = \frac{1}{T^{\frac{3}{2}}} \frac{1}{n^3} e^{x_0} \int_{x_0}^{\infty} \frac{e^{-x}}{x} \, dx \tag{1.11}$$

and

$$x_0 = \frac{\chi_n}{kT},$$

χ_n is the ionization energy for the nth level, k is Boltzmann's constant, and T is the electron temperature. Cillié has tabulated $M(n, T)$ for levels up to the 10th. From these and equation (1.11), the quantities α and α' were computed for the same electron temperature as Cillié's by summation of the quantities $M(n, T)$. The values are given in Table 1.10. The contributions by levels higher than the 10th were not negligible and Zanstra allowed for them by replacing the summation for those levels by the integral

$$\int_{n=1}^{\infty} M(n, T) \, dn.$$

For temperatures 1,000–50,000° K. it amounts to 20, 11, 8, 6, and 4 per cent. respectively for the values of temperatures given in Table 1.10, of the total for all levels.

TABLE 1.10

	$T: 1,000° K.$	$5,000° K.$	$10,000° K.$	$20,000° K.$	$50,000° K.$
$10^{14}\alpha$	218	74	46	27	13 5
$10^{14}\alpha'$	122	33	17·7	9·4	3 6

Recombination of this type has been recorded spectroscopically by observation of the Balmer series limit continuum in electrodeless discharges [76, 77] in sparks [78] and arcs [61].

Recombination processes in the upper atmosphere have been extensively studied by Massey and his collaborators. For example, Bates, Buckingham, Massey, and Unwin [80] have worked out values of $\sum\limits_{n}^{\infty} Q^n$ in hydrogen (radiative electron-ion recombination) where n is the state in question and Q the cross-section given by

$$\alpha = vQ, \tag{1.12}$$

where v is the electron velocity.

For small values of the electron energy ϵ, where $\epsilon = \frac{1}{2}mv^2$ in eV, Q^0 varies as $1/\epsilon$ and correspondingly $\alpha_0 = \dfrac{5 \cdot 1 \times 10^{-14}}{\sqrt{V}}$ cm.^{3}/ion-sec. Table 1 11 gives some values.

TABLE 1.11

Electron energy (eV)	.	0 28	0 13	0 060	0 034
$\sum\limits_{1}^{\infty} Q^n \times 10^{21}$ (cm.^{3})	. .	23	53 7	119	272

Data for different values of n are given in the paper [80], where the discrepancy between the above values for hydrogen and the experimental values of Kenty and Mohler (references given above), i.e. $\alpha \sim 2 \times 10^{-10}$ cm.^{3}/ion-sec., is pointed out [54]. As an example, it is found that for an electron energy of 2 eV, $\alpha \sim 10^{-12}$ cm ^{3}/ion-sec.

Massey and Bates [81] give values for the radiative recombination coefficient in atomic oxygen for various electron temperatures (250–8,000° K.). The contributions from the ground state and from excited states are tabulated separately, the total at 2,000° K. is $0 \cdot 89 \times 10^{-12}$ cm.^{3}/ion-sec , which is of the same order as that found for hydrogen. Massey and Bates [81] extend the discussion of attachment and detachment processes in oxygen, with particular reference to the upper atmosphere [82]

It is possible to calculate the cross-sections Q_e for radiative electron-ion recombination to the ground state from the cross-sections Q_α for photo-ionization, using Milne's [87] relation, which states that

$$Q_a = \frac{m^2v^2c^2}{2h^2\nu^2} Q_e, \tag{1.13}$$

where m is the electron mass, v is the velocity of the electron after photoionization or before recombination, c is the velocity of light, h is

Planck's constant and ν is the frequency of the absorbed or emitted radiation. If $h\nu \equiv$ ionization potential, then $v = 0$. Using Bates's values for Q_a (Table 1.4) the data of Table 1.12 may be derived.

TABLE 1.12

Atom	Recombination cross-section Q_e (cm ^{2})		
	0 1 eV electrons	0 5 eV electrons	2·0 eV electrons
N	$8\,0 \times 10^{-22}$	$1\,9 \times 10^{-22}$	$5\,7 \times 10^{-23}$
O	$1\,6 \times 10^{-22}$	$3\,4 \times 10^{-22}$	$1\,0 \times 10^{-23}$
Ne	$5\,3 \times 10^{-22}$	$1\,1 \times 10^{-22}$	$3\,1 \times 10^{-23}$
He	$9\,2 \times 10^{-22}$	$1\,9 \times 10^{-22}$	$5\,4 \times 10^{-22}$
H	$2\,6 \times 10^{-20}$	$5\,5 \times 10^{-21}$	$1\,8 \times 10^{-21}$

There are many published researches on the volume recombination of negative and positive ions, i.e. in gases where electron attachment can, if field and temperature conditions are favourable, be appreciable. The results show that α is about 10^{-7} to 10^{-6} cm.^{3}/ion-sec. at atmospheric pressure in most of the common gases such as air, CO_2, O_2, etc. At 1 atm. and room temperature, Compton and Langmuir [63] give values of α varying only between 0 85 and $1·71 \times 10^{-6}$ cm.^{3}/ion-sec. for air, CO_2, O_2, H_2, H_2O, SO_2, N_2O, and CO. More recent results for air and O_2 are given by Sayers [64] and Gardner [65], who studied also the effects of initial recombination. Sayers found α in air to increase from 1×10^{-6} cm.^{3}/ion-sec. at 50 mm. Hg to a maximum of about $2·5 \times 10^{-6}$ cm ^{3}/ion-sec. at about 1 atm., thereafter falling but fitting well by extrapolation to Machler's results [66] for high-pressure air. At about 15 atm. α is $0·5 \times 10^{-6}$ cm.^{3}/ion-sec. Gardner found α in pure O_2 to increase from 1 to 2×10^{-6} cm.^{3}/ion-sec. for a pressure increase from about 100 to 750 mm. Hg, and to fall at 1 atm. from 4×10^{-6} at $200°$ K. to 1×10^{-6} cm ^{3}/ion-sec. at $450°$ K.

The early theoretical work of Langevin [69] and Thomson [70] on recombination processes is in general supported by the work of Sayers and others. With the exception of those of Sayers, Gardner, and a few other workers, the experimental values of α, for the above-mentioned volume recombination processes, were very scanty until the new data of Biondi and Brown [56, 126] and others [127] became available. Indeed for this reason the use of recombination coefficients in many cases of discharge physics must be limited by the fact that α is known, in general, to little better than an order of magnitude. Many individual researches such as those of Mohler [50, 58], have been carried out with great care, but uncertainties generally arise when extrapolations to other conditions are attempted.

The method used by Gardner [65], Sayers [64], and others is, very briefly, to produce ionization between two plates by a pulse of X-rays. The ions are then swept to the plates, by the application of an electric field and the total ionization at various times after the cessation of the ionization process can then be measured. Allowance is made for diffusion.

Mohler's probe method consisted of measuring the concentrations with a normal Langmuir probe at various times after the arc discharge terminals had been short-circuited by a commutator. A spectroscopic method is referred to briefly by Zanstra [75]. Jaffe's experimental and theoretical work on columnar recombination [67] is outstanding.

Attachment and detachment

Early experiments [5] show that the mobility of negative ions is generally approximately equal to that of the positive ions. When a negative ion loses an electron, the latter then acquires a much larger mobility and this will be the value applicable to negative ions in gases where attachment is negligible. For example, Túxen [106] has observed the following negative ions O^-, O_2^-, NO_2^-, NO_3^-, OH^-, and H^-, but did not detect N^-, N_2^-, He^-, Ne^-, or A^-.

The stability of negative ions is measured by the electron affinity of the neutral particle in question, and this is also the energy necessary to detach an electron from the ion [19]. A positive electron affinity indicates a stable negative ion and the higher the electron affinity, the greater the stability of the ion. Some electron affinities are given in Table 1.13 [19].

TABLE 1.13

Atom	Electron affinity eV
H	0 75
He	−0 53
N	−0 6
O	2·2
F	3 94†
Ne	−1 20
Cl	3 8
A	−1 0

† Recent values tend to be lower than this [222]

It is clear from Table 1.13 that in experiments where negative ions are to be avoided, the gases used should be free of O, F, or Cl. The other halogens also have very high electron affinities. The avoidance of water vapour is particularly important in such cases, not only as a gaseous contaminant but as adsorbed layers in the apparatus.

Because of the insufficiency of the data it is not possible to tabulate the probabilities of the various, and sometimes complicated, attachment and detachment cross-sections as a function of electron energies for the various gases, or as mean cross-sections as functions of E/p. An excellent review of collision processes involving negative ions is due to Massey [19] (see also Loeb [3]).

Experiments with negative ions are often carried out with streams of electrons which acquire their terminal energies of drift in an electric field E in a gas. If A is the chance of attachment per electron per cm. in the field direction, then the loss of electron current in a distance dx due to this cause is

$$dI = -IA\, dx.$$

In moving this distance dx, parallel to the field, the electron moves through a greater distance than dx, i.e. through $c_e\, dx/u_e$ where c_e and u_e are the random and drift velocities. If Q_a is the effective cross-section for attachment,

$$A = NQ_a \frac{c_e}{u_e}. \tag{1.14}$$

If h is the probability of attachment per collision and λ is the electron mean free path then

$$A = \frac{hc_e}{\lambda u_e}, \tag{1 15}$$

c_e and u_e are related [105] by the expression

$$u_e = 0\ 75\frac{Ee\lambda}{mc_e}, \tag{1.16}$$

u_e is discussed elsewhere (pp. 33–41) and may be found from Ramsauer cross-sections. In the above treatment, mean values of electron velocities are implied, and these are functions of E/p [86, 99].

Simple capture can be either purely radiative of the two-body type or of the three-body type Kinetic energy ($\frac{1}{2}mv^2 + eV_a$) must be dissipated in both cases, where $\frac{1}{2}mv^2$ is the electron energy and V_a is the electron affinity. Cross-sections for radiative capture are shown in Fig. 1.14 for O. The collision processes in molecular gases are of various types [19]. The cross-sections for various processes in O_2 [107] are given in Table 1.14 at arbitrary electron energies. The last reaction shows the production of an ion pair by dissociation [19, 108 to 112]

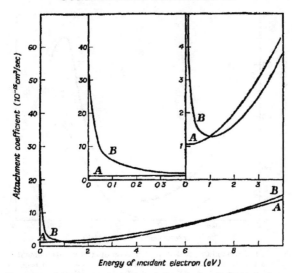

FIG 1 14   Calculated attachment coefficients for electrons in atomic oxygen on the assumption of two different values of the polarizability parameter p. A, $p = 5\ 7$ atomic units, B, $p = 18\ 3$ atomic units.

TABLE 1 14

Reaction	Cross-section (cm^2)
$O_2 + e \rightarrow O + O^-$	8×10^{-19}
$O_2 + e \rightarrow O' + O^-$	$1\ 3 \times 10^{-20}$
$O_2 + e \rightarrow O^+ + O^- + e$	$2\ 2 \times 10^{-19}$

Some results due to Bradbury [113] are given in Fig. 1.15. Here h is plotted against mean electron energy but the latter can be replaced by E/p values from Townsend [105] or from Druyvesteyn and Penning [86] It is suggested [19] that the low energy peak seems doubtful. Bradbury [115] and Bradbury and Tatel [116] have also studied attachment in NO, CO, NH_3, HCl, Cl_2, CO_2, N_2O, SO_2, H_2S, and H_2O, see also work by Bailey and Duncanson and Bailey and Rudd [101]. The results of Healey and Kirkpatrick are shown in Fig. 1.15 [101]. The data for O_2 have been obtained by Healey and Kirkpatrick [114]. It is apparent that in O_2, attachment is small for $E/p > 20$ [117]. The value of h in the molecular halogens is ~ 10 times that for O_2 [101].

The average number of collisions δ which an electron must make before attaching itself to a neutral atom is given for various gases in Table 1.15 [94, p. 97].

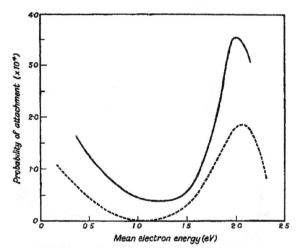

Fɪɢ 1 15 Probability (by 10^4) per collision of electron attachment to O_2 as a function of the mean electron energy ——— Observed by Bradbury - - - - Observed by Healey and Kirkpatrick.

Table 1 15

Gas		δ
Noble gases⎫ H_2 and N_2⎭	.	∞
CO	. .	. 1 6 × 10^8
NH_3	. .	. 9 9 × 10^7
N_2O	. .	. 6 1 × 10^5
Air	. .	. 2 0 × 10^5
O_2, H_2O		4 0 × 10^4
Cl_2	. . .	2 1 × 10^2

Table 1.15 is widely quoted in the literature, but it cannot be used as more than a general guide to the relative probabilities of attachment, since these vary with electron energies, etc., often in a complicated way. Massey's book [19] should be used as a source of more accurate, detailed, and modern data.

The detachment of electrons from negative ions has been relatively little studied, because of the difficulty of obtaining large concentrations of negative ions. Massey [19] shows that the maximum absorption cross-section for photoelectric detachment in Cl is 2×10^{-17} cm.² when the energy of the ejected electron is ∼ 8 eV. Massey and Smith have studied the ionization of Cl⁻ by electrons [19].

Loeb [3] has carried out experiments on detachment in O_2 and found that it occurred for $E/p > 90$ V/cm./mm. Hg, although the exact process in question does not appear certain. This type of experiment is often of importance in, for example, negative point corona (see Chapter III). Thus Weissler and Mohr have shown, in corona studies with CCl_2F_2/air mixtures that Cl^- ions appear to be stable even at $E/p > 200$.

The probable importance of negative ion formation in breakdown in gases such as CCl_4, CCl_2F_2 has been discussed by Warren *et al.* [110, 111, 112] and other investigators [118, 119, 120]. The paper by Warren *et al.* [110] gives further references. The mechanism of operation of T.R. switches, widely used in radar practice, involves the formation of negative ions [121].

Mobilities

The mobility k of a particle is given by

$$k = \frac{u}{E}, \tag{1.17}$$

where u is the particle drift velocity. Early work had shown that k is independent of E/p over a wide range of E/p values, provided that the speed of drift in the field is appreciably less than the velocity of thermal agitation.

k is related to the diffusion coefficient D for a given ion by

$$k = \frac{NeD}{p}, \tag{1.18}$$

where N is the number of molecules/cm.^{3} at pressure p. D is given by definition for uniaxial flow by

$$\frac{\partial n}{\partial t} = -DA\frac{\partial n}{\partial x}, \tag{1 19}$$

where n is the ion concentration, A is the cross-section of the region of flow, and the net diffusion is from regions of high to low n (see p. 41).

Using kinetic theory methods Langevin [95] obtained the following expression for the mobility of an ion of mass m moving through a gas consisting of molecules of mass M:

$$k = 0.815\frac{e}{M}\frac{\lambda}{C}\sqrt{\frac{m+M}{m}},$$

where C is the r.m.s. velocity of the molecules and λ is the mean free path of the ion. On the assumption of thermal equilibrium between ions and molecules the expression may also be written in the form

$$k = 0.815\frac{e}{m}\frac{\lambda}{C_1}\sqrt{\frac{m+M}{M}},$$

where C_1 is the r.m.s. velocity of the ion. For an electron, $M \gg m$, and the expression reduces to

$$k = 0 \cdot 815 \frac{e}{m} \frac{\lambda}{C_1}.$$

The above expressions are deduced on the assumption of elastic collisions between ions and molecules when considered as solid spheres.

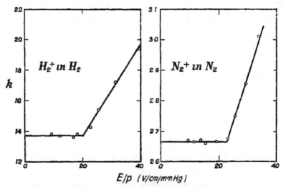

FIG. 1 16 (a) and (b) Variation of mobility of H_2^+ in H_2 and N_2^+ in N_2 as a function of E/p. k is measured at 1 atmosphere pressure and room temperature.

More exact theoretical relations, which take into account attractive and repulsive forces between ions and molecules, have been deduced by Langevin [95], Compton and Langmuir [63], Hassé and Cook [218, 219], and others. A full discussion of these later theories is given by Loeb [3]. Recent theoretical expressions for the mobilities of electrons are given in equations (1.21) and (1.22).

Modern data on the mobility of positive ions are given by Tyndall [91]. They require modification if used for high values of E/p, as the work of Mitchell and Ridler [91], and Huxley [224] has shown. Generally, the independence of k on E/p disappears at $E/p \sim 20$ V/cm./mm. Hg and data for H_2 and N_2 (by Mitchell, Ridler, and Munson [91] at Bristol) are given in Figs. 1.16 (a) and (b). Tyndall's book [91] gives an exhaustive account of the Bristol work on mobilities, particularly of ions of alkali metals, and a brief review of theoretical work. Massey's recent review [213] deals with the effects of various types of atom-atom collisions on mobilities (see also the work of Hasted [211], and of Sena [125]).

The effect of temperature variation on the mobilities of positive ions

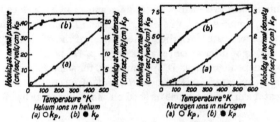

FIGS. 1.17 and 1 18. Variation of mobility of He$^+$ (He$_2^+$?) in He
and N$_2^+$ in N$_2$ with temperature

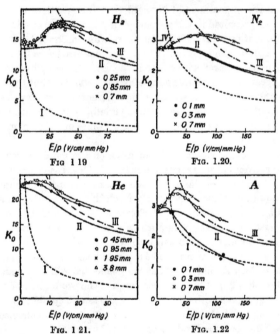

FIG. 1 19.   FIG. 1.20.

FIG. 1 21.   FIG. 1.22

FIGS. 1 19–22 Hershey's data on mobilities of K$^+$ positive ions.
I, II, III are theoretical curves [3]   IV is experimental curve
given by Mitchell and Ridler [91]

is shown in Figs. 1.17 and 1.18 for He and N$_2$. k_p is the mobility as
measured at 18° C. and normal pressure [91, p. 61] and k_ρ is the
mobility at normal density. The following relation holds at a temperature T° K.

$$k_p = \frac{291}{T} k_\rho.$$

(1.20)

Data obtained by Hershey [92, 93] are shown in Figs. 1.19 to 1 22; Hershey's K_0 denotes the value of k_p (equation 1.20). The variation of K_0 with pressure, even over the limited range of the experiments (0·1–3·8 mm. Hg) is of interest since, as Loeb [3, p. 76] pointed out, K_0 or k should be independent of p. It appears that, for some reason, when the mean free paths are long the measurement of k_p is affected. Recent experimental work with a pulse method on the rare gases is by Hornbeck [212].

The effect of impurities in the gases studied is often important. Mitchell and Ridler [91] found that 0·02 per cent. Hg vapour in N_2 suppressed the appearance of N_2^+ and that the addition of $\sim 0·8$ per cent. H_2 to N_2 resulted in the formation of a large number of NH_3^+ ions ($k = 3·22$). A study of $Hg/Xe/N_2$ mixtures is also described by the above authors. Tyndall and Powell [91, p 38] measured $k = 17$ in highly purified He as compared with $k = 5$ for a particular sample of unpurified He. It appears that the most simple explanation, i.e. that the effect of impurities is marked if their ionization potentials are less than that of the parent gas, is not sufficient. Tyndall [91, p. 48] suggests that metastable atoms may be important, and points out that more experimental data are required.

The results of the vast amount of work on the mobility of negative ions and electrons is not so easily summarized. One reason is that a study of negative ions is complicated by the attachment and detachment reactions that may occur, particularly in electro-negative gases such as O_2 and Cl_2. Since attachment coefficients vary so widely among the common gases, the effect of impurities such as O_2 or Cl_2 in A or Ne is serious in cases where attachment is to be avoided.

In cases where detachment can be neglected, the mobilities of negative and positive ions should be about the same according to Langevin's theory [3 (p. 39), 95]. Cluster formation, i.e. the production of complex molecular aggregates, often complicates matters [3 (p. 39), 91].

In gases where electrons can exist without becoming attached, their mobilities may be measured. The latter, as would be expected, are much larger than the values found for positive ions or massive negative ions. Recent data are given in Figs. 1.23–1.26. (Note that

$$K_e = k_e \frac{p}{760} \frac{273}{T}, \text{ and that the mean drift velocity } v = k_e E.)$$

Early theoretical studies of electron mobilities, or drift velocities v, have been made by Langevin, and pioneer experiments by Townsend. Some of the most recent experimental work is by Nielsen and Bradbury

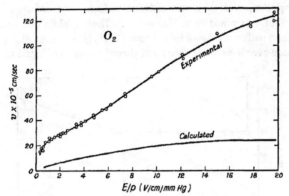

FIG 1 23. Electron drift velocities in oxygen. (Bradbury and Nielsen)

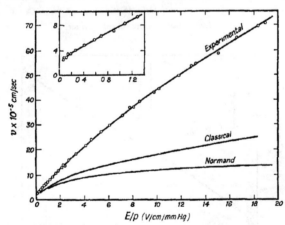

FIG. 1.24. Electron drift velocities in nitrogen. The curves marked 'Classical' and 'Normand' were derived using respectively kinetic theory and Ramsauer cross-sections for elastic electron collisions.

[96, 97], and their expression for k_e (modified from Compton's original expression [3]) reads

$$k_e = \frac{10^{-10}\frac{1}{p}\frac{1}{\sigma_1^2}T^{\frac{1}{4}}}{\left[1+\left\{1+1\cdot 15\times 10^{-31}\left(\frac{E}{p}\right)^2\left(\frac{1}{\sigma_1^4 f}\right)\right\}^{\frac{1}{2}}\right]^{\frac{1}{2}}} \qquad (1.21)$$

with k_e in cm./sec./V/cm., E in V/cm., and p in mm. Hg. f is the fraction of the electron energy lost per collision (see Loeb [3], pp. 186–7 and this work, pp. 3–4) and σ_1 is the collision radius. The electron energy distribution is assumed to be Maxwellian. Thus, Nielsen and Bradbury determined drift velocities v for electrons in H_2 ($p \sim 7$ mm. Hg) purified by passage over hot copper gauze and through a series of liquid air traps.

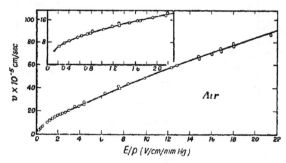

FIG 1.25. Electron drift velocities in air.

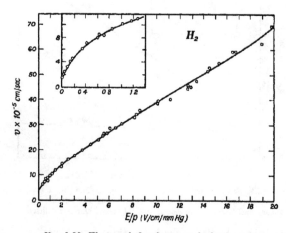

FIG 1 26. Electron drift velocities in hydrogen

No massive negative ions could be detected. The results are given in Fig 1.27, where the drift velocity is shown as a function of E/p.

It is of interest to compare the values of drift velocity found by direct measurement with the values derivable from equation (1.21), and such a comparison is given in Fig. 1.27. Curve A is derived by taking the kinetic theory value of $\sigma_1 = 1 \cdot 045 \times 10^{-8}$ cm.^{2} and $f = 2m/M$. Curve B

is obtained by the use of Normand's values [98] for the Ramsauer cross-sections in hydrogen. The discrepancy is still great and Nielsen and Bradbury suggest that it may be due to inadequate knowledge of f, since it appears that the average fraction of its energy which an electron can transfer to a molecule must be greater than that indicated by classical momentum transfer. This seems reasonable if excitation of vibrational

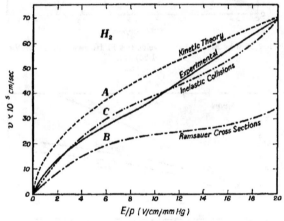

FIG 1.27. Drift velocities of electrons in hydrogen as a function of E/p

levels (possible in molecular gases) is considered as a means of energy transfer in addition to the elastic collisions. Nielsen and Bradbury suggest that f be increased sixteen-fold, at $E/p = 20$, on the basis of Ramien's work on collisions of slow electrons in hydrogen, and the results then lie on curve C (Fig. 1.27), an excellent agreement which the above authors consider may be partly fortuitous.

For these reasons it was clear that experiments on monatomic gases would be of great interest. These were carried out by Nielsen and Bradbury, with results shown in Fig. 1.28. Taking the results for He, it is found that use of Normand's Ramsauer cross-sections enables good agreement with experiment to be obtained. Above $E/p = 2.5$, however, the curves depart and it may be inferred that at this value of E/p there are sufficient electrons in the high energy tail of the energy distribution to cause excitation, so that the inelastic collisions begin to be significant in determining the average value of f. Nielsen [214] then proceeded to assess the importance of this effect on the assumption that the electron

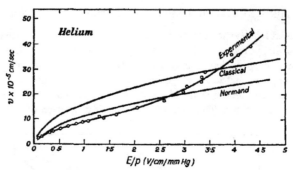

Fig 1.28 (a) Electron drift velocities in He (see caption of Fig 1 24)

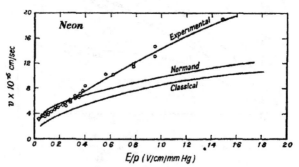

Fig 1 28 (b) Electron drift velocities in Ne (see caption of Fig. 1 24)

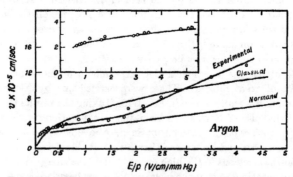

Fig. 1.28 (c). Electron drift velocities in A (see caption of Fig. 1.24).

energy distribution was Maxwellian, and by use of the data found by Townsend and his school on the average terminal energy of the electrons as a function of E/p Some of Nielsen's data are given in Table 1.16.

TABLE 1.16

Gas	Deviation observed at E/p	Electron energy in electron volts (at r m s velocity)	Electron energy in electron volts (at most probable velocity)	Energy in eV which is exceeded by fastest 5% of electrons	First excitation potential in electron volts
He	2 25	4 25	2 83	16–17	19 8
Ne	0 4	4 3	2 86	16–17	16·6
A	1·75	11 1	2 33	41	11 6

Thus in Ne the effect of inelastic collisions at the point of deviation should be just appreciable. In He it should be less important but in A it should be very marked and should have been noticed at lower values of E/p. Therefore Nielsen's simple interpretation seemed inadequate. Allen [100] later tested his derivation of the electron energy distributions in He, Ne, and A against Nielsen's experimental values of drift velocity, with fairly good agreement. This treatment involves the use of an arbitrary upper limit to the energy distribution.

The experimental values of drift velocity found some thirteen years before the work of Nielsen and Bradbury by Townsend, Bailey, and their collaborators show in general remarkable agreement with the later work. The data are tabulated most conveniently by Healey and Reed [101].

The procedure adopted above, in order to find a formula to fit observed values of electron velocities, was (a) to assume a Maxwellian distribution for electron energies and then (b) to correct this by allowing in various ways for inelastic collisions. This procedure can clearly be made in one step by redetermining the distribution and making allowance a priori for inelastic collisions. Thus, where the electronic mean free path λ_e is independent of electron velocity and $\rho(\epsilon)$ is the energy distribution of the electrons, the mean velocity of drift [86] is

$$u = \frac{2eE\lambda_e}{3(2m)^{\frac{1}{2}}} \int_0^\infty \frac{\rho(\epsilon)}{\epsilon^{\frac{1}{2}}} \, d\epsilon \Big/ \int_0^\infty \rho(\epsilon) \, d\epsilon. \tag{1.22}$$

The form of the distribution function $\rho(\epsilon)$ is discussed on p. 43.

Townsend [99] showed that the following relation holds for electrons·

$$T_e = \frac{eED}{ku_e}. \tag{1.23}$$

T_e is the electron temperature and is based on a Maxwellian distribution, which need not necessarily apply. E is the field strength, D is the diffusion coefficient (defined later, see p. 41), and k and u_e are Boltzmann's constant and the drift velocity respectively. Townsend also

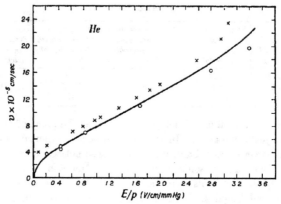

FIG. 1.29. Allen's values of electron drift velocities in He.

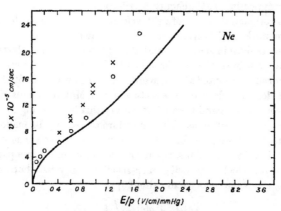

FIG 1 30. Allen's values of electron drift velocities in Ne.

showed that D/u_e could be measured, as could u_e alone by a different arrangement, and therefore D may be calculated. Allis and Allen [102] and Llewellyn Jones [103] have studied the theory of Townsend's methods in the light of various distributions [86, 104, 3 (chap. v)].

Allen's calculations (curves) and experimental data (points) are illustrated in Figs. 1.29–1.31. These data refer to low currents, where interaction effects are negligible. Allen concluded that electron temperatures and drift velocities may be calculated [99, 102] without any adjustable parameters and give excellent agreement with experiment for low

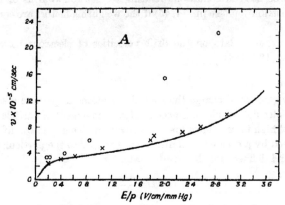

Fig 1.31. Allen's values of electron drift velocities in A.

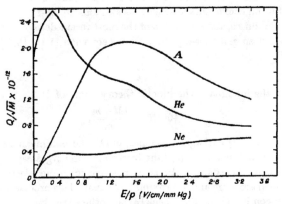

Fig. 1.32   Allen's cross-section data for some of the rare gases

E/p. The measurements in argon reveal an unusual mode of variation of T_e with E/p. Allen also gives curves of $Q/\sqrt{M}$ (Fig. 1.32), where Q is the electron collision cross-section and M the molecular weight. The electron mean free path is given by $\lambda_e = 1/NQ$, if N is the number of molecules/cm.[3] at the given pressure, so that mean free paths, for the limited range of E/p values, may be deduced from Allen's results. Similar data have been given by Townsend and his collaborators [105] It must be emphasized however that the derivation of mean free paths from the simple relation $\lambda_e = 1/NQ$ using values for Q as given, for example, in Figs. 1.1. to 1.8, can only be used in conditions relating to electron swarms of known energy distribution (see pp. 44, 45) if the weighting factors are correctly found.

Recently, new data on the drift velocities of electrons have been published [122, 124].

Diffusion

In any gaseous discharge the spatial variations in particle (electron, ion, or molecule) populations necessitates a movement of particles from regions of high to regions of low concentration. This process, diffusion, is governed by a coefficient of diffusion D proper to the particular conditions and defined, for the steady state, by

$$n = -DA \frac{dN}{dx} dt, \tag{1 24}$$

where n is the net number of particles diffusing in time dt and A is the area across which the flow takes place. N is the particle density.

The diffusion coefficients for electrons and ions are best derived from values of mobilities, and a selection of the most recent data on the latter has already been given (see. p. 31). From equation (1.18) [3, 7], since $p = NkT$,

$$\frac{D}{\bar{k}} = \frac{kT}{e}, \tag{1.25}$$

where $\bar{k}$ is the mobility. The kinetic theory value of D is

$$D = \tfrac{1}{3} 0.815 \lambda c \sqrt{\frac{M+m}{m}}, \tag{1.26}$$

c is the r.m.s. velocity of the gas molecules, and λ is the ionic mean free path.

Ambipolar diffusion is important in certain discharges especially in plasmas where the concentrations of positive ions and electrons (or negative ions) are large and approximately equal. This process has been studied by von Engel and Steenbeck [5] and others [86, 94]. The following is a treatment given by Loeb [3] based on that of von Engel and Steenbeck.

The electrons in a plasma tend to diffuse more slowly, owing to the Coulomb forces, than they would in the absence of positive ions. In fact, the average velocity of the positive ions $\bar{v}_+$ equals that of the negative ions $\bar{v}_-$, and we can therefore write both equal to an average ion velocity v.

The flux of particles of either sign is

$$N_\perp v_\pm = -D_\pm \frac{dN_\pm}{dx} + N_\pm k_\pm E, \qquad (1.27)$$

where E is the field, x is the direction of the concentration gradient, and k is the relevant ionic mobility. It is generally justifiable, at least to the first order, to put $N_+ = N_- = N$ and, eliminating E,

$$\bar{v} = \frac{D_+ k_- + D_- k_+}{k_+ + k_-} \frac{1}{N} \frac{dN}{dx}. \qquad (1.28)$$

By definition of D_a, the ambipolar coefficient of diffusion, we have, for unit area,

$$N\bar{v} = D_a \frac{dN}{dx}. \qquad (1.29)$$

Hence, from equations (1.28) and (1.29),

$$D_a = \frac{D_+ k_- + D_- k_+}{k_+ + k_-}, \qquad (1.30)$$

and, since $\qquad \dfrac{k_+}{D_+} = \dfrac{e}{kT_+}$ and $\dfrac{k_-}{D_-} = \dfrac{e}{kT_-},$

it follows that in cases where $T_e = T_- \gg T_+$ and where $k_e = k_- \gg k_+$ we have

$$D_a = D_+ \frac{T_e}{T_+} = D_- \frac{k_+}{k_e} = D_- \frac{k_+}{k_-}. \qquad (1.31)$$

D_a can then be found from the data given elsewhere in this chapter or in the original papers [123].

Biondi and Brown [57] have studied ambipolar diffusion in He, using a u.h.f resonant-cavity technique. They found, for example, that the product $D_a p$, i e. ambipolar diffusion coefficient × pressure, which should be constant for given conditions of temperature, etc., was 540 cm.²/sec./mm. Hg for an average particle energy of 0·039 eV in a case where the electrons and ions were in equilibrium with the gas, so that

$$D_a = 2 \frac{D_+ k_-}{k_- + k_+} \simeq 2D_+. \qquad (1.32)$$

At constant pressure, the gas density varies inversely as the gas temperature, and therefore $\qquad D_a \propto T^2.$ $\qquad\qquad (1.33)$

This relation was checked at a pressure of 8·5 mm. Hg for T^2 varying from about 9 to 16×10^4 (° K.)², over which range D_a varied from about 65 to 110 cm./sec. Further work has since been carried out [126, 127, 128].

Townsend has shown that the concentration of diffusing particles at a point distant r from a source at time t is

$$n = \left(\frac{n_0}{4Dt}\right)^{\frac{3}{2}} e^{-r^2/4Dt}, \tag{1.34}$$

where n_0 is the concentration at the source. Cobine [94], using the above formula of Townsend, has derived the mean square distance $\bar{r}$ of diffusing particles from the source [129] as

$$\bar{r}^2 = \int_V \frac{nr^2}{N} \, dv, \tag{1 35}$$

where dv is the element of volume at distance r from the source. n is the concentration at dv and N is the total number of diffusing particles in the volume V. It is found, for the 3-dimensional case, that

$$\bar{r}^2 = 6Dt + \text{constant}. \tag{1.36}$$

The factor 6 requires to be replaced by 2 or 4 for the 1- or 2-dimensional cases respectively

Diffusion processes are of importance in streamer discharge and spark-channel studies (see Chapters IV, VI, and X)

Energy distributions

The importance of the distribution of electron energies in discharges is great, because electron collisions are often the main cause of excitation and ionization. Only in certain types of experiment in which mono-energetic beams of electrons are used, for example, in Ramsauer's apparatus [88, 89, 90], can the effects of a finite spread in electron energies be ignored In gaseous discharges the electrons possess an energy spectrum, which if based on the Maxwell distribution determines the electron temperature according to the relation. mean electron energy $= \frac{3}{2}kT_e$, where k is Boltzmann's constant.

More generally [104], if the electron energy distribution is $\rho(\epsilon)$, where ϵ is the electron energy, the mean energy is

$$\bar{\epsilon} = \int_0^\infty \epsilon \rho(\epsilon) \, d\epsilon \Big/ \int_0^\infty \rho(\epsilon) \, d\epsilon. \tag{1 37}$$

The distribution results from the statistical equilibrium between the energies gained in the field and lost at collisions, provided the dimensions of the discharge are sufficiently great and the electrical conditions stable (the effects of transient disturbances and non-isothermal plasmas are discussed on p. 49). For these reasons, it is often convenient, as a special case, to study the positive column, where the axial electric field is sensibly uniform over appreciable distances [104] Early work with probes has shown that the distribution in positive columns is generally Maxwellian. The following treatment summarizes the more important results described by Druyvesteyn and Penning [86].

The Maxwellian distribution is

$$\rho_m(\epsilon) = C_m \epsilon^{\frac{1}{2}} e^{-\epsilon/kT}, \tag{1.38}$$

where C_m is a constant. The distribution for the case of small currents, constant field E, and elastic collisions only, where the electron energy is much greater than the atomic energy, is

$$\rho(\epsilon) = C\epsilon^{\frac{1}{2}} \exp[-3\delta\epsilon^2/2\lambda^2 e^2 E^2]$$

$$= C\epsilon^{\frac{1}{2}} \exp\left[-0.55 \frac{\epsilon^2}{\bar{\epsilon}^2}\right], \tag{1.39}$$

where $\bar{\epsilon} = \sqrt{2}\,\Gamma(\tfrac{5}{4})\lambda e E/\Gamma(\tfrac{3}{4})(3\delta)^{\frac{1}{4}}$ and $\bar{v} = \bar{\epsilon}/\epsilon = 0\,605\lambda_{\mu_0} E/\delta^{\frac{1}{4}}p_0$,

$C = (3\delta/2\lambda^2 e^2 E^2)^{\frac{3}{4}}. 2ne/\Gamma(\tfrac{3}{4})$ and $\delta = 2m/M$.

λ is the electron mean free path, Γ is the gamma function. For a given gas, $\rho = f(E/p_0)$ where p_0 is the pressure reduced to $0°$ C. These two distributions are shown in Fig 1.33. If, as a further refinement, λ is a function of ϵ, and not, as hitherto assumed, independent of ϵ, then

$$\rho(\epsilon) = C\epsilon^{\frac{1}{2}} \exp\left[\frac{3}{-e^2 E^2} \int_0^{\epsilon} \frac{\delta\epsilon}{\lambda^2}\, d\epsilon\right]. \tag{1 39a}$$

This gives a Maxwellian distribution if δ/λ^2 is a constant. If the energy of the electrons is not very large compared with that of the atoms, of temperature T, then for constant λ [135]

$$\rho(\epsilon) = C\epsilon^{\frac{1}{2}} e^{-\epsilon/kT} \left(\epsilon + \frac{\lambda^2 e^2 E^2}{3\delta kT}\right)^{\lambda^2 e^2 E^2/3\delta k^2 T^2}. \tag{1.40}$$

The next step is to introduce inelastic collisions producing excitation and ionization. There are two general cases if excitation only is considered. (1) If the energy loss is small compared with the electron energy (e.g. if the vibrational levels of the ground state only can be excited, as in molecular gases). Then equation (1 39a) applies if

$$\delta\epsilon = \frac{2m\epsilon}{M} + \sum_n K_{v_n}\epsilon_{v_n}, \tag{1.41}$$

where K_{v_n} is the probability of excitation of the nth vibrational level, of energy ϵ_{v_n}. (2) If the energy loss in one exciting collision is large, as with monatomic gases, then equation (1.39a) may be used if $\epsilon < \epsilon_h$, where ϵ_h is the lowest excitation energy. In the case where $\epsilon > \epsilon_h$ approximations

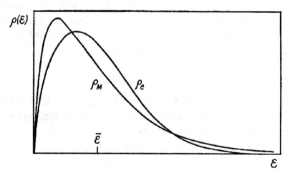

FIG. 1.33   Electron energy distributions ρ_m is Maxwellian and
ρ_e is given by equation (1 39)   The curves are normalized for mean
energy ($\bar{\epsilon}$) and electron concentration.

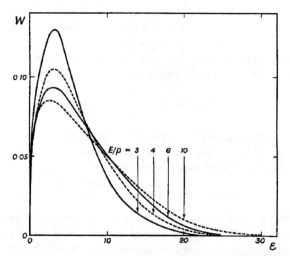

FIG 1 34   Energy distributions of electrons in helium (Smit) for
various values of E/p in V/cm /mm Hg

have to be made and equations have been deduced by Druyvesteyn
[131] for Ne and by Smit [132] for He. Smit included also the effect of
ionizing collisions. Smit's results are given in Fig 1.34 where the dashed
lines correspond to a simplified approximate distribution given by

$$\rho_2(\epsilon) = C_2 \exp\left\{\frac{-[3K_h(\epsilon_2)]^{\frac{1}{2}}\epsilon}{\lambda e E}\right\}, \tag{1.42}$$

where K_h is the experimental value of the probability for excitation at an energy ϵ_2 between the excitation energy ϵ_h and the ionization energy ϵ_i,

$$\epsilon_2 = \epsilon_h + 2(\epsilon_i - \epsilon_h)/3. \tag{1.43}$$

Smit's work has recently been extended, again for He, by Dunlop and Emeléus [217] to higher electron energies, i.e. up to 35 eV.

If, contrary to the assumptions made in deducing the above distributions, the electron and ion concentrations (N_e and N_i) are high and still closely equal, so that the field is still low, it can be shown that electron/ion collisions affect the distribution very little. Electron interaction may be important [133, 134] and in this case an arbitrary initial distribution function becomes almost Maxwellian when an electron has travelled a distance S given by

$$S = C\bar{\epsilon}^2/N_e e^4. \tag{1.44}$$

The values of C given by different authors vary. According to Druyvesteyn and Penning [86] only the effects due to exciting collisions have so far been studied with electron interaction. These authors introduce a quantity B [86], given by

$$B = \frac{3\pi n e^2}{\lambda E^2} \log_e \left[\frac{\epsilon + \frac{1}{2}\bar{\epsilon}}{e^2(\pi n)^{\frac{1}{3}}} \right]. \tag{1.45}$$

When $B > 5$, the interaction is important and the distribution almost Maxwellian, but when $B < 0.1$, the interaction may be neglected. Cahn [133, 134] has given a comprehensive treatment of electron interaction effects.

It is difficult to determine $\bar{\epsilon}$ in cases where inelastic collisions are important. Figs 1.35 and 1.36 show data, partly deduced by calculation from the Smit and Druyvesteyn distributions and partly by experiment [99]. The agreement is as satisfactory as could be expected. If the ion density is large, say $10^9/\mathrm{cm}^3$ or more [3, p. 233], as in a positive column of high current density, the electron interaction is so important that the electron energy distribution is Maxwellian. Then the mean energy can be calculated relatively simply. Mierdel [136] found the results of curve c, Fig. 1.36, for Ne.

Townsend's method for determining $\bar{\epsilon}$ gave results of great value, as may be seen from Figs. 1.35 and 1.36, but is indirect. The probe method of Langmuir and Mott-Smith [137] is the most widely used method for determining the velocity distribution of electrons and ions, although it fails at pressures greater than a few cm. Hg [138, 139]. The basis of this method is that a small wire is inserted in the plasma and the currents (+ or −) flowing to it, as its potential (− or +) relative to the discharge tube anode is altered, are measured. From the observed current-voltage characteristic of the probe certain properties of the plasma may be deduced.

If correctly used [3] the probe method gives ion and electron temperatures, concentrations of electrons and positive ions, and the space

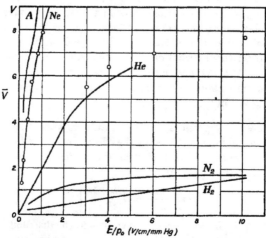

Fig. 1 35. Mean energy ($\bar{V}$) as a function of E/p_0 for different gases at small currents. The curves are according to Townsend, the circles are values calculated by Druyvesteyn (Ne) and Smit (He). p_0 is the gas pressure in mm. Hg corrected to 0° C.

potential. A major assumption is that the distribution is Maxwellian, in which case the probe characteristics are easily interpreted. Druyvesteyn [140], however, showed that the electron velocity distribution, even if not Maxwellian, could be given in terms of probe potential v_s and of probe electron current i_-, in the negative probe voltage region, by

$$\rho(v) = \rho\left(\sqrt{\frac{2eV}{m}}\right) = \frac{4m}{Al^2} V \frac{\partial^2 i}{\partial v_s^2}, \qquad (1.46)$$

where A is the probe area and V is the difference between the space potential and probe potential. Emeléus proposed a convenient method for measuring directly the second derivative (equation 1.46) which was used by Sloane and McGregor [141] and later by van Gorcum [142].

Loeb [3, pp. 251–6] and others [143, 144] discuss in some detail the sources of error that may be encountered in probe measurements, and a good recent review of probe methods is given by Francis and Jenkins [104]. The work of Emeléus and his school is pre-eminent in this field.

It has been indicated in the earlier sections on recombination and photoionization that spectroscopic methods of measuring electron temperatures are available. In cases where an excitation temperature† exists, and where it can reasonably be assumed to equal the electron temperature [145, 146, 147, 148], measurements of the former give

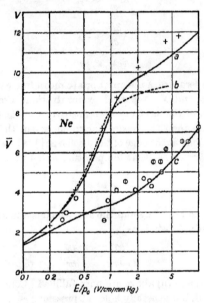

FIG. 1.36. Mean energy $(\bar{V})$ of the electrons as a function of E/p_0 in Ne. a (Druyvesteyn) and b (Allen) calculated for small current. c (Mierdel) calculated for large current $+$ experimental results for small current (Townsend), ○ for large current (⊖ Seeliger and Hirchert, ○ Druyvesteyn, ⊕ Holdt, as cited by Mierdel). p_0 is the gas pressure in mm. Hg corrected to 0° C.

information on the latter (Williams *et al.* [148] give references to the earlier work of Ornstein and his school).

It should be emphasized here that, despite the above references the existence of a single excitation temperature for a wide range of excitation energies in a discharge still appears to be an open question [149, 150, 151, 152].

† The temperature corresponding to the distribution of excited atom populations in the various excited states (see Chapter X).

Ornstein and Brinkman [153] describing conditions in arc discharges, point out that, on the basis of the Lorentz theory of electrons in metals, the departure from a Maxwellian distribution for the particles in a discharge is negligible if

(i)
$$\lambda \frac{\partial T_g}{\partial r} \ll T_g, \tag{1.47}$$

(ii)
$$\lambda \frac{\partial N}{\partial r} \ll N, \tag{1.48}$$

(iii)
$$Ee\lambda_e \ll \tfrac{1}{2}mv^2. \tag{1.49}$$

T_g is the gas temperature, λ the appropriate mean free path, r the radius of the discharge, and N the particle concentration. Consequently the energy variation of the electrons and ions over the electron mean free path is much less than the mean kinetic energy of the electrons. As $\partial T_g/\partial r \sim 10^4 \, °\text{K./cm}$ at the edge of the discharge [153] for $T_g \sim 4{,}000° \, \text{K.}$ and since $\lambda \sim 10^{-4}$ cm. at that temperature and 1 atm., then

$$\lambda \frac{\partial T_g}{\partial r} \sim 1° \ll T_g.$$

For constant pressure throughout the arc, which is a justifiable assumption (see, however, Alfvén [154]),

$$-\frac{1}{T_g} \frac{\partial T_g}{\partial r} = \frac{1}{N} \frac{\partial N}{\partial r}, \tag{1.50}$$

and conditions (i) and (ii) above are both fulfilled Condition (iii) is not so readily fulfilled, but generally holds at pressures ~ 1 atm or greater.

In electrical breakdown processes, it is important to determine the properties of a transient or non-isothermal plasma [155]. Weizel and Rompe [145] have recently analysed the transient equilibria set up in the various stages of spark discharges, and a brief account of their work will now be given.

First consider the electron population. A small disturbance of a Maxwellian distribution, assumed to apply *a priori*, will last only for a time [155]

$$\Delta\tau = \frac{S_e}{v}. \tag{1.51}$$

S_e is the so-called relaxation distance and v is the electron velocity. For typical arc conditions $v \sim 10^{-8}$ cm /sec. and $S_e \sim 10^{-5}$ cm., so that $\Delta\tau \sim 10^{-13}$ sec. Thus, for ordinary electrical conditions, a Maxwellian distribution of electron energies will generally apply.

Consider next the excitation and ionization processes. The variation of ion or electron density N_e is given by

$$\frac{dN_e}{dt} = N_e(Z_1 - Z_2),$$

where Z_1 and Z_2 are the numbers of ionizing and recombination collisions, respectively, per electron per unit time. We also have

$$Z_1 = N_0 v Q_i \quad \text{and} \quad Z_2 = N_e^2 v Q_c,$$

where N_0 is the concentration of neutral particles, Q_i is the ionization cross-section, N_e is the ion density, and Q_c is the recombination cross-section. Weizel and Rompe state that they ignore electron/ion (2-body) recombination, since a state of thermal equilibrium is considered and 3-body recombinations are operative. (This assumption must be stressed as thermal equilibrium need not necessarily apply in rapidly changing conditions.)

Thus
$$\frac{dN_e}{dt} = v N_e N_0 Q_i - v N_e^3 Q_c.$$

If we write $N_e = u^{-\frac{1}{2}}$, then $u - u_\infty = (u_0 - u_\infty)e^{-2vN_0 Q_i t}$, where $u = u_0$ at $t = 0$ and u_∞ is the equilibrium value of u.

The relaxation time is

$$\Delta \tau_i = \frac{1}{2 v N_0 Q_i} = \frac{1}{2Z_1} = \frac{\tau_i}{2}, \tag{1.52}$$

where τ_i is the time elapsing between successive electron collisions. After this time the disturbance has fallen to $1/e$ of its initial value. In ordinary arc plasmas $\Delta \tau_i \sim 10^{-12}$ sec so that the ionization attains its equilibrium almost as quickly as the electron energy distribution.

The population of excited atoms may be studied in a similar manner and

$$\frac{dN_a}{dt} = N_e Z_1 - N_a Z_2,$$

in which now Z_1 is the number of excitation processes caused by electron impact per electron per sec. Z_2 is the number of collisions of the second kind per excited atom per sec. Spontaneous de-excitation is neglected since thermal equilibrium is assumed to be virtually established. Again,

$$Z_1 = N_0 v Q_a \quad \text{and} \quad Z_2 = N_e v Q_a'.N_0$$

where N_a is the concentration of excited atoms, and Q_a and Q_a' are respectively the excitation cross-section and the cross-section for collisions of the second kind. The relaxation time for this process is

$$\Delta \tau_a = \frac{1}{Z_2}.$$

For equilibrium conditions we have

$$Z_2 = N_e Z_1 / N_a, \qquad \Delta\tau_a = \frac{N_a}{N_e Z_1}, \tag{1.53}$$

from which

$$\frac{\Delta\tau_a}{\Delta\tau_i} = \frac{2N_a Q_i}{N_e Q_a}. \tag{1.54}$$

Generally $\quad Q_i \sim \frac{1}{10} Q_a \quad$ and $\quad \dfrac{N_a}{N_e} \sim 10, \quad$ so $\Delta\tau_a \sim \Delta\tau_i.$ (1.55)

It is thus deduced, on the basis of the assumptions made, that the electron, ion, and excited atom populations are generally in equilibrium with each other for all electrical changes likely to be encountered.

It should be repeated that neither radiative recombination nor the spontaneous return of excited atoms to their ground states are considered, since in general these processes are not comparable in importance with 3-body recombination processes or collisions of the second kind   These assumptions require further justification for low pressure discharges.

The interchange of energy among atoms is of importance, and if an atom has kinetic energy greater by $\Delta\epsilon$ than the mean value ϵ it will on the average transfer $\Delta\epsilon/2$ to an equally heavy atom with which it collides. Equilibrium is therefore quickly reached and as an approximation the relaxation time may be taken as λ/v (where λ is the mean free path), i.e. approximately 10^{-8}–10^{-10} sec. For elastic electron-atom collisions, the electron will transfer approximately $2m/M$ of its excess energy, and the corresponding relaxation time $\Delta\tau_0$ is longer than for atom-atom collisions

We have

$$\Delta\tau_0 = \frac{M}{2N_0 \, v Q \sqrt{m}}, \tag{1.56}$$

where Q is the elastic collision cross-section †Since, for example, generally $v \sim 10^8$ cm /sec., $N_0 Q \sim 10^3$, and for Hg $M/m \sim 4 \times 10^5$ it follows that $\Delta\tau_0 \sim 4 \times 10^{-6}$ sec. For the extreme case of H_2, where $M/m \sim 2,000$, $\Delta\tau_0 \sim 2 \times 10^{-8}$ sec. The values deduced in the above manner by Weizel and Rompe for heavy atoms are given in Table 1.17 for 1 atm.

Non-stationary arc discharges (or some spark channels, where conditions tend to approach those of an arc plasma) may be divided roughly into two groups. The first group comprises discharges in which the translation of the atoms achieves equilibrium with the electron population, for example arcs operated at frequencies $< 10^5$ c./s. More rapidly

† The validity of this simple formula depends on the relative values of the electron and gas temperatures at the end of the relaxation period [223] (see p. 4; equation (1.4) should be used to give a more accurate form of equation (1 56)).

TABLE 1.17

Process	Relaxation time in sec
Maxwell distribution for electron energies . . .	$\Delta_\tau \sim 10^{-13}$
Ionization	$\Delta_{\tau_i} \sim 10^{-12}$
Excitation	$\Delta_{\tau_a} \sim 10^{-11}$
Energy transfer from electrons to atoms .	$\Delta_{\tau_0} \sim 10^{-5}-10^{-6}$
Maxwell distribution of atoms .	$\sim 10^{-8}-10^{-10}$

occurring processes (the second group) may prevent this equilibrium. The electron temperature will still tend to govern the translation of electrons and the ionization and excitation processes, whilst the gas temperature (differing now from the electron temperature) determines the mean energy of the atoms. The initial processes of spark discharges and spark channels lasting $\sim 10^{-5}$ sec. or less fall into this category, but even here if the discharge lasts $\gg 10^{-12}$ sec. an electron temperature in the Maxwellian sense still exists. The properties of spark channels are discussed further in Chapter X, but it is necessary to state that very little relevant experimental evidence exists for transient discharges.

A discussion of decaying plasmas is given in Chapter X where further references to time-varying electron and excitation temperatures are given.

Townsend's first ionization coefficient

In his early studies of the variation of current between parallel plane electrodes in a gas as a function of the applied electric field, Townsend [3, 105, 156] observed that, with increasing voltage gradient, the current at first increases as shown diagrammatically in Fig. 1 37 to a nearly constant value i_0 which corresponds to the photocurrent produced at the cathode, for example, by external ultra-violet illumination. At higher voltage gradients, in the region bc of Fig. 1 37, the current increases above the value i_0 at a rate which increases rapidly with the gradient. This amplification of the current was ascribed by Townsend to the ionization of the gas by electron collision. Photoelectrons leaving the cathode gain sufficient energy in the applied field to cause ionization when they collide with the gas atoms or molecules. The additional electrons so formed also gain energy in the field and produce further electrons.

Townsend introduced a coefficient α to define the number of electrons produced in the path of a single electron travelling a distance of 1 cm.

in the direction of the field. The current i flowing in the gap is then given by

$$i = i_0 e^{\alpha d}, \tag{1.57}$$

where d cm. is the gap length.

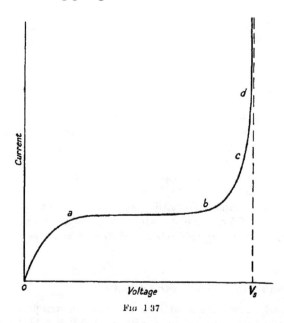

Fig 1 37

Some investigators [86] have introduced a coefficient η in place of α. The coefficient η denotes the number of electrons produced per volt potential difference passed through by an electron Consequently

$$\alpha = \eta E,$$

where E is the voltage gradient in the gap in V/cm., and equation (1.57) becomes

$$i = i_0 e^{\eta V},$$

where V volts is the potential difference across the gap. While there are certain advantages to be gained by the adoption of the coefficient η rather than α the latter coefficient is now so widely used that it will be adopted in general throughout this book.

In order to determine α the current i flowing between two parallel plane electrodes is measured as a function of the voltage gradient. Then the value of α for any particular value of E can be calculated from equation (1.57) Full details of the technique used in the measurements is given by Loeb [3]. For accurate results the primary photocurrent i_0 must be adjusted to a sufficiently low value so that it does not exceed more than about 10^{-14} A/cm.^{2}, otherwise space charge distortion of the electric field in the gap occurs and modifies the resultant value of α [158].

Numerous measurements of α in various gases have been carried out by Townsend and his colleagues from 1900 onwards, and subsequently by other workers The investigations show that α/p is a function of E/p, in accordance with theoretical considerations (see p. 60). Some of the more recent measurements for a number of gases are listed below:

Air	Paavola [159], Sanders [160], Masch [161], Llewellyn Jones and Parker [220].
Nitrogen	Ayres [162], Masch [161], Posin [163], Bowls [164].
Hydrogen	Ayres [162], Hale [165], Llewellyn Jones [166]
Argon, neon, and argon-neon mixtures	Ayres [162], Townsend and McCallum [167], Kruithof and Penning [168], Glotov [169], Huxford [170], Engstrom and Huxford [171].
Krypton, xenon	Kruithof [172].
Benzene, toluene, etc	Badareu and Valeriu [173, 221]
Mercury vapour	Badareu and Bratescu [174].
Sulphur hexafluoride	Hochberg and Sandberg [175]

A full discussion of the results and of the various effects observed in the different gases has been given by Loeb [3], and only a brief account of some of the more notable features is attempted here.

Fig. 1.38 shows some curves obtained by Kruithof and Penning [168] for neon and neon-argon mixtures, together with a curve for neon by Townsend and McCallum [167]. The values for α in pure neon or pure argon are markedly less than in neon containing a small admixture of argon. This effect may be explained by the ionization of argon atoms (ionization potential 15·8 V) by metastable neon atoms (excitation potential 16·53 V). The probability for this process is high so that slight quantities of argon present in the neon have considerable influence. The effect is so pronounced that clearly in all measurements of α care must be taken to define precisely the gas conditions.

Curves of α/p in nitrogen, as obtained by Bowls [164], are given in Fig. 1 39. These curves again show the large effect caused by impurities such as mercury vapour or sodium in the gas. The curve labelled Pt is that obtained with platinum electrodes in pure nitrogen. The curve labelled Hg is that obtained when mercury vapour is present, at room temperature. The curve labelled Na is that with sodium present. At the higher values of E/p the value of α/p is increased by about 25 per cent. by the presence of sodium and about 15 per cent by the mercury. Such results may again be explained by considerations similar to those for the

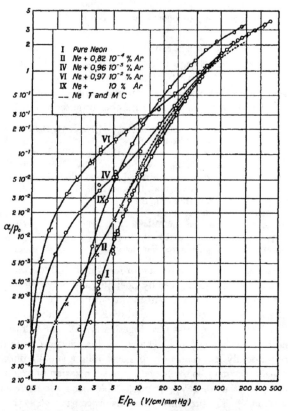

FIG 1 38. Values of α/p_0 in pure neon, and in mixtures of neon and argon, as a function of E/p_0 p_0 is the gas pressure in mm. Hg corrected to 0° C.

neon-argon mixtures, and Kaplan [176] has shown that there are many excited metastable states in nitrogen that could ionize mercury.

Observations by Hale [165] for hydrogen confirm that the presence of sodium in the discharge chamber can affect the measured values of α. In Fig. 1.40 the curve marked Pt is that obtained for pure hydrogen with clean platinum electrodes. The curve marked Na corresponds to

hydrogen with sodium present Curve T obtained in Townsend's laboratory crosses Hale's Pt curve at $E/p \sim 300$. According to Loeb [3], the
influence of sodium is caused by a volatile sodium compound.

In the later studies by Hale [165] α has been measured in hydrogen for
cathodes of nickel and aluminium. The results obtained are the same
as those for a platinum cathode. The variation of α/p with E/p has been

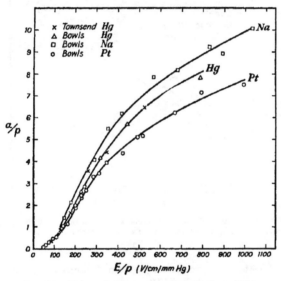

Fig. 1.39. Values of α/p in nitrogen as a function of E/p

investigated for higher values of E/p than those used previously, and it
is found that α/p declines with increasing E/p for $900 < E/p < 1400$,
but that for $E/p > 1400$ α/p increases again. The decline in α/p is
attributed by Hale [165] to a decrease in the cross-section for ionization
by electrons at higher electron energies. The subsequent rise in α/p is
attributed to the failure of electrons to attain their terminal energy
in the gap at the gas pressures used, the probability of ionization
being greater at energies below the maximum terminal energy under
these conditions, but the physical mechanisms involved are not
apparent.

The various results obtained by Kruithof and Penning [168], Bowls
[164], and Hale [165], drawing attention to the important influence of

impurities on the value of α, make it apparent that in previous investigations the measured values of α may not be correctly applicable to the pure gas. In many of the earlier studies mercury vapour is a probable contaminant because of the use of mercury diffusion pumps with possibly inadequate cold traps.

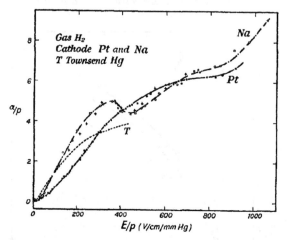

FIG 1.40 Values of α/p as a function of E/p Two sodium surfaces were used and the corresponding results are shown as circles and crosses

Tables 1 18 to 1.21 give recently obtained values for α/p in various gases. E is in V/cm., p in mm. Hg.

TABLE 1.18

Values of α/p in Pure Hydrogen (Hale [165])

E/p	α/p	E/p	α/p	E/p	α/p	E/p	α/p
22 7	0 041	130 4	0 995	252 6	2 757	452 0	4 750
24 0	0 040	136 2	1 085	271 4	2 928	481 0	4 870
30 6	0 123	144 7	1 119	277 9	3 071	516 0	5 210
32 3	0 129	146 0	1 220	299 0	3 200	536 0	5 290
45 5	0 172	153 8	1 157	306 4	3 268	600 0	5 810
52 3	0 238	164 7	1 358	328 4	3 820	663 5	6 050
61 9	0 334	175 8	1 690	337 4	3·870	694 5	6 120
68 6	0 392	209 6	2 056	362 0	4 010	787·8	6 260
74 7	0 444	219 7	2 245	392 6	4 290	833 5	6 180
85 0	0 578	228 0	2 380	424 0	4 463	867 0	6 300
100·0	0 785	240 0	2 540	448 0	4 736	916 7	6 700

TABLE 1.19

Values of α/p in Pure Nitrogen (Bowls [164]) and in Mercury-contaminated Nitrogen (Posin [163])

E/p	α/p Bowls	α/p Posin	E/p	α/p Bowls	α/p Posin
20	.	0 000087	195	1 898	.
26	.	0 000258	196		2 522
30		0 000911	215	2 215	3 07
35		0 00305	250	2 68	3 50
40	.	0 00734	290	3 378	
45	.	0 01353	320	3 58	
59	0 1189	0 0934	350	3 96	4 93
65	0 2128		440	4 372	
78	0 297	0 2830	450		6 00
94	0 4047		500	5 163	6 23
100		0 7000	660	6 203	
115	0 612		750		8 78
120		0 9500	800	7 19	
140	0 961	1 400	1,000	7 58	10 8
160	1 196		.	.	
166		2 020	..	..	..

TABLE 1.20

Values of α/p in Mercury-contaminated Air (Sanders [160]; Masch [161])

E/p	α/p Sanders	α/p Masch	E/p	α/p Sanders	α/p Masch
20	0 00003	.	50	0 0554	0 057
22	0 00005		60	0 127	0 130
24	0 00013	.	70	0 224	0 235
26	0 00023		80	0 340	0 365
28	0 00043	.	90	0 491	0 51
30	0 00091		100	0 637	0 68
31	0 00136	0 00152	110	0 806	0 85
32	0 00201	0 00204	120	1 007	1 05
33	0 00305	0 00309	130	1 236	1 23
34	0 00459	0 0044	140	1 477	1·40
35	0 00605	0 0059	150	1 602	1 60
36	0 00820	0 0076	160	1 758	1 83
40	0 0167	0 0168			

TABLE 1·21

Values of α/p in the Rare Gases (Druyvesteyn and Penning [157])

E/p	He	Ne	A	Kr	Xe	Ne+0·01% A
0·5					•	0·00075
0·7	•				•	0·0062
1·0	•			•		0·0194
1·5					•	0·042
2·0		0·00062		•	•	0·061
2·5		0·00115		•	•	0·079
3·0		0·00192		•		0·094
4		0·0044				0·122
5		0·0080	0·00025	0·00015		0·145
6		0·0132	0·00054	0·00036	0·00006	0·167
8		0·0290	0·00208	0·00144	0·00016	0·204
10		0·050	0·0057	0·0041	0·0006	0·240
12		0·074	0·0132	0·0095	0·0017	0·273
15	0·121	0·115	0·0315	0·024	0·0054	0·320
20	0·190	0·192	0·076	0·068	0·022	0·402
25	0·256	0·275	0·14	0·135	0·0525	0·482
30	0·330	0·360	0·225	0·213	0·102	0·564
40	0·464	0·540	0·432	0·416	0·248	0·724
50	0·590	0·710	0·675	0·660	0·440	0·875
70	0·812	1·03	1·195	1·195	0·925	1·160
100	1·10	1·49	1·98	2·04	1·80	1·53
150		2·10	3·26	3·48	3·33	2·19
200	..	2·60	4·44	4·84	4·94	2·68
300		3·21	6·39	7·08	7·92	3·33
500	•		9·05	10·45	12·50	..
1,000	•		12·3	14·7	19·8	•
2,000	•		16·8		25·4	

An approximate theoretical expression for the relation between α, the electric field E, and the gas pressure p has been deduced by Townsend [105] from consideration of the following simple assumptions. After an electron has travelled a free path λ in the direction of the field E, the energy gained is given by $Ee\lambda$. If $Ee\lambda \geqslant eV_i$, where V_i is the ionizing potential of the gas, the electron will ionize the gas. The chance that an electron will ionize is then governed by the probability of occurrence of free paths $\lambda \geqslant \lambda_i = V_i/E$.

From kinetic theory, the number n of electrons that have free paths of length greater than i is given by

$$n = n_0 \exp\left[-\frac{\lambda_i}{\lambda_m}\right] = n_0 \exp\left[-\frac{V_i}{\lambda_m E}\right], \qquad (1.58)$$

where n_0 is the number of electrons starting out on free paths and λ_m is the mean free path.

The number α of ionizing collisions per cm. of path, which is equal to the number of free paths multiplied by the chance of a free path being of more than ionizing length λ_i, can then be expressed as

$$\alpha = \frac{1}{\lambda_m} \exp\left[-\frac{V_i}{\lambda_m E}\right]. \tag{1 59}$$

As $1/\lambda_m = Ap$, where A is a constant, (1.59) can be written

$$\frac{\alpha}{p} = A \exp\left[-\frac{AV_i p}{E}\right],$$

or

$$\frac{\alpha}{p} = A \exp\left[-\frac{Bp}{E}\right]. \tag{1.60}$$

Consequently, $\alpha/p = f(E/p)$, as found experimentally.

Equation (1 60) has been derived from a number of simplifying assumptions which are known to be inexact, as, for instance, that the probability of ionization is zero for energies less than eV_i and is unity for energies greater than eV_i. However, by judicious choice of the constants A and B, equation (1.60) may be made to give reasonable agreement with experiment over a range of values of E/p. Appropriate values of A and B, as given by von Engel and Steenbeck [5], together with the range of E/p for which equation (1.60) is then valid, are listed in Table 1.22.

TABLE 1.22

Gas	A	B	Range of E/p
Air . . .	14 6	365	150–160
Argon .	13 6	235	100–600
Carbon dioxide	20 0	466	500–1,000
Hydrogen	5 0	130	150–400
Helium . .	2 8	34	20–150

A more exact theoretical calculation of α must include consideration of the probability $P(V)$ that an electron with energy V will ionize a molecule by collision in unit length of random path at unit pressure, and of the energy distribution function $F(V)$, where $F(V) \, dV$ is the fraction of electrons per unit volume with energy of random motion between V and $V+dV$. Then

$$\frac{\alpha}{p} = W^{-1} \int_I^\infty v F(V) P(V) \, dV, \tag{1.61}$$

where W is the mean drift velocity of the electrons at unit pressure in the field E, v is the speed of an electron with energy V, and I is the ionization energy of the molecule. The mean value of V is given by

$$V_a = \int_0^\infty V F(V)\, dV. \qquad (1.62)$$

α/p, W, $P(V)$, I, and V_a can be measured experimentally.

From a comparison of the experimental values of α/p with those predicted from equation (1 61), for various forms of $F(V)$, Eméleus, Lunt, and Meek [177] have shown that reasonable agreement is obtained in air, nitrogen, hydrogen, and oxygen when $F(V)$ is assumed to have the Maxwellian form

$$c V^{\frac{1}{2}} \exp\left(-\frac{3V}{2\bar{V}_a}\right), \qquad (1.63\,\mathrm{a})$$

where c is a constant. In a subsequent theoretical analysis by Deas and Eméleus [157] calculations have also been made to determine the value of α/p when $F(V)$ is assumed to correspond to the case of elastic collisions between electrons and molecules, namely

$$C V^{\frac{1}{2}} \exp\left[-\frac{0.55 V^2}{V_a^2}\right] \qquad (1.63\,\mathrm{b})$$

where C is a constant.

If it is assumed that

$$P(V) = b(V-I), \qquad (1.64)$$

where b is a constant, and that $F(V)$ is Maxwellian, as given by equation (1.63 a), then equation (1.61) for α/p becomes

$$\left(\frac{\alpha}{p}\right)_{\mathrm{max}} = 5.48 \times 10^7 b\, W^{-1} V_a^{\frac{1}{2}}\left(1+\frac{4V_a}{3}\right)\exp\left(-\frac{3I}{2V_m}\right). \qquad (1.65)$$

Alternatively, if $F(V)$ corresponds to equation (1.63 b) for elastic collisions

$$\left[\frac{\alpha}{p}\right]_{\mathrm{el}} = 6.72 \times 10^7 b\, W^{-1} V_a^{\frac{1}{2}}\left[1-2\pi^{-\frac{1}{2}}\int_0^{y_i} \exp(-y^2)\, dy\right], \qquad (1.66)$$

where

$$y_i = \left(\frac{I}{V_a}\right) 0.55^{\frac{1}{2}}.$$

A comparison between the results for α/p calculated from equations (1 65) and (1.66) and Sanders's experimental values in air [160] is given in Table 1.23. For the lower values of E/p the results given by $(\alpha/p)_{\mathrm{el}}$ diverge appreciably more from the experimental values than those given by $(\alpha/p)_{\mathrm{max}}$, but for larger values of E/p the ratio of the two calculated values becomes progressively less. Consequently, the choice of the form of $F(V)$ becomes less critical for large values of E/p than for small values. A similar tendency is recorded also for hydrogen and nitrogen, as shown in Tables 1 24 and 1.25, where the experimental values given are those by Hale [165], Posin [163], and Bowls [164].

TABLE 1.23

Values of α/p for Air

E/p	V_a	(α/p) (Sanders)	$(\alpha/p)_{max}$	$(\alpha/p)_{el}$
20	2	$3\ 4 \times 10^{-5}$	$2\ 2 \times 10^{-4}$	$6\ 7 \times 10^{-17}$
36	3	$8\ 2 \times 10^{-3}$	$1\ 1 \times 10^{-2}$	$8\ 6 \times 10^{-8}$
55	4	$9\ 0 \times 10^{-2}$	$7\ 4 \times 10^{-2}$	$1\ 5 \times 10^{-4}$
75	5	$2\ 8 \times 10^{-1}$	$2\ 4 \times 10^{-1}$	$5\ 4 \times 10^{-3}$
100	6	$6\ 4 \times 10^{-1}$	$5\ 3 \times 10^{-1}$	$4\ 1 \times 10^{-2}$
130	7	$1\ 24$	$9\ 2 \times 10^{-1}$	$1\ 5 \times 10^{-1}$
163	8	$1\ 65$	$1\ 37$	$3\ 4 \times 10^{-1}$
239	10		$2\ 45$	$1\ 04$
455	14		$4\ 91$	$3\ 06$
595	16		$6\ 14$	$4\ 19$

TABLE 1 24

Values of α/p for Hydrogen

E/p	V_a	(α/p) (Hale)	$(\alpha/p)_{max}$	$(\alpha/p)_{el}$
13	2		$4\ 6 \times 10^{-4}$	$1\ 04 \times 10^{-16}$
31 5	4	$8\ 5 \times 10^{-2}$	$1\ 1 \times 10^{-1}$	$2\ 3 \times 10^{-4}$
57 5	6	$2\ 8 \times 10^{-1}$	$4\ 9 \times 10^{-1}$	$4\ 5 \times 10^{-2}$
88	8	$6\ 25 \times 10^{-1}$	$9\ 6 \times 10^{-1}$	$2\ 6 \times 10^{-1}$
123	10	$9\ 5 \times 10^{-1}$	$1\ 42$	$6\ 3 \times 10^{-1}$
203	14	$2\ 00$	$2\ 22$	$1\ 42$

TABLE 1 25

Values of α/p for Nitrogen

E/p	V_a	(α/p) (Posin)	(α/p) (Bowls)	$(\alpha/p)_{max}$	$(\alpha/p)_{el}$
20	2 2	$8\ 7 \times 10^{-5}$	.	$1\ 1 \times 10^{-3}$	$3\ 9 \times 10^{-13}$
30	2 67	$9\ 1 \times 10^{-4}$		$6\ 2 \times 10^{-3}$	$4\ 2 \times 10^{-8}$
50	4 0	$3\ 3 \times 10^{-2}$		$1\ 1 \times 10^{-1}$	$3\ 3 \times 10^{-4}$
86	6 0	$3\ 8 \times 10^{-1}$	$3\ 3 \times 10^{-1}$	$7\ 5 \times 10^{-1}$	$6\ 9 \times 10^{-2}$
106	7 0	$7\ 3 \times 10^{-1}$	$5\ 1 \times 10^{-1}$	$1\ 26$	$2\ 3 \times 10^{-1}$
160	9 0	$1\ 79$	$1\ 20$	$2\ 56$	$9\ 3 \times 10^{-1}$

From the above results in air, hydrogen, and nitrogen, Eméleus, Lunt, and Meek [177] conclude that the experimental values of α/p in these three gases are compatible with the assumption that the energy distribution function $F(V)$ has the Maxwellian form, but that discrimination between various possible forms of $F(V)$ becomes less certain at the higher values of E/p.

Calculations for argon and neon give less satisfactory agreement with experiment, the discrepancy being between 2 and 4 orders of magnitude for the Maxwellian form of $F(V)$. Evidently in these gases the rate of diminution of $F(V)$ with increase in V above V_a is extremely rapid. Calculations have been made therefore by Deas and Eméleus[157] on the assumption that $F(V)$ has the form

$$A_n V^{\frac{1}{2}} \exp[-R_n V]^n \qquad (1.67)$$

with n ranging between 3 and 10, and it is found that agreement between theory and experiment, for α/p in neon, can only be obtained by taking a particular value of n for each value of E/p. Closer agreement with experiment in the case of neon has been obtained in a theoretical analysis by Druyvesteyn [131] who has calculated α for values of E/p in the range 5 to 30, from consideration of a velocity distribution of electrons, taking account of elastic, exciting, and ionizing collisions, and assuming that both the excitation and ionization probabilities are linear functions of the electron energy. Druyvesteyn's results are given in Table 1.26, where they are compared with the experimental values obtained by Townsend and McCallum [167] and by Glotov [169]. The theory has been further extended by Druyvesteyn and Penning [86] who give results also for argon in reasonable agreement with experiment.

TABLE 1.26

Values of α/p in Neon

E/p	$(\alpha/p)_{exp}$ (*Townsend and McCallum*)	$(\alpha/p)_{exp}$ (*Glotov*)	$(\alpha/p)_{calc}$ (*Druyvesteyn*)	$(\alpha/p)_{calc}$ (*Glotov*)
7 5	0 030	0 031	0 029	0 024
10	0 058	0 060	0 066	0 053
20	0 220	0 238	0 23	0 22
30	0 39	0 41	0 41	0 40

A calculation of α/p in helium has been made by Dunlop [178] using equation (1.61) with results as given in Table 1.27. The distribution function $F(V)$ is assumed to be of the form given by Smit [132]. The values adopted for W are close to those calculated by Smit [86] but are rather less than those obtained directly or by extrapolation from the experiments of Bradbury and Nielsen [96, 97].

TABLE 1.27

E/p	3	4	5	6	8	10
W	17 5	23·5	30 2	37 5	52 0	68 0
α/p	7×10^{-5}	10×10^{-3}	$4 2 \times 10^{-3}$	$7 7 \times 10^{-3}$	$2·4 \times 10^{-2}$	$4 6 \times 10^{-2}$

A theoretical evaluation of α in pure neon has also been made by Glotov [179] using an expression derived by Moralev [180], namely

$$\frac{\alpha}{p} = \frac{u_t p}{0.44\lambda_0^2 E} P_t, \tag{1.68}$$

where u_t is the energy corresponding to the most probable electron velocity in volts, λ_0 is the mean free path of an electron for a neon pressure of 1 mm. Hg and is assumed to be 0·08 cm., and P_t is the probability of ionization of a neon atom in a collision u_t is a function of E/p and can be determined if the probability of excitation of neon atoms is known. In Glotov's calculations u_t is computed from consideration of the probability of excitation of the first metastable level of neon. P_t is computed from a formula given by Kollath [181]. Resultant values for α/p as calculated for various values of E/p from equation (1.68) are given in Table 1.26.

Glotov [179] has extended his calculations to the case of neon-argon mixtures and has computed α on the assumption that it contains three components corresponding respectively to the normal ionization coefficient in neon, the normal ionization coefficient in argon, and an effective ionization coefficient determined by the ionization of argon atoms by metastable neon atoms. Reasonable agreement is again obtained between theory and experiment, the maximum discrepancy amounting to 25 per cent.

Another derivation for α/p, taking into account the energy loss of electrons on impact, and on the assumption of a Maxwellian distribution for electron velocities and an ionization probability function of the form given in equation (1.64), has been made by von Engel and Steenbeck [5] who show that

$$\frac{\alpha}{p} = \frac{600aV_i}{2^{\frac{1}{2}}\pi^{\frac{1}{2}}f^{\frac{1}{2}}}\left(1 + \frac{eV_i}{2kT_e}\right)\exp\left[-\frac{2^{\frac{1}{2}}}{\pi^{\frac{1}{2}}}\frac{f^{\frac{1}{2}}V_i}{(E/p)\lambda_{e0}}\right], \tag{1.69}$$

where a is related to the electron mean free path λ_e and to the quantity b of equation (1.64) by the equation $a\lambda_e = b$. f is the fractional loss of energy on electron collision, T_e is the absolute temperature of the electrons, and λ_{e0} is the electron mean free path at 1 mm. Hg. At high values of E/p in some gases $\frac{eV_i}{2kT_e} \ll 1$ and equation (1.69) then reduces to the Townsend form for α/p as given in equation (1.60). According to Loeb [3], this condition obtains in air for $E/p > 200$ only.

In uniform fields the magnitude of the current flowing, when amplified by electron multiplication alone is given by

$$i = i_0 \exp[\alpha d]. \qquad (1\ 70)$$

In non-uniform fields it is usual to rewrite this expression in the form

$$i = i_0 \exp\left[\int_0^d \alpha \, dx \right] \qquad (1.71)$$

and, if the field distribution is known, the value of the integral may be calculated on the assumption that the value of α at any point in the gap, where the gradient is E, is the same as that determined for the same gradient E in a uniform field. This procedure is frequently adopted in theoretical analyses of non-uniform fields, as for instance between co-axial cylinders. However, experiments by Fisher and Weissler [182], Morton [183], and Johnson [184] have shown that the above method leads to appreciable inaccuracies in the case of highly divergent fields. In studies of the current flowing in hydrogen at near atmospheric pressure between confocal paraboloids Fisher and Weissler [182] observed that the multiplication was higher than that expected from the Townsend equation when the field changed by more than 2 per cent. over an electron mean free path. A similar result was obtained by Morton [183] who investigated currents in hydrogen, at pressures below 5 mm Hg, between coaxial cylinders. The reason for these observations is attributed to the fact that in a highly divergent field the electrons do not reach their terminal energy distribution. Morton has also shown, in a detailed theoretical treatment of the problem, that the ionization does not depend on the field at which the ionizing impact occurs but that the field influences the result only in that it affects the energy distribution. In later investigations by Johnson [184] the equivalent value of α, denoted by α', in a divergent field has been determined for both air and hydrogen, and the results are given in Tables 1.28 and 1.29 where they are compared with the normal value of α corresponding to a uniform field in nitrogen and hydrogen. This work is of interest in the study of corona discharges in non-uniform fields (see Chapter III).

A microwave method for measuring the average energy of electrons in a gas in the presence of a D.C. field has been developed by Varnerin and Brown [185] who have determined the ratio between the electron diffusion coefficient D and the mobility μ. By solving for the A.C. and D.C. distribution functions it is possible to compare quantitatively the effective A.C. ionization coefficient ζ (see p. 385) and the Townsend coefficient η in hydrogen. Values of η as calculated from the observed

TABLE 1.28

Comparison of values of α'/p (the apparent α/p) for air with Bowls's values of α/p for nitrogen [164]

E/p	α'/p	α/p
60	0 127	0 17
80	0 33	0 28
100	0 60	0 46
200	2 53	2 00
400	5 15	4 35
600	9 8	5 80
800	12 2	7 2
1,000	12 0	7 6

TABLE 1 29

Comparison of values of α'/p (the apparent α/p) for hydrogen with Hale's values of α/p for hydrogen [165]

E/p	α'/p	α/p
30	0 057	0 080
45	0 266	0·170
60	0 52	0 31
100	1 06	0 71
200	3 60	1 95
400	6 00	4 30
600	6 00	5 80

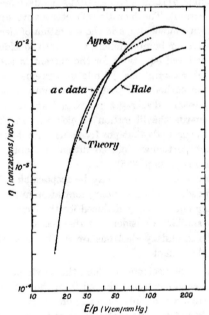

Fig 1.41. Comparison of theoretical and experimental values for η in hydrogen.

values of D/μ and ζ are plotted as a smooth curve in Fig. 1.41, which includes also the curves given by Ayres [162] and Hale [165]. The dotted curve is that estimated from theoretical considerations of the average energy as calculated from the distribution functions.

Townsend's second ionization coefficient γ

The electrons leaving the cathode in a discharge gap cause ionization of the gas molecules according to the Townsend α-mechanism. Other inelastic collisions also occur in which (1) molecules are excited to a level from which they decay with the emission of one or more photons, or (2) molecules are excited to a metastable level. Consequently, positive ions, photons, and metastable atoms are available in the gas to act as agents for the production of secondary emission of electrons either in the gas or at the cathode.

In his early measurements of the current flowing in parallel plate gaps at low pressures Townsend [3, 105, 156] observed that if the voltage was raised sufficiently the current increased at a more rapid rate than that given by equation (1.57) and therefore that secondary mechanisms must be affecting the current. The liberation of electrons in the gas by collisions of positive ions with molecules, and the liberation of electrons from the cathode by positive ion bombardment, were both considered by Townsend [105], who deduced equations for the current in the self-sustained discharge on such assumptions. While in agreement that secondary emission from the cathode by positive ion bombardment is a major factor in low pressure discharges many investigators disagree with Townsend as regards the liberation of electrons in the gas by this mechanism, on the grounds that positive ions cannot acquire sufficient energy under the particular field conditions to ionize gas atoms on collision [3, 186, 187] (see p 285)

Although various mechanisms may be responsible, either singly or jointly, for the production of secondary ionization in the discharge gap, the corresponding equations, as deduced for the current flowing in the gap, are closely similar. Consider first the case for a maintained discharge where the secondary electrons are produced at the cathode by positive ion bombardment.

Let n = number of electrons reaching the anode per sec.,

$\quad n_0$ = number of electrons released from the cathode by external means, e.g. by ultraviolet illumination,

$\quad n_+$ = number of electrons released from the cathode by positive ion bombardment,

$\quad \gamma$ = number of electrons released from the cathode per incident positive ion.

Then
$$n = (n_0 + n_+)e^{\alpha d},$$
$$n_+ = \gamma[n - (n_0 + n_+)].$$

Eliminating n_+, $n = n_0 \dfrac{e^{\alpha d}}{1 - \gamma(e^{\alpha d} - 1)}$

or $i = i_0 \dfrac{e^{\alpha d}}{1 - \gamma(e^{\alpha d} - 1)}.$ (1.72)

Clearly, for $\gamma = 0$, equation (1.72) reduces to the form given by equation (1.57).

In his original derivation, Townsend [3, 105, 156] introduced a coefficient β to represent the number of ion pairs produced by a positive ion in travelling 1 cm. in the direction of the field. It may then be shown that

$$ i = i_0 \frac{(\alpha - \beta)e^{(\alpha - \beta)d}}{\alpha - \beta e^{(\alpha - \beta)d}}. $$ (1.73)

A similar expression for i can also be deduced for the case where secondary emission from the cathode is caused by photon impact [3]. Then

$$ i = i_0 \frac{\alpha e^{\alpha d}}{\alpha - \theta \eta g e^{(\alpha - \mu)d}}, $$ (1.74)

where θ is the number of photons produced by an electron per cm. in the field direction, η is the fraction of the photons which produce electrons that succeed in leaving the cathode, g is a geometrical factor giving the fraction of photons created in the gas that reach the cathode, μ is the absorption coefficient of the photons in the gas.

Because of the similarity between the equations (1.72), (1.73), and (1.74), it is now usual in the consideration of the Townsend theory of the spark, to express the various secondary ionization effects by the single ionization coefficient γ, denoting the number of secondary electrons per positive ion, with the realization that this may represent one or more of several mechanisms [3].

Determination of the value of γ as a function of E/p can be made from equation (1.72) by measurement of the current in the gap for various gas pressures, field strengths, and gap lengths. In a number of investigations [3, 168, 188, 189, 190] γ is calculated from equation (1.72) from a knowledge of the breakdown voltage and curves of α/p as a function of E/p. It has been shown by various investigators, e.g. Huxford [170] for caesium cathodes in argon and Engstrom [189] for barium electrodes in argon, that the values so obtained are in agreement with those determined from direct experiment in conditions where the Townsend mechanism is known to be operative. As in the measurements of α/p care must be taken to use only low current densities in order to avoid space-charge distortion of the field and consequent falsification of the results [158].

Some curves showing the variation of γ with E/p in several rare gases [86], are given in Fig. 1.42. Curves for several cathode materials in argon, determined by several investigators [86, 168, 188, 191, 192] are summarized in Fig. 1.43. With a given cathode metal the value of γ

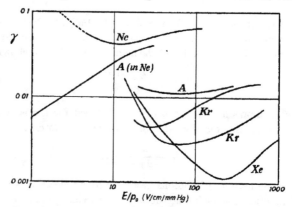

FIG. 1.42. Values of γ for copper in the rare gases. p_0 is the gas pressure in mm Hg reduced to $0°$ C.

depends markedly on the surface condition, as illustrated by the two curves for copper in Fig. 1.43, the lower curve being recorded after treatment of the cathode by a glow discharge, and also by the two curves for krypton given in Fig. 1.42.

Certain conclusions are drawn by Druyvesteyn and Penning [86] for the results in the rare gases, though it is pointed out that knowledge is still inexact on this subject. Direct experiments in vacuum show that helium ions [193, 194], neon ions [196], and caesium ions [195] are able to liberate electrons from metals, the efficiency increasing with increasing ionization potential of the gas and with decreasing work function of the cathode [86]. If γ_p denotes that part of γ which is produced by positive ion bombardment of the cathode, then the value of γ_p found in the above experiments for neon and argon is of the same order of magnitude as the value of γ in Fig 1.42, and it is therefore concluded by Druyvesteyn and Penning [86] that the secondary mechanism is caused principally in these gases, for the particular conditions investigated, by the liberation of electrons from the cathode by positive ion bombardment.

Druyvesteyn and Penning [86] point out that at high values of E/p

γ should have the same value as in vacuum ($p \to 0$), but that it should decrease gradually with decreasing E/p as a consequence of the increased loss of electrons by diffusion back to the cathode. This is observed in the various curves obtained, as for instance in neon at $E/p > 20$. At lower values of E/p, γ increases again, and this is attributed by Kruithof and

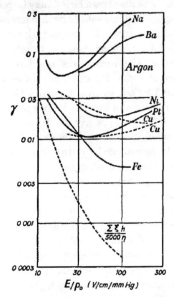

Fig 1.43 Values of γ for A with different cathode materials, according to Kruithof and Penning (Cu), Penning and Addink (Fe), Schofer (Ni and Ba), Ehrenkranz (Pt and Na) The proportion of the number of excitations to the number of ionizations is roughly equal to $Z\xi_h/\eta$ p_0 is the gas pressure in mm Hg reduced to 0° C

Penning [197] to the liberation of electrons from the cathode by photons or by metastable atoms. The importance of photons in causing the emission of electrons from the cathode in discharges in the rare gases is also stressed by Kenty [198]

Studies of the relative influence of positive ions and of metastables in a discharge in argon, with activated caesium cathodes have been made by Huxford [170], and later by Engstrom and Huxford [171]. The separation of the various factors governing secondary emission processes is not readily achieved by study of the static discharge, and Engstrom and Huxford [171] have proposed a new approach to the problem by the study of the time lags in transient discharges Their original analysis has been extended by Newton [199] who has assumed that secondary emission may be caused at the cathode by the impact of photons, ions, and metastable atoms, and has shown how it may be possible to separate the effects of these agents experimentally by measurement of the growth of current in transient discharges. The method is based on the differences in the times required for photons, ions, and metastables to produce emission at the cathode. If a discharge is considered in a gap of 1 cm., in a gas at a pressure of 1 mm. Hg and a field of 50 V/cm., an electron takes a time of about 10^{-8} sec to cross the gap The time taken for an excited molecule to radiate and for the photons to reach the cathode is also of the order of 10^{-8} sec A positive ion drifting under the influence of the field takes about 10^{-6} sec. to reach the cathode, while a metastable, which can only reach the cathode by diffusion, takes a time of the order of 10^{-3} sec Therefore transient discharges should exhibit three relatively distinct stages in their growth, governed by these differences in time required for secondary emission to be produced at the cathode by these various processes, and Newton has calculated the steplike nature of the resultant current in the discharge on this basis So far only the third phase, corresponding to the first arrival of metastables at the cathode, has been studied experimentally, by Engstrom and Huxford [171], who have shown that in the Townsend discharge in argon with an activated caesium cathode the metastable atoms are highly efficient in liberating secondary electrons, the minimum efficiency being about 0 4 electrons per incident metastable. The relative contribution made by the metastables to the total value of γ decreases as E/p is increased.

The action of metastable atoms either in the gas or at the cathode cannot always be a contributing factor to the production of secondary electrons, as there are many gases in which metastables do not occur. Also, in a single pure gas containing metastables, ionization will not be produced by the action of metastables as the energy of the metastable atom is less than the ionization energy. However, the mechanism is undoubtedly effective in mixtures of gases, such as argon and neon, where ionization of A atoms ($V_i = 15\ 68$ volts) can be caused by the action of Ne metastables ($V_m = 16 \cdot 53$ volts).

Experiments have been made by Costa [200] to examine the influence of photons on secondary emission from the cathode for discharges in air and in hydrogen In his measurements the radiation from a discharge is allowed to fall on the cathode of a second discharge gap in the same gas, and from the resultant current recorded in the second gap it is deduced that in air at 0·72 mm. Hg about 50 per cent. of the secondary emission can be attributed to the liberation of electrons by photon impact on the cathode. When the gas pressure is raised to 7·1 mm. Hg the proportion falls to 20 per cent , and it is concluded that the reduction is caused by the greater absorption of photons in the gas at higher pressure. In similar experiments in hydrogen the proportion of the secondary emission caused by photons is higher than in air. Other experimental evidence in support of the importance of photons causing secondary emission at the cathode in discharges in air and in hydrogen is given by Schade [201], and by Hale [165], but Llewellyn Jones [166] has concluded that, on the basis of his measurements in hydrogen for several cathode materials, the secondary emission from the cathode is not caused by photons except for relatively low values of E/p (< 28).

Values of γ in nitrogen and in hydrogen, as determined by Bowls [164] and by Hale [165] respectively, are shown in the curves of Fig. 1.44 and Fig. 1.45. The three curves shown for nitrogen correspond to the results obtained with a clean platinum cathode in uncontaminated gas, then with the same cathode when the cathode and the gas are contaminated with mercury vapour, and finally with a sodium cathode. The measurements show the pronounced effect of the presence of mercury vapour on the value of γ, and it is probable that, in many of the previous investigations of γ in various gases, this factor may have influenced the results obtained. The various peaks in the γ curves given by Bowls [164] and by Hale [165] have only been explained tentatively. Loeb [3] interprets the peak in the value of γ at $E/p \sim 100$ for the clean platinum cathode as caused by photons reaching the cathode from the gas, these photons being produced as a consequence of the ionization proceeding simultaneously in the gas by electron collisions according to the Townsend α-mechanism. The presence of mercury vapour in the gas causes an increased absorption of the photons, so that the number reaching the cathode falls to a negligible value and hence the peak in the curve for γ disappears. Because of the low work function for sodium it might have been expected that γ would be detectable at much lower values of E/p than for platinum, and also that greater values for γ would be obtained. However, it appears that a volatile sodium compound causes an enhanced absorption of photons in the gas. A similar interpretation is given for Hale's results [165], but the situation is more complex as several peaks occur at certain favourable values of E/p. Similar results to those

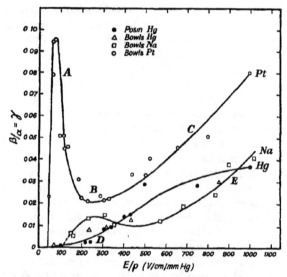

FIG 1 44 Values of the ratio β/α as a function of E/p in nitrogen

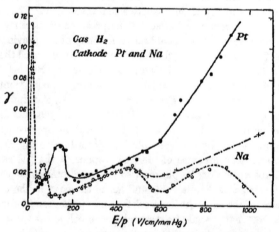

FIG. 1 45. Values of γ as a function of E/p in hydrogen.

obtained with a platinum cathode have also been recorded by Hale [165] for nickel and aluminium cathodes. The curves for γ as a function of E/p have peaks at $E/p = 131$ for both nickel and aluminium as compared with the peak observed at $E/p = 120$ for platinum. The maximum value of γ is $0 \cdot 115$ for aluminium and $0 \cdot 095$ for nickel.

As shown in the curves of Fig. 1.44 for nitrogen the value of γ is very small for $E/p < 60$ approximately. Similarly in the other gases, at reduced values of E/p the value of γ becomes undetectable. Low values of γ for tungsten in oxygen [203] and for water drops in air [202] have been measured. However, as discharges occur in gases at atmospheric and higher pressures for still lower values of E/p (for instance the breakdown of a 10 cm. gap in air at atmospheric pressure takes place at a field strength such that $E/p \sim 35$), it is evident that other discharge mechanisms may be operative. In such cases the liberation of electrons in the gas by photoionization appears to become effective [25 to 32 and 204 to 211], and is a necessary feature of recent theories of the spark mechanism for the breakdown of gaps of 1 cm. or more at atmospheric and higher pressures. (See Chapter VI.)

The work of Oliphant and others, on the secondary emission of electrons due to positive ion bombardment of surfaces *in vacuo* is of considerable importance [193, 194]. Many other papers on similar work have appeared (references are given in the review papers [86, 104]), and the results obtained give equivalent cross-sections for the secondary emission process. The relation between these results and those obtained with positive ion swarms in gaseous discharges is similar to the relation between ionization cross-sections and α (see p. 60). Thomson and Thomson [215] give a summary of work up to 1932 and review the work carried out with high energy positive ions (10 keV or more).

REFERENCES QUOTED IN CHAPTER I

1. N. F. MOTT and H. S. W. MASSEY, *Theory of Atomic Collisions*. Oxford, 1949.
2. F. L. ARNOT, *Collision Processes in Gases*. Methuen, London, 1933.
3. L. B. LOEB, *Fundamental Processes of Electrical Discharge in Gases*. Wiley, New York, 1939.
4. R. SEELIGER, *Einfuhrung in die Physik der Gasentladungen*, Barth, Leipzig, 1934.
5. A. VON ENGEL and M. STEENBECK, *Elektrische Gasentladungen*. Springer, Berlin, 1934.
6. W. HOPWOOD, *Proc. Phys. Soc. B*, **62** (1949), 657.
7. K. K. DARROW, *Electrical Phenomena in Gases*. Williams and Wilkins, Baltimore, 1932.
8. A. E RUARK and H. C. UREY, *Atoms, Molecules and Quanta*. McGraw-Hill, New York, 1930.
9. A. C. G. MITCHELL and M. W. ZEMANSKY, *Resonance Radiation and Excited Atoms*. Cambridge, 1934.
10. W. HEITLER, *Quantum Theory of Radiation*. Oxford, 1944.

11. A. A. KRUITHOF and F. M. PENNING, *Physica*, **4** (1937), 430.

12 W. BLEAKNEY, *Phys. Rev.* **35** (1930), 139, 1180.

13. W. BLEAKNEY, ibid. **36** (1930), 1303.

14. R. A. MILLIKAN and I. S. BOWEN, ibid. **23** (1924), 1.

15. B. EDLÉN, *Z. Phys.* **100** (1936), 726.

16. D. R. BATES, *Mon Not. Roy. Astr. Soc.* **100** (1939), 25.

17. D. R. BATES and H. S. W. MASSEY, *Proc. Roy Soc.* A, **187** (1946), 261.

18 SU-SHU HUANG, *Astrophys. J.* **108** (1948), 354.

19. H. S. W. MASSEY, *Negative Ions.* Cambridge, 1950.

20. J. P. VINTI, *Phys Rev* **44** (1933), 524.

21. J. A. WHEELER, ibid. **43** (1933), 258.

22. R W DITCHBURN, *Proc. Roy. Soc.* A, **117** (1928), 486.

23. R W. DITCHBURN and J. HARDING, ibid. **157** (1936), 66.

24. E. O LAWRENCE and N E. EDLEFSEN, *Phys. Rev.* **34** (1929), 233.

25. E. G SCHNEIDER, *J. Opt. Soc. Amer.* **30** (1940), 128.

26. P. D. FOOTE and F. L MOHLER, *Phys Rev* **26** (1925), 195.

27. F. L. MOHLER and C BOECKNER, *J. Res. Nat. Bur. Stand. Wash.* **2** (1929), 489.

28. E. O LAWRENCE and N E. EDLEFSEN, *Phys. Rev.* **34** (1929), 1056.

29. R. GEBALLE, ibid **66** (1944), 316

30. H. RAETHER, *Z. Phys.* **110** (1938), 611.

31. H. GREINER, ibid. **81** (1933), 543.

32 W. CHRISTOPH, *Ann. Phys Lpz.* **30** (1937), 446.

33. A. A. JAFFE, J D. CRAGGS, and C. BALAKRISHNAN, *Proc. Phys. Soc.* B, **62** (1949), 39.

34. C. G. CILLIÉ, *Mon. Not. Roy. Astr. Soc.* **92** (1932), 820.

35 H A. KRAMERS, *Phil. Mag.* **46** (1923), 837.

36. C. G. CILLIÉ, *Mon. Not. Roy. Astr. Soc.* **96** (1936), 771.

37 J. D. CRAGGS and W. HOPWOOD, *Proc. Phys. Soc.* **59** (1947), 771.

38 R. LADENBURG and C C. VAN VOORHIS, *Phys Rev.* **43** (1933), 315.

39. F. L. MOHLER, *Proc. Nat Acad Sci Wash.* **12** (1926), 494.

40. K. H. KINGDON, *Phys. Rev.* **21** (1923), 408

41. J J. HOPFIELD, ibid. **20** (1924), 573

42. H. D. SMYTH, ibid. **25** (1925), 452.

43. F. L. MOHLER, ibid **28** (1926), 46.

44 R LADENBURG, C C. VAN VOORHIS, and J. C BOYCE, ibid. **40** (1932), 1018.

45. W. M. PRESTON, ibid **57** (1940), 887.

46. E. DERSHEM and M. SCHEIN, ibid. **37** (1931), 1238.

47. F. HOLWECK, *De la lumière aux rayons X.* Presses Universitaires, Paris, 1927.

48. R. W. LARENZ, *Z. Phys.* **129** (1951), 343.

49. P. RAVENHILL and J. D. CRAGGS, in course of preparation.

50. F. L MOHLER, *J. Res Nat Bur. Stand Wash.* **19** (1937), 559.

51 C KENTY, *Phys. Rev.* **32** (1928), 624.

52 H. W. WEBB and D. SINCLAIR, ibid. **37** (1931), 182.

53. L. HAYNER, *Z. Phys* **35** (1925), 365.

54 D. R. BATES and H. S W. MASSEY, *Proc. Roy. Soc.* A, **192** (1947), 1.

55. P. DANDURAND and R. B. HOLT, *Phys. Rev.* **82** (1949), 278.

56. M. A. BIONDI and S. C. BROWN, ibid. **76** (1949), 1697.

57. M. A. BIONDI and S. C. BROWN, ibid **75** (1949), 1700.

58. F. L. MOHLER, ibid. **31** (1928), 187.

59. P. M. Morse and E. C. G. Stueckelberg, *Phys. Rev.* **35** (1930), 116.
60. H. Zanstra, *Proc. Roy. Soc.* A, **186** (1946), 236
61. H Edels and J. D. Craggs, *Proc. Phys. Soc. Lond.* A, **64** (1951), 562.
62. D. R. Bates, *Phys Rev.* **77** (1950), 718.
63. K. T. Compton and I. Langmuir, *Rev Mod. Phys.* **2** (1930), 128.
64. J. Sayers, *Proc. Roy. Soc.* A, **169** (1938), 83.
65. M. E. Gardner, *Phys. Rev.* **53** (1938), 75.
66. W. Machler, *Z. Phys.* **104** (1936), 1.
67. G. Jaffe, *Ann. Phys. Lpz.* **1** (1929), 977.
68. R. L. F. Boyd, *Proc. Phys. Soc. Lond.* A, **63** (1950), 543.
69. P. Langevin, *Ann. Chim. Phys.* **28** (1903), 289, 433.
70. J. J. Thomson, *Phil. Mag.* **47** (1924), 337.
71. B. T. McClure, R A. Johnson, and R B. Holt, *Phys. Rev.* **79** (1950), 232.
72. E C G. Stueckelberg and P. M. Morse, ibid. **36** (1930), 16.
73. Lord Rayleigh, *Proc. Roy. Soc.* A, **183** (1944), 26.
74. J. D. Craggs and J M. Meek, ibid. **186** (1946), 241.
75. H. Zanstra, *Astrophys. J.* **65** (1927), 50.
76. D. Chalonge and Ny Tsi Zé, *J. Phys. Radium,* **1** (1930), 416.
77. G. Herzberg, *Ann Phys. Lpz* **84** (1927), 565.
78. J. D. Craggs and W. Hopwood, *Proc. Phys Soc.* **59** (1947), 755.
79. H. Edels and J. D. Craggs, *Proc. Phys. Soc. Lond.* A, **64** (1951), 574
80. D R. Bates, H. A. Buckingham, H. S. W. Massey, and J J. Unwin. *Proc. Roy Soc.* A, **170** (1939), 322
81. H. S. W. Massey and D. R Bates, *Rep. Phys. Soc. Progr Phys* **9** (1942–3), 62.
82. J. Sayers, ibid. 52.
83. A. M. Cravath, *Phys. Rev* **36** (1930), 248.
84. L. D. Smullin and C. G. Montgomery, *Microwave Duplexers.* McGraw-Hill, New York, 1948.
85. J. D. Craggs, W. Hopwood, and J M. Meek, *J. Appl. Phys.* **18** (1947), 919.
86. M. J Druyvesteyn and F M. Penning, *Rev. Mod Phys.* **12** (1940), 87.
87. E. A. Milne, *Phil. Mag.* **47** (1924), 209,
88. C. Ramsauer and R Kollath, *Ann. Phys Lpz.* **3** (1929), 536.
89. R. B. Brode, *Rev. Mod Phys.* **5** (1933), 257.
90. C. Ramsauer and R. Kollath, *Ann. Phys. Lpz.* **12** (1932), 529, 837.
91. A. M. Tyndall, *Mobility of Positive Ions.* Cambridge, 1938.
92. A. V. Hershey, *Phys. Rev.* **56** (1939), 908.
93. A. V. Hershey, ibid. 916.
94. J. D. Cobine, *Gaseous Conductors.* McGraw-Hill, New York, 1941
95. P. Langevin, *Ann. Chim. Phys.* **8** (1905), 238
96. R. A. Nielsen and N. E. Bradbury, *Phys. Rev* **49** (1936), 388
97. R. A. Nielsen and N E. Bradbury, ibid. **51** (1937), 69.
98. C. E. Normand, ibid **35** (1930), 1217.
99. J. S. Townsend, *J. Frankl Inst.* **200** (1925), 563.
100. H. W. Allen, *Phys. Rev.* **52** (1937), 707.
101. R. H. Healey and J. W. Reed, *The Behaviour of Slow Electrons in Gases.* Amalgamated Wireless of Australasia, Sydney, 1941.
102. W. P. Allis and H. W. Allen, *Phys Rev.* **52** (1937), 703.

103. F. LLEWELLYN JONES, *Proc. Phys. Soc. Lond.* **61** (1944), 239.
104. V. J. FRANCIS and H. G. JENKINS, *Rep Phys. Soc. Progr. Phys.* **7** (1940), 230.
105. J. S. TOWNSEND, *Electrons in Gases.* Hutchinson, 1947.
106. O. TUXEN, *Z. Phys.* **103** (1936), 463.
107. W. W. LOZIER, *Phys. Rev* **42** (1934), 518.
108. H. D. HAGSTRUM and J. T. TATE, ibid. **59** (1941), 354.
109. R. H. VOUGHT, ibid. **71** (1947), 93.
110. J. W. WARREN, W. HOPWOOD, and J. D. CRAGGS, *Proc. Phys. Soc. Lond.* B, **63** (1950), 180.
111. J. W. WARREN and J D CRAGGS, in course of preparation.
112. J. W. WARREN and J. D. CRAGGS, in course of preparation.
113. N. E. BRADBURY, *Phys. Rev.* **44** (1933), 883.
114. R. H. HEALEY and C. B. KIRKPATRICK, unpublished (quoted in reference 101).
115. N. E. BRADBURY, *J. Chem. Phys.* **2** (1934), 827, 840.
116. N. E. BRADBURY and J. TATEL, ibid. (1934), 835.
117. W. DE GROOT and F M. PENNING, *Ned. T. Natuurkde.* **11** (1945), 156.
118. B. HOCHBERG and E SANDBERG, *C R. Acad. Sci. U.R.S.S.* **53** (1946), 511.
119. E. E. CHARLTON and F. S. COOPER, *G. E. Rev.* **40** (1937), 438.
120. H C. POLLOCK and F. S. COOPER, *Phys. Rev.* **56** (1939), 170.
121. H. MARGENAU, F. L. MCMILLAN, I. H. DEARNLEY, C. H. PEARSALL, and C G MONTGOMERY, ibid. **70** (1946), 349.
122 B. ROSSI and H. S. STAUB, *Ionisation Chambers and Counters.* McGraw-Hill, New York, 1949.
123. P. SCHULZ, *Ann. Phys Lpz.* **6** (1947), 318.
124 E. D. KLEMA and J. S. ALLEN, *Phys. Rev.* **77** (1950), 661.
125. L. SENA, *J. Phys. U.S.S.R.* **10** (1946), 179.
126. M. A. BIONDI, *Phys. Rev.* **79** (1950), 733.
127. R. B. HOLT, J. M. RICHARDSON, R. HOWLAND, and B. T. MCCLURE, ibid. **77** (1950), 239.
128. R. A. JOHNSON, B. T. MCCLURE, and R. B. HOLT, ibid. **80** (1950), 376.
129. W. R. HARPER, *Proc Camb. Phil. Soc.* **31** (1935), 430
130. L. B. LOEB and J. M. MEEK, *The Mechanism of the Electric Spark.* Stanford, 1941.
131. M. J. DRUYVESTEYN, *Physica,* **3** (1936), 65.
132. J. A. SMIT, ibid. (1937), 543.
133. J. H CAHN, *Phys Rev.* **75** (1949), 293.
134. J. H. CAHN, ibid (1949), 838.
135. B DAVYDOV, *Phys. Zeits. Sowjetunion,* **12** (1937), 269.
136. G. MIERDEL, *Wiss. Veroff. Siemens-Werk,* **17** (1938), 71.
137. I. LANGMUIR and H. MOTT-SMITH, *Phys. Rev.* **28** (1928), 727.
138. R C. MASON, ibid. **51** (1937), 28
139. T. JONSSON, *Trans. Chalmers Univ. Tech. Gothenburg* (No. 65), 1948.
140. M. J. DRUYVESTEYN, *Z. Phys.* **64** (1930), 790.
141. R. H. SLOANE and E. J. R. MCGREGOR, *Phil. Mag.* **18** (1934), 193.
142. A. H VAN GORCUM, *Physica,* **3** (1936), 207.
143. R L. F. BOYD, *Nature,* London, **165** (1950), 228.
144. R. L. F. BOYD, *Proc. Roy. Soc.* A, **201** (1950), 329.

145. W WEIZEL and R. ROMPE, *Theorie Elektrische Lichtbogen und Funken.* Barth, Leipzig, 1949.
146. H. JAGODZINSKI, *Z. Phys.* **120** (1943), 318.
147. R. LADENBURG, *Rev Mod. Phys.* **5** (1933), 243.
148. G. C. WILLIAMS, J. D. CRAGGS, and W. HOPWOOD, *Proc. Phys. Soc.* B, **62** (1949), 49.
149. C. J. BRAUDO, J. D. CRAGGS, and G C. WILLIAMS, *Spectrochim Acta,* **3** (1949), 546.
150. B. T. BARNES and E. Q ADAMS, *Phys. Rev.* **53** (1938), 545.
151. B. T. BARNES and E. Q. ADAMS, ibid. 556
152. H. EDELS, *E.R.A. Report,* L/T 230 (1950).
153. L. S. ORNSTEIN and H. BRINKMAN, *Physica,* **1** (1934), 797.
154. H. ALFVÉN, *Cosmical Electrodynamics,* Oxford, 1950.
155. R. ROMPE and M. STEENBECK, *Ergebnisse der exakt Naturwiss.* **18** (1939), 25.
156. J. S. TOWNSEND, *Electricity in Gases,* Oxford, 1914
157. H. D. DEAS and K. G. EMELÉUS, *Phil Mag* **40** (1949), 460.
158. R N. VARNEY, H J WHITE, L. B. LOEB, and D Q POSIN, *Phys. Rev.* **48** (1935), 818.
159. M. PAAVOLA, *Archiv f. Elek.* **22** (1929), 443
160. F. H. SANDERS, *Phys. Rev.* **41** (1932), 667; **44** (1933), 1020.
161. K MASCH, *Arch. Elektrotech.* **26** (1932), 589.
162. T. L R. AYRES, *Phil Mag* **45** (1923), 353.
163 D Q POSIN, *Phys Rev* **50** (1936), 650.
164. W. E. BOWLS, ibid **53** (1938), 293
165. D. H. HALE, ibid. **54** (1938), 241, **55** (1939), 815, **56** (1939), 1149.
166. F. LLEWELLYN JONES, *Phil. Mag* **28** (1939), 192
167 J. S. TOWNSEND, and S. P. MCCALLUM, ibid. **5** (1928), 695; **6** (1928), 857.
168. A. A. KRUITHOF and F. M. PENNING, *Physica,* **3** (1936), 515; **4** (1937), 430.
169. I. I. GLOTOV, *Phys. Zeits. Sowjetunion,* **12** (1937), 256.
170. W. S. HUXFORD, *Phys. Rev.* **55** (1939), 754.
171. R. W. ENGSTROM and W. S. HUXFORD, ibid. **58** (1940), 67.
172 A. A. KRUITHOF, *Physica,* **7** (1940), 519.
173. E. BADAREU and M. VALERIU, *Bull. Soc Roumaine de Phys.* **42** (1941), 9, **43** (1942), 35.
174. E. BADAREU and G. G. BRATESCU, ibid. **45** (1944), 9.
175. B. HOCHBERG and E SANDBERG, *J. Tech. Phys. U.S.S.R.* **12** (1942), 65.
176. J. KAPLAN, *Phys. Rev.* **55** (1939), 111.
177. K. G. EMELÉUS, R W. LUNT, and C. A. MEEK, *Proc. Roy. Soc.* A, **156** (1936), 394.
178. S. H. DUNLOP, *Nature,* **164** (1949), 452.
179. I. I. GLOTOV, *Phys Zeits. Sowjetunion,* **13** (1938), 84
180 S. K. MORALEV, *J. Exp. Theor. Phys. U.S.S R.* **7** (1937), 764.
181. R. KOLLATH, *Phys. Zeits.* **31** (1930), 981.
182. L. H. FISHER and G. L. WEISSLER, *Phys. Rev.* **66** (1944), 95.
183. P. L. MORTON, ibid. **70** (1946), 358.
184. G. W. JOHNSON, ibid **73** (1948), 284.
185. L. J. VARNERIN and S. C. BROWN, ibid. **79** (1950), 946.
186. G. HOLST and E. OOSTERHUIS, *Phil. Mag.* **46** (1923), 1117.

187. R. N. Varney, L B Loeb, and W. R. Hazeltine, ibid. **29** (1940), 379.
188. R Schofer, *Z. Phys.* **110** (1938), 21.
189. R. W. Engstrom, *Phys. Rev.* **55** (1939), 239.
190 D. H. Hale and W. S. Huxford, *J. Applied Physics*, **18** (1947), 586.
191. F. Ehrenkranz, *Phys. Rev.* **55** (1939), 219.
192. F. M. Penning and C. C. J. Addink, *Physica*, 1 (1934), 1007.
193. M. L E. Oliphant, *Proc Roy Soc.* A, **124** (1929), 228.
194 M. L. E. Oliphant and P. B. Moon, ibid (1929), 388
195. J. Kock, *Z. Phys.* **100** (1936), 700.
196 F. M. Penning, *Proc. Roy. Acad. Amsterdam*, **31** (1928), 14; **33** (1930), 841.
197. A. A. Kruithof and F. M. Penning, *Physica*, **5** (1938), 203.
198. K. Kenty, *Phys. Rev* **43** (1933), 181.
199. R. R. Newton, ibid. **73** (1948), 570.
200. H. Costa, *Z. Phys* **113** (1939), 531; **116** (1940), 508.
201. R. Schade, ibid. **111** (1938), 437.
202. W. N. English, *Phys Rev.* **74** (1948), 179.
203. J. D Craggs and J. M. Meek, *Proc. Phys. Soc.* **61** (1948), 327.
204. A N. Cravath, *Phys Rev.* **47** (1935), 254.
205. G. Déchène, *J. Phys. Radium*, **7** (1936), 533.
206. G. L. Weissler, *Phys. Rev.* **63** (1943), 96
207. K. Masch, *Z Phys.* **79** (1932), 672.
208. L. B. Loeb and R. A. Wijsman, *J. Appl. Phys.* **19** (1948), 797.
209. L. B. Loeb, *Rev Mod Phys.* **20** (1948), 151.
210 L. B. Loeb, *Phys. Rev.* **74** (1948), 210.
211. J. B. Hasted, *Proc. Roy. Soc.* A, **205** (1951), 421.
212. J A. Hornbeck, *Phys Rev.* **80** (1950), 297.
213 H. S W. Massey, *Rep. Phys. Soc. Progr. Phys.* **12** (1948–9), 248.
214. R. A. Nielsen, *Phys. Rev.* **50** (1936), 950.
215. J. J Thomson and G. P. Thomson, *Conduction of Electricity through Gases*. Cambridge University Press, 1933.
216. H. Trumpy, *Z. Phys.* **42** (1927), 327 and **44** (1927), 1575.
217. S. H. Dunlop and K G. Eméléus, *Brit. J. Appl. Phys* **2** (1951), 163.
218. H. R. Hassé and W R. Cook, *Proc Roy. Soc* A, **125** (1929), 196.
219 H. R. Hassé and W. R. Cook, *Phil. Mag.* **12** (1931), 554.
220. F. Llewellyn Jones and A. B. Parker, *Nature*, **165** (1950), 960.
221. M. Valeriu-Petrescu, *Bull. Soc Roumaine de Phys.* **44** (1943), 3.
222. R. H. Johnston, *J Chem. Phys.* **19** (1951), 1391.
223. R. D. Craig and J. D. Craggs, in course of preparation.
224. L. G. H. Huxley, *Phil. Mag.* **5** (1928), 721.

II

BREAKDOWN AT LOW GAS PRESSURES

TOWNSEND CRITERION FOR A SPARK AND PASCHEN'S LAW

It has been shown in Chapter I that the current flowing in a uniform field can be expressed by the relation

$$i = i_0 \frac{e^{\alpha d}}{1 - \gamma(e^{\alpha d} - 1)}, \tag{2.1}$$

where γ may represent one or more of several secondary mechanisms The value of $\gamma(e^{\alpha d} - 1)$ is zero at low voltage gradients but increases as the voltage gradient is raised until eventually

$$\gamma(e^{\alpha d} - 1) = 1. \tag{2.2}$$

The denominator of equation (2.1) then becomes zero and the current i as defined by equation (2.1) is then indeterminate. According to the theory of the spark as suggested originally by Townsend [1, 2] this condition defines the onset of the spark. In general at the value of α corresponding to breakdown $e^{\alpha d} \gg 1$ and therefore equation (2.2) can be written in the form

$$\gamma e^{\alpha d} = 1 \tag{2.3}$$

The significance of the Townsend sparking criterion has been discussed by various authors following a modified interpretation proposed by Holst and Oosterhuis [3]. The subject is treated in a recent paper by Loeb [4] who summarizes the present views in the following manner:

(a) For $\gamma e^{\alpha d} < 1$ the discharge current i is given by (2 1) and is not self-maintained, i.e. the current ceases if the primary current i_0 is reduced to zero by the removal of the external source of irradiation producing it.

(b) For $\gamma e^{\alpha d} = 1$ the number $e^{\alpha d}$ of ion pairs produced in the gap by the passage of one electron avalanche is sufficiently large that the resultant positive ions, on bombarding the cathode, are able to release one secondary electron and so cause a repetition of the process. The discharge is then self-maintaining and can continue in the absence of the source producing i_0, so that the criterion $\gamma e^{\alpha d} = 1$ can be said to define the sparking threshold.

(c) For $\gamma e^{\alpha d} > 1$, the ionization produced by successive avalanches is cumulative. The spark discharge grows the more rapidly the greater the amount by which $\gamma e^{\alpha d}$ exceeds unity.

As both the ionization in the gas by electron collision and the secondary emission of electrons from the cathode by the γ mechanism are chance phenomena, the quantities $e^{\alpha d}$ and γ will each fluctuate about mean values. The product $\gamma e^{\alpha d}$ will then vary for individual avalanches. Consequently if V_s is the voltage corresponding to $\gamma e^{\alpha d} = 1$ on the average, there is a possibility that breakdown may occur at a voltage V slightly less than V_s as the result of the occurrence of an exceptional

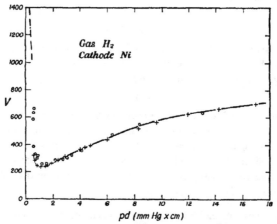

Fig. 2 1. The breakdown voltage curve for a nickel cathode in hydrogen. The crosses denote the measured values, the circles denote the calculated values

avalanche or a train of avalanches for which, individually, $\gamma e^{\alpha d} > 1$. However, because of the large rate of change of the product $\gamma e^{\alpha d}$ as a function of voltage gradient, the sparking threshold is found to be relatively sharply defined and there are few data concerning the shape of the sparking probability curves.

The Townsend criterion of equation (2.3) enables the breakdown voltage of a gap to be determined by reference to the appropriate curves relating α/p and γ with E/p. Comparison between the calculated breakdown voltage and the experimentally determined breakdown voltage gives close agreement for short gaps at reduced gas pressures, where breakdown occurs at a voltage less than 10 kV or so.

An example of the agreement obtained is shown in Fig. 2.1, which gives Hale's results for the breakdown of hydrogen [5]. The crosses correspond to the measured sparking voltages, the circles to those calculated from Hale's measured values of α and γ.

The Townsend theory of the spark enables an explanation to be made of the experimental observation known as Paschen's law, which states that if the length of a discharge gap and the gas pressure are altered in such a way that their product is unchanged the magnitude of the breakdown voltage remains constant [6]. As shown in Chapter I the ionization coefficients may be expressed as functions of the field strength and gas pressure·

$$\alpha/p = F_1(E/p), \qquad \gamma = F_2(E/p).$$

Insertion of these functions in equation (2.3) yields

$$F_2(E/p)\exp[pd\ F_1(E/p)] = 1. \tag{2.4}$$

If we write $Ed = V$, the voltage across a uniform field gap, equation (2.4) can be rewritten as

$$F_2(V/pd)\exp[pd\ F_1(V/pd)] = 1. \tag{2.5}$$

From equation (2.5) it is clear that for a given value of pd there is a particular value of V, and hence

$$V = F(pd). \tag{2.6}$$

The breakdown voltages of uniform field gaps in gases at different pressures can therefore be plotted in the forms of curves relating the voltage with the product of gap length and gas pressure, and this procedure is frequently adopted in this chapter. However, deviations from Paschen's law have been reported and are referred to on p. 96 et seq.

Paschen's law in the form given in equation (2.6) obtains for a constant gas temperature. The effect of the gas temperature can be introduced by consideration again of the quantities α and γ which depend on the mean free path of the ionizing particles. In the derivation of these quantities in Chapter I the mean free path was stated to be inversely proportional to the gas pressure on the assumption of constant gas temperature. If, however, the gas temperature and the gas pressure are both varied, the mean free path now changes in accordance with the ratio of temperature to gas pressure, i.e inversely as the gas density. Consequently when account is taken of variations of both temperature and pressure it is necessary to consider curves relating α/ρ and γ with E/ρ rather than α/p and γ with E/p. Under such conditions Paschen's law may be written in the more general form

$$V = f(\rho d), \tag{2.7}$$

which becomes of the same form as equation (2.6) for constant temperature.

The accuracy of equation (2.7) stating that the breakdown voltage in a uniform field is a function only of the product of the gas density and the gap length has been confirmed by several investigators [7, 8, 9, 10]. In measurements by Bowker [8] the breakdown of gaps of 0·4 cm. to 1·0 cm. between spheres of 2 cm. diameter in hydrogen and in nitrogen has been studied for temperatures up to 860° C. For the range of gas densities covered, between about 0·2 and 1·2 times that of the gas at 0° C. and 760 mm. Hg, the sparking voltage in both gases is found to be independent of the gas temperature and the gas pressure provided that the gas density is constant.

Breakdown voltage characteristics in uniform fields

Typical curves [11] relating the breakdown voltage with the product of gas pressure times gap length in several gases are compared in Fig. 2.2, where p_0 is the gas pressure in mm. Hg reduced to 0° C. The breakdown voltage for each gas decreases as $p_0 d$ is reduced, reaches a minimum at a value of $p_0 d$ of the order of 1 to 10 mm. Hg × cm., and then increases Differences are recorded between the breakdown curves for the various gases, as is to be expected from consideration of the breakdown criterion of equation (2.3) and the curves relating α/p and γ with E/p as given in Chapter I. The fact that α appears in the exponential term of the breakdown criterion implies that the magnitude of α has a greater influence than that of γ on the magnitude of the breakdown voltage.

The breakdown between parallel plates in neon has been investigated by Townsend and McCallum [12], Penning [13], and Penning and Addink [14]. The results obtained by the last two investigators, as given in Fig 2 3, show that small amounts of argon in the neon have a large effect on the breakdown voltage. This is explained [15] by the presence of metastable neon atoms, the excitation potential of which exceeds the ionization potential of argon atoms (see Chapter I) The influence of metastable neon atoms in ionizing the argon can be shown by diminishing their concentration either by irradiation with light from a neon discharge or by the addition of a third gas, when the breakdown voltage is observed to increase. Fig. 2.4 gives some curves obtained by Penning [13] showing the effect of illumination from a neon discharge on the breakdown voltage curves for neon containing 0·002 per cent. of argon. The same illumination causes a negligible effect on the breakdown voltage curves for pure neon. Fig. 2 5 shows the breakdown voltage curves as a function of argon content for several intensities of illumination [16].

According to Penning and Addink [14] small admixtures of all gases

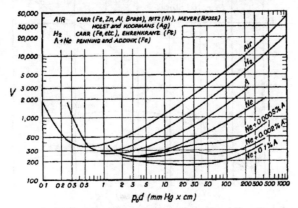

FIG 2 2 Typical breakdown voltage curves for different gases between parallel plate electrodes. p_0 is the gas pressure in mm. Hg corrected to 0° C.

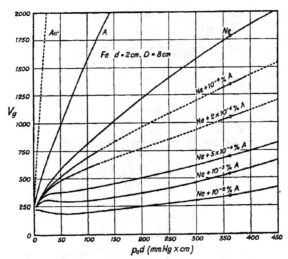

FIG. 2 3. Breakdown voltage curves in neon-argon mixtures between large parallel plates at 2 cm. spacing p_0 is the gas pressure in mm. Hg corrected to 0° C.

except helium decrease, as far as is known, the breakdown voltage of neon. In the case of helium the ionization potential is higher than that of neon.

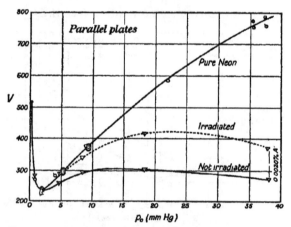

Fig 2 4 Influence of illumination from a neon discharge on the breakdown voltage curve for neon containing 0 002 per cent of argon. p_0 is the gas pressure in mm Hg corrected to 0° C.

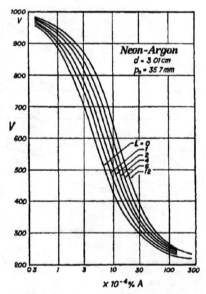

Fig. 2.5. Breakdown voltage curves of neon-argon mixtures for different intensities of illumination.

The breakdown voltage varies with the cathode material, the variation becoming particularly noticeable at the lower values of the product pd near the minimum breakdown voltage. Some curves given by Jacobs and La Rocque[17] for the breakdown of argon with cathodes of barium, magnesium, and aluminium are shown in Fig 2.6. Similar results in argon have been recorded by several other workers. Penning and

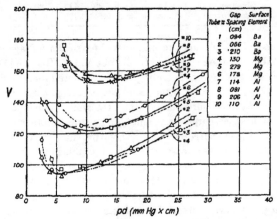

FIG 2 6. Breakdown voltage curves in argon for barium, magnesium, and aluminium cathodes.

Addink [14] observed a minimum breakdown voltage of about 270 V for an iron cathode The results for nickel and platinum as obtained by Ehrenkranz [18] and Schofer [19] respectively are about the same, at 200 V approximately. With a sodium cathode Ehrenkranz [18] measured a minimum breakdown voltage of roughly 90 V, a similar value being obtained by Heymann [20] for potassium The large variations observed are accounted for by the differences in the coefficient γ for the cathodes concerned, and confirm that the emission of electrons from the cathode as the result either of bombardment by positive ions, by radiation from the gas, or by other processes (see Chapter I) must be an important factor in the spark mechanism for the conditions investigated.

Studies of the minimum breakdown voltages in the other inert gases have been made by Jacobs and La Rocque [21] for barium, magnesium, and aluminium cathodes. The results are given in Table 2.1. A minimum breakdown voltage as low as 64 V has been detected by Cueilleron [22] for the breakdown in neon between caesium-coated electrodes, the gas pressure being 26 mm. Hg.

TABLE 2.1

Minimum Breakdown Voltages

	He	Ne	A	Kr	Xe
	(volts)				
Barium .	157	129	94	104	83
Magnesium	160	150	123	115	120
Aluminium	189	160	154	135	150

A method of calculating the minimum breakdown voltage of cold-cathode tubes has been suggested by Jacobs and La Rocque [17]. The calculated values of γ corresponding to the measured minimum breakdown voltages, for several cathodes in argon, are plotted against the values of the cathode work functions and the minimum breakdown voltages. With these two curves an attempt is made to estimate the minimum breakdown voltage in argon for any cathode provided the work function is known. On this basis a reasonable value is obtained for the minimum breakdown voltage of a gap in argon with a sodium cathode. The method has been criticized by Hale and Huxford [23] who point out that the agreement between the calculated and measured values of the minimum breakdown voltages for a sodium cathode in argon may be fortuitous. For instance Bowls [24] found γ in nitrogen to be appreciably lower for a sodium cathode than for a platinum cathode, but Ehrenkranz [18] recorded lower minimum breakdown voltages with a sodium cathode than those with a platinum cathode. Hale and Huxford [23] also point out that the minimum breakdown voltages for nickel and platinum cathodes of work functions 5·01 and 6·30 V respectively, are 195 V in each case, but with an iron cathode of work function 4·72 V the minimum breakdown voltage is 265 V. It can therefore be concluded that while, in general, materials with low work functions show high values of γ there are exceptions to this rule and any predictions of minimum breakdown voltages based on the work functions of the cathode may be greatly in error.

An unusual form of breakdown voltage characteristic has been recorded by Penning [25] for discharges in helium at values of pd below that corresponding to the minimum breakdown voltage. The curve obtained is shown in Fig. 2.7 and implies that for a given value of pd there may be three breakdown voltages. The form of the curve has been explained by Penning [25] by consideration of the ionization probability of electrons in helium. As the gap length is reduced for a given gas pressure, there is a decrease in the number of atoms between the electrodes and in the number of ionizing collisions. Consequently in order for the breakdown criterion $\gamma e^{\alpha d} = 1$ to be satisfied the value of $e^{\alpha d}$ must be maintained, if γ remains constant, and therefore α must be increased.

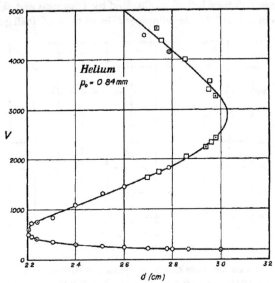

FIG. 2 7 Breakdown voltage characteristic for helium at values of pd below that corresponding to the minimum breakdown voltage.

However, the probability of ionization by an electron does not increase continuously with increasing electron velocity but reaches a maximum at ~ 100 V. When the voltage gradient is such that the number of ionizing collisions is at a maximum a decrease in pd cannot be compensated by increasing the voltage across the gap. Therefore no sparks occur for values of pd below a certain value and the breakdown curve bends to the right as shown in Fig 2 7 at a voltage of about 600 V. For larger values of d the required value of $e^{\alpha d}$ can be obtained for two voltage gradients on either side of the maximum in the ionizing probability curve, the same value of α being produced by both gradients. The situation, however, is made more complicated by the fact that γ varies also with the gradient and it is also possible that at sufficiently high velocities ionization by positive ions may occur in the gas. These effects cause breakdown to occur again at a third voltage and the breakdown curve therefore bends back again for voltages above about 3,000 V, as shown in Fig. 2.7.

The breakdown in helium has also been investigated by McCallum and Klatzow [26] and Townsend and McCallum [27]. The breakdown voltage curve in pure helium as obtained by Townsend and McCallum [27] is given in Fig. 2 8. The influence of the addition of small quantities

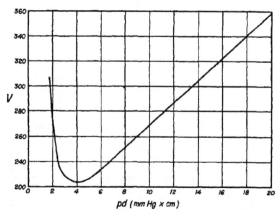

FIG 2.8. Breakdown voltage curve for helium.

of other gases on the breakdown of helium is discussed by McCallum and Klatzow [26], who give the curves of Fig 2.9 showing the currents flowing between parallel plates in pure helium, pure argon, and helium containing 0·025 per cent of argon. The respective values of E/p used for the three curves are 50 in helium, 200 in argon, and 15 in the mixture. If the curves had been plotted for the same value of E/p in all three cases the current in the mixture would have been even more markedly in excess of those measured in the pure gases. McCallum and Klatzow explain the observed effect as being caused by the action of direct collisions of electrons with argon atoms and disagree with the interpretation as given by Penning [16] in terms of metastable helium atoms.

The dependence of the breakdown voltage of a 0·35 cm. gap in hydrogen at pressures between 1 and 520 mm. Hg for six different cathode materials has been studied by Llewellyn Jones and Henderson [28]. The results near the minimum breakdown voltage are shown in Fig. 2.10. Except in the case of the nickel electrode, which has the highest work function, the minimum sparking voltage increases with the work function of the cathode and is lowered by the presence of impurities or oxidation

The results are interpreted by Llewellyn Jones and Henderson [28] as demonstrating that electron emission from the cathode must be a factor in the breakdown process, the influence of oxidation or impurities being such as to enhance this emission   The anomalous results obtained with nickel are possibly due to a similar effect to that noted by Farns-

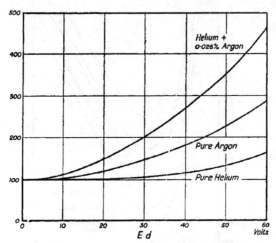

FIG. 2 9. Conductivity between parallel plates

worth [29], who found that the exposure of nickel to hydrogen after degassing caused a marked increase in the secondary emission under electron impact.

With increasing values of pd the percentage differences between the extreme values of the breakdown voltages for the various cathodes decreases but it is still detectable at pd = 200 mm. Hg × cm. Llewellyn Jones and Henderson [28] consider that at pressures greater than atmospheric the difference in breakdown voltage becomes small enough to fall within the experimental error, and the breakdown voltage would then appear to be independent of the cathode material, at least for the normal types of metals used.

The mechanism of the spark in hydrogen has been discussed in a paper by Llewellyn Jones [30], who has analysed the breakdown voltages to determine the value of γ at breakdown from the relation $\gamma e^{\alpha d} = 1$ and has made certain deductions concerning the relative importance of three secondary processes of ionization, namely the liberation of electrons from the cathode by positive ion bombardment, the liberation of electrons from the cathode by photoelectric effect, and the liberation of electrons in the gas caused by collision of positive ions with the gas molecules. From a comparison of the values of γ obtained with the different metals Llewellyn Jones concludes that with aluminium the emission of electrons from the cathode must contribute at least 50 per

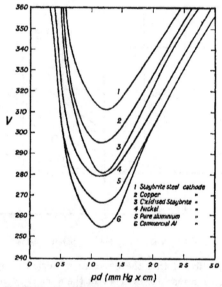

FIG 2 10   Breakdown voltage curves in hydrogen
for different cathode materials

cent. of all the secondary ionization. The corresponding minimum fractions for nickel and copper are 41 per cent. and 28 per cent. respectively. These values are determined on the assumption that there is no electron emission from a steel cathode. If electron emission does occur from steel the proportions for the other cathodes is raised. For instance if the emission from steel is one-half that from aluminium then all the secondary emission with the aluminium cathode must be due to the emission of electrons from the cathode. The relative contributions of positive ion bombardment and of photoelectric effect have been considered by Llewellyn Jones who concludes that the former process predominates over a wide range of values of E/p from 28 to 300 V/cm./mm. Hg, while the influence of radiation from the gas is important only for breakdown at higher gas pressures when $E/p < 28$.

Breakdown measurements in hydrogen have been studied by Ehrenkranz [18] who used both platinum and sodium cathodes. The breakdown voltage is lower with a sodium cathode than with a platinum
cathode, the difference being about 10 per cent. at $pd = 300$ mm. Hg × cm.
and increasing with decreasing pd. The minimum breakdown voltage
is approximately 300 V with platinum and 180 V with sodium. For

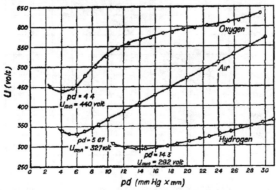

FIG. 2.11. Breakdown voltage curves in oxygen, air, and hydrogen

values of $pd > 50$ the curve obtained by Ehrenkranz [18] with a platinum cathode is closely the same as that given by Llewellyn Jones and
Henderson [28] for a steel cathode, but for lower values of pd the results
with platinum are slightly higher.

The breakdown curve in hydrogen as recorded by Fricke [31] is given
in Figs 2.11 and 2.12. For higher values of pd the curve agrees closely
with those of Llewellyn Jones and Henderson [28], Ehrenkranz [18],
and Hale [5, 32] but there is an appreciable difference near the minimum
of the curve. The minimum breakdown voltage is 292 V and occurs at
$pd = 1.43$. Sparking voltages in mixtures of hydrogen and nitrogen as
obtained by Frey [33] are given in Fig. 2.13.

The breakdown voltage characteristic in nitrogen for platinum and
sodium electrodes as measured by Ehrenkranz [18] is given in Fig. 2.14.
The pressure range is from 0.5 to 600 mm. Hg. A lowering of the sparking
voltage in the presence of sodium is to be expected because of the higher
values of α under such conditions as noted by Bowls [24]. However, the
lowering is larger than that expected on this basis and may be associated
with the high pre-breakdown currents observed by Ehrenkranz when
a sodium cathode was used. A comparison between the experimental

results of Ehrenkranz and the theoretical breakdown curves computed
from Bowls's [24] and Posin's [34] values for α and γ are given in Fig. 2.15.
Figs. 2.11 and 2.12 include curves for oxygen obtained by Fricke [31]

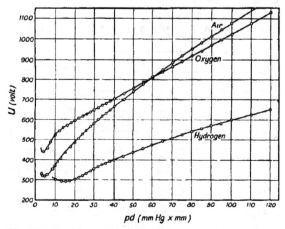

Fig. 2.12. Breakdown voltage curves in oxygen, air, and hydrogen.

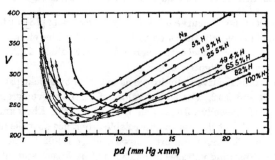

Fig. 2.13.   Breakdown voltage curves in hydrogen-nitrogen
mixtures.

and curves for air obtained by Meyer [35]. The curves for air approximate
to those for nitrogen for values of $pd > 100$ mm. Hg × cm. approximately.
The influence of magnetic fields on the breakdown voltages for discharges in air was studied by Meyer [35] who found that the minimum
breakdown voltage was increased from 327 V to 470 V by the application
of a magnetic field of 1885 gauss.

Measurements of the breakdown voltage in deuterium for pressures

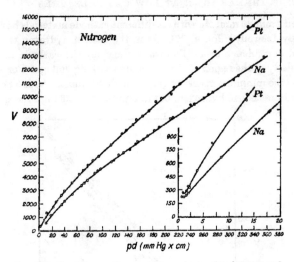

FIG. 2 14. Breakdown voltage curves in nitrogen for platinum and sodium cathodes.

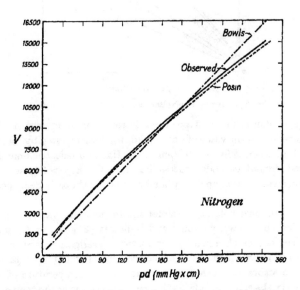

FIG. 2.15. Comparison between calculated and measured breakdown voltage curves in nitrogen.

between 1 and 26 mm. Hg with six different cathode materials have been made by Llewellyn Jones [36]. As in the case of hydrogen there is a marked dependence of breakdown voltage on the cathode material, particularly in the region of the minimum sparking voltage. For a given cathode the minimum sparking voltage in deuterium is about 3 per cent. greater than that in hydrogen, but at a pressure of 26 mm. the sparking

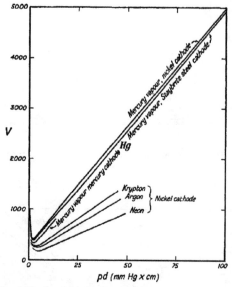

FIG. 2 16   Breakdown voltage curves in mercury vapour.

voltage in deuterium is 3 per cent. lower than in hydrogen From considerations of the value of γ at breakdown it is suggested by Llewellyn Jones that when $E/p \sim 150$ V/cm /mm Hg, photoelectric emission is the predominant cathode process, but that for $E/p > 250$ secondary emission is caused principally by bombardment of the cathode by positive ions.

The breakdown voltage characteristic for mercury vapour has been determined by Llewellyn Jones and Galloway [37] and by Grigorovici [38]. The curves obtained by the former investigations are shown in Fig. 2.16 and give the variation of the breakdown voltage with pd, where p is the pressure at which a permanent gas at a temperature of 15° C. would have the same density as the mercury vapour at the temperature $T°$ C. so that the curves may then be directly compared with those for permanent gases. The difference between the curves for nickel and for steel cathodes is small. The minimum sparking voltages recorded for clean electrodes were 400 V for the nickel cathode and 380 V for the

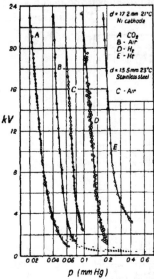

FIG. 2.17. Breakdown voltage curves in air, CO₂, H₂, and He at low gas pressures.

steel cathode. With the cathodes covered by a film of mercury the minimum sparking voltage was 305 V for both nickel and steel. Similar results were obtained by Grigorovici [38] who records minimum sparking voltages of 520 V for a clean iron cathode and of 330 V for the same cathode covered with mercury.

For the measurement of breakdown voltages at values of pd less than that corresponding to the minimum breakdown voltage difficulties arise in that the breakdown does not occur most readily across the shortest gap between the electrode but prefers to take a longer path [40]. Particular attention must therefore be given to the design of the electrode system. In the arrangement used by Quinn [39] the electrodes consist of two coaxial nickel cylinders each with hemispherical ends, the gap being slightly greater than the difference between the radii of the cylinder. Some results obtained by Quinn [39] for the breakdown of a 1·72 cm. gap in several gases are given in Fig. 2.17. The dotted curves near the minimum have been computed by Quinn from the earlier data of Carr [41], who used voltages up to 1,000 V only. The lack of agreement is understandable as Carr used brass electrodes with rubber insulation so that the system was not outgassed. Quinn's results in air are in reasonable agreement with those obtained by Cerwin [42] as given in Fig. 2.18, which shows the variation of the breakdown voltage of air between plane nickel electrodes for gaps of 2 to 10 mm. with voltages up to 80 kV.

Deviations from Paschen's law

In experiments on neon Townsend and McCallum [43] observed that the sparking voltage with a large spacing between the plate electrodes was larger than that with a small spacing for a constant value of pressure times gap length. Later investigation by McCallum and Klatzow [44] confirmed this result for neon and showed that the same effect occurred in argon but not in helium. The values of breakdown voltage obtained in these gases for a constant value of $pd = 7·4$ mm. Hg $\times$ cm. and with

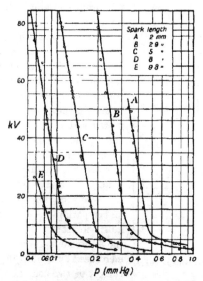

FIG. 2 18. Breakdown voltage curves in air as a function of gas pressure for different gap lengths

an electrode plate diameter of $D = 3·5$ cm. are given in Table 2.2. The variation of breakdown voltage in neon and argon is explained by the investigators as being caused by the loss of electrons by diffusion from the discharge region, the loss being the greater with increased spacing for a given electrode diameter. The fact that no change occurs in helium is explained by the lower diffusion rate in this gas as shown by measurements of the photocurrents in the three gases.

Similar results in the inert gases have been recorded by Buttner [45]. At $pd = 20$ mm. Hg $\times$ cm. the breakdown voltage in neon is 365 V for $p = 5$ mm. Hg and is 355 V for $p = 15$ mm Hg. In neon containing 6×10^{-3} per cent. of argon, at the same value of pd the breakdown voltage is 300 V for $p = 5$ mm. Hg and is 255 V for $p = 15$ mm. Hg. Heymann [20] has also observed an increase in the breakdown voltage

H

TABLE 2.2

d cm.	p mm. Hg	D/d	Breakdown voltage		
			He	Ne	A
0 3	24·7	11 7	242	204	328
0·4	18 5	8·7	242	206	338
0 5	14 8	7 0	242	210	343
0 6	12 3	5 8	242	214	347
0 7	10 6	5 0	242	217	352
0·8	9 2	4 4	242	220	358

for constant pd as the gap length is increased for discharges in argon and argon-neon mixtures. Some curves obtained by Heymann are given in Fig. 2.19. The variations are again explained as caused by a decrease in

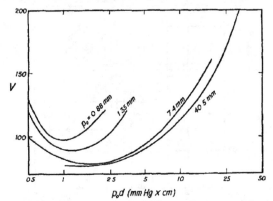

FIG 2 19. Breakdown voltage curves in an argon-neon mixture
(93·4 per cent. Ne, 6·6 per cent. A)

the effective α in the gas because of the loss of electrons and positive ions from the discharge gap to the walls of the container. The loss factor per unit potential difference is shown theoretically to be inversely proportional to the voltage gradient and electrode radius, when the electrodes are enclosed by a close-fitting glass tube, but the theoretical analysis is not fully substantiated by the experimental results.

Deviations from Paschen's law have been noted in other gases. For instance in measurements by Fricke [31] the breakdown voltage in hydrogen at $pd = 10$ mm. Hg × cm is 612 V at $p = 5·2$ mm. Hg and is 587 V at $p = 21·3$ mm. Hg. Small deviations of this order have been overlooked in the previous measurements on air by Carr [41] and other investigators, but re-examination of these earlier publications shows that such deviations were in fact recorded.

A careful analysis has been carried out by Slepian and Mason [46] to check the applicability of Paschen's law For this purpose they assume that the general relation between the breakdown voltage and the pressure and gap length is of the form

$$V = f(p^m d^n),\qquad (2.8)$$

which may be written as $p^m d^n = F(V)$.

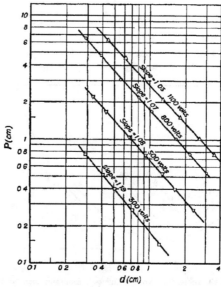

FIG. 2.20. Curves relating gas pressure with gap length for given breakdown voltages of hydrogen in a uniform field.

Then for any particular value of V,

$$m \log p + n \log d = \text{constant}.\qquad (2.9)$$

A curve relating $\log p$ with $\log d$ should then be a straight line of slope $-n/m$. If Paschen's law holds, $m = n$ and the slope of the line should be -1. This method of presenting the data gives a more precise comparison of the experimental results than that obtained by inspection of the points when plotted on a curve relating V with pd. In Fig. 2.20 the results given by Fricke [31] for the breakdown of hydrogen in a uniform field are plotted according to the above procedure relating $\log p$ with $\log d$ for various values of breakdown voltage The slopes of the resultant curves are not constant but vary from $-1\cdot03$ to $-1\cdot19$ and show therefore the occurrence of departures from Paschen's law. It is possible that these deviations can again be explained by the fact that the diameter of the electrodes may have been insufficiently large in comparison with the distance between them.

The electric breakdown in carbon dioxide, from low pressures through the critical point into the liquid state, has been investigated by Young [132] Paschen's law is verified for low gas densities, of 0 015 gm /cm.³ At higher gas densities small departures from Paschen's law are observed for long gaps and large departures for small gaps Simultaneously the scatter of the measured breakdown voltages becomes independent of illumination and the breakdown field dependent on the cathode material. Measurements of the pre-breakdown currents give values for the Townsend ionization coefficient α as well as for the field emission constants. For small gaps the pre-breakdown currents are higher than those predicted by the normal field emission equation, indicating that some additional process, possibly space-charge modification of the field, may become effective at the shorter gaps The transition from the gaseous to the liquid state does not produce a discontinuity in the breakdown voltage curve.

Breakdown in non-uniform fields

In non-uniform fields such as those between coaxial cylinders the value of α varies across the gap, and the electron multiplication in an avalanche is governed by the integral of α over the path travelled. The Townsend breakdown criterion can then be written

$$\gamma\left[\exp\left(\int_{R_2}^{R_1}\alpha\,dr\right)-1\right] = 1, \tag{2.10}$$

where R_1 and R_2 are the radii of the inner and outer cylinders Equation (2.10) may be rewritten as

$$\exp\left(\int_{R_2}^{R_1}\alpha\,dr\right) = 1+1/\gamma \tag{2 11}$$

or

$$\int_{R_2}^{R_1}\alpha\,dr = \log_e(1+1/\gamma). \tag{2.12}$$

Numerous measurements have been made by various investigators, including Huxley [47], Bruce [48], Boulind [49], Penning [13], and Craggs and Meek [50], of the breakdown of gases in non-uniform fields between coaxial cylinders. The curves given in Figs 2.21 to 2.25 show the results obtained for the breakdown voltage as a function of gas pressure in

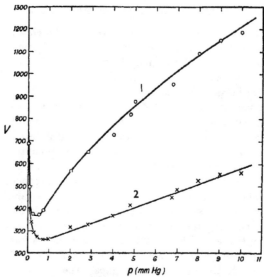

Fig 2 21 Breakdown voltage curves for nitrogen between a wire
and a coaxial cylinder (radii 0 083 and 2 3 cm respectively)
Curve 1 refers to a positive wire, curve 2 to a negative wire.

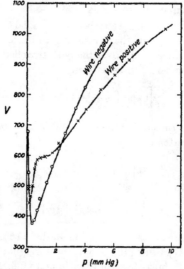

Fig 2 22 Breakdown voltage curves for oxygen between a wire and a
coaxial cylinder (radii 0 158 and 2 3 cm respectively)

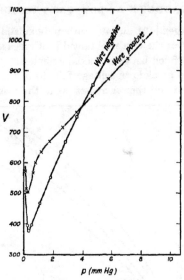

Fig 2.23. Breakdown voltage curves for air between a wire and a coaxial cylinder (radii 0 158 and 2 3 cm. respectively).

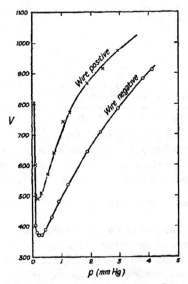

Fig 2 24 Breakdown voltage curves for carbon dioxide between a wire and a coaxial cylinder (radii 0 158 and 2 3 cm. respectively)

nitrogen [47], oxygen [49], air [49], carbon dioxide [49], and hydrogen
[48]. Fig. 2.26 gives the curves obtained by Craggs and Meek [50] for
the breakdown of freon between coaxial cylinders.

In all cases the breakdown voltage falls to a minimum as the gas
pressure is reduced and then increases, as in the case of uniform field

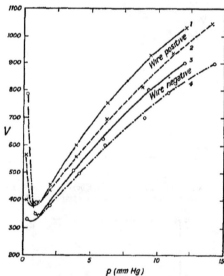

FIG. 2 25   Breakdown voltage curves in hydrogen.
Curves 1 and 3 refer to a wire 3 16 mm in diameter,
curves 2 and 4 to a wire 1 65 mm in diameter   Curves
1 and 2 refer to a positive wire, curves 3 and 4 to a
negative wire   (Radius of outer cylinder is 2 3 cm)

breakdown. The fact that the breakdown voltage is generally lower
when the inner cylinder is negative than when it is positive is usually
explained by consideration of the higher field at the cathode in the
former case so that γ is greater and therefore a lower value is needed of

$$\exp\left(\int_{R_2}^{R_1} \alpha \, dr\right)$$ in order to satisfy the breakdown criterion (2.11). Exceptions

occur in the case of oxygen and air, as Boulind and others have shown.
The reasons for these exceptions are not altogether clear, but as pointed
out by Craggs and Meek [50] it is probably not entirely due to the high
electron attachment coefficient in oxygen, as has been suggested,
because in the measurements with freon, where the attachment coefficient is also high (in the presence of free halogens when freon is partly

dissociated), the negative breakdown curve lies below the positive curve
as shown in Fig 2 26. It is possible, however, that the work function of
the cathode may be appreciably affected in the presence of oxygen either
as an adsorbed oxygen layer or by oxidation, and calculations based on
the experiments by Craggs and Meek indicate that the value of γ at the
onset of negative breakdown is much lower in oxygen than in the other
gases.

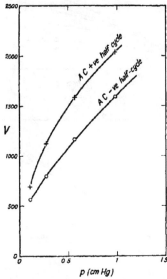

Fig. 2 26   Breakdown voltage curves
for freon (CCl$_2$F$_2$) between a wire and a
coaxial cylinder (radii 0 063 and 1 25 cm
respectively).

Curves showing the breakdown voltage between cylinders of radii
0 087 and 2 3 cm. in neon and in neon-argon mixtures as recorded by
Penning [13] are given in Fig. 2 27. The curves show that small admixtures of argon to the neon cause a large reduction in the breakdown
voltage at a given gas pressure because of ionization of argon atoms by
metastable neon atoms. The reduction is greater for a positive wire than
for a negative wire, as shown by the curves of Fig 2.28, which gives the
variation of breakdown voltage for a particular pair of electrodes in the
gas mixture at $p = 37 \cdot 6$ mm. Hg when the percentage of argon is
increased from about 10^{-3} per cent. Druyvesteyn and Penning [11]
explain the observed behaviour in the following way. With a positive

wire A, as illustrated in Fig 2.29, the number of ionizations and excitations caused by an electron avalanche increases rapidly with decreasing

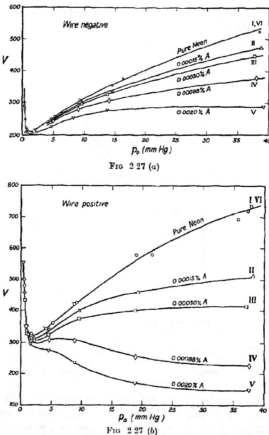

Fig 2 27 (a)

Fig 2 27 (b)

Breakdown voltage curves for a coaxial wire and cylinder, of radii
0 087 and 2 3 cm respectively, in neon and neon-argon mixtures
The numbers I to VI denote the order in which the curves were
obtained experimentally p_0 is the gas pressure in mm. Hg
corrected to 0° C

distance measured from the wire and therefore the bulk of the metastable neon atoms lie in a small cylinder of radius R_c surrounding the wire. The metastable atoms travel a certain distance before ionizing an argon atom, and consequently many of the ions formed by this process will lie outside the cylinder R_e. The electrons so produced have then to travel across the whole potential difference of the cylinder R_e and so give rise to many more new electrons than would be the case if the initial electrons

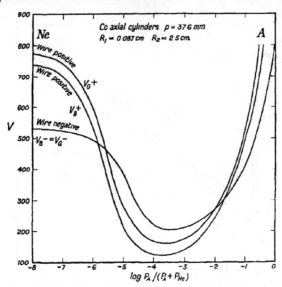

FIG. 2.28. Breakdown voltage V_B and starting voltage of the glow discharge V_G for neon-argon mixtures between coaxial cylinders at $p = 37\,6$ mm Hg.

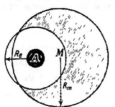

FIG. 2 29. Ion formation by metastable atoms around a wire A as anode.

had started from the original position of the metastable atoms. This leads to a considerable reduction in the value of the breakdown voltage V_B when the wire is positive but does not do so when the wire is negative. The effect is so marked that, as shown in Figs. 2.27 and 2.28, the positive breakdown voltage falls below the negative breakdown voltage.

Curves for the breakdown of helium and argon are given in Figs. 2.30 and 2.31. The helium contains some neon (< 1 per cent.) but, as Penning [13] points out, the effect of this impurity is negligible as the ionization

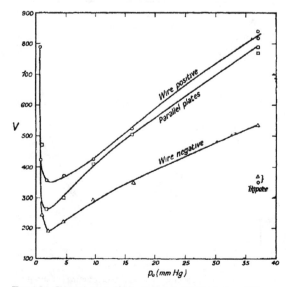

Fig. 2 30. Breakdown voltage characteristics for helium between parallel plates and between a coaxial wire and cylinder (radii 0 087 and 2 3 cm). p_0 is the gas pressure in mm. Hg corrected to 0° C.

potential for neon is higher than the excitation potential of metastable helium atoms. In the case of argon a small amount of mercury, for which the ionization potential is less than the excitation potential of metastable argon atoms, caused a slight reduction in the breakdown voltage.

Direct comparison of the results obtained by various investigators of the breakdown characteristics is not always possible because of the different diameters of electrodes used. However, as shown by Townsend and subsequent workers [2, 47], it is permissible to compare curves relating $R_1 E_1$ and $R_1 p$ where R_1 is the radius of the inner cylinder, E_1 is the voltage gradient at the inner cylinder when breakdown occurs, and p is the gas pressure. The reason for the relation between $R_1 E$ and $R_1 p$ is apparent from the following analysis [11].

When the breakdown voltage V_B is applied between the coaxial cylinders the field E at radius r is given by

$$E = \frac{V_B}{r \log_e(R_2/R_1)}. \tag{2.13}$$

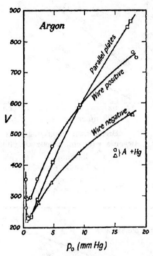

FIG. 2.31 Breakdown voltage characteristics for argon between parallel plates and between a coaxial wire and cylinder (radii 0 087 and 2 3 cm.) p_0 is the gas pressure in mm Hg corrected to 0° C

It follows that

$$dr = \frac{R_1 E_1}{E^2} dE$$

and, if we write $\alpha = \eta E$ (see p. 53) the integral of α in equation (2.12) can be expressed as

$$\int_{R_1}^{R_1} \alpha \, dr = \int_{E_1}^{E_1} R_1 E_1 \eta \frac{dE}{E}.$$

The breakdown criterion of equation (2 12) then becomes

$$R_1 E_1 \int_{E_2}^{E_1} \eta \frac{dE}{E} = \log_e(1+1/\gamma)$$

or

$$\frac{E_1}{p} R_1 p \int_{E_2/p}^{E_1/p} \frac{\eta}{E/p} \, d(E/p) = \log_e(1+1/\gamma). \tag{2.14}$$

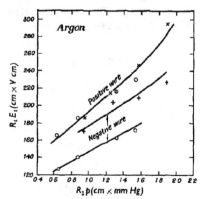

FIG. 2 32   Breakdown characteristics for argon between a wire and a coaxial cylinder.

Positive wire $\left\{ \begin{array}{l} \odot \text{ Penning} \\ \times \text{ Craggs and Meek} \end{array} \right.$

Negative wire $\left\{ \begin{array}{l} \odot \text{ Penning} \\ + \text{ Craggs and Meek} \end{array} \right.$

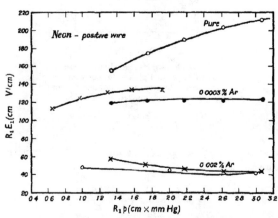

FIG 2 33. Breakdown characteristics for neon between a positive wire and a coaxial cylinder

$\left. \begin{array}{l} \odot \\ \bullet \\ \odot \end{array} \right\}$ Penning. $\times$ Craggs and Meek.
                                                          $\bigcirc$ Huxley

When $R_2 \gg R_1$ the value of E_2/p at the outer cylinder is so small that there is little error in replacing it by 0 as the limit in the integral in equation (2.14). It then follows from consideration of equation (2.14), and of the fact that both η and γ are functions of E/p, that for any particular value of pR_1 there is a particular value of E_1/p to give a solution of equation (2.14) and similarly a particular value of $E_1 R_1$.

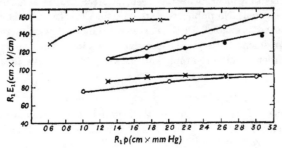

FIG. 2.34. Breakdown characteristics for neon between a negative wire and a coaxial cylinder.

○ Pure
● 0 0003% A } Penning.   × Craggs and Meek.
✖ 0 002% A   ○ Huxley.

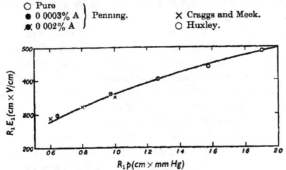

FIG. 2.35. Breakdown characteristics for oxygen between a positive wire and a coaxial cylinder.
⊙ Craggs and Meek.
× Boulind.

Consequently the breakdown voltage is determined by the field strength at the wire surface and is independent of the radius of the outer cylinder [51]. This deduction is applicable, however, to a wire cathode only. For a wire anode the value of γ depends on E_2/p and not on E_1/p and therefore the value of R_2 may be expected to have an influence.

From the above analysis it is then evident that the results obtained for cylinders of different radii may be compared if they are plotted in the form of curves relating $R_1 E_1$ with $R_1 p$. Figs. 2.32 to 2.35 show several such curves as obtained by various investigators for neon, argon, and oxygen [13, 47, 49, 50] The curves in argon as recorded by Penning [13] and by Craggs and Meek [50] are in close agreement for a positive wire but differ noticeably for a negative wire probably because of the difference in wire materials used and the consequent change in γ. In Penning's measurements a nickel wire was used, whereas Craggs and Meek employed tungsten. No published data on γ for these two metals in argon appear to be available. In the curves for the other gases there is also closer agreement between the positive breakdown voltages recorded by the different investigators than between the negative breakdown voltages.

Time lags in breakdown

The time which elapses between the instant of application of a voltage to a gap and the occurrence of breakdown is known as the time lag, and consists of two separate parts:

1. The statistical time lag caused by the need for an electron to appear in the gap during the period of application of the voltage in order to initiate the discharge.

2. The formative time lag corresponding to the time required for the discharge, once it has been initiated, to develop across the gap.

The statistical time lag is clearly a variable quantity which depends on the amount of pre-ionization, or irradiation, of the gap   Such irradiation may be produced by ultraviolet light, radioactive materials, or other methods. A gap subjected to an impulse voltage may break down if the peak voltage reaches the D.C. breakdown value provided that the irradiation is sufficient to cause an electron to be present to initiate the breakdown process. With less irradiation present the voltage must be maintained above the D.C. breakdown value for a longer period before an electron appears. Therefore, to cause breakdown of a gap with an impulse voltage of defined wave-shape, increasing values of peak voltage are required as the amount of irradiation is decreased so that the voltage is in excess of the D.C. breakdown value for increasingly long time intervals.

von Laue [52] has calculated the statistical time lag in terms of the number P_0 of primary electrons formed per sec in the gap by irradiation and the probability W that a primary electron gives rise to an avalanche of sufficient multiplication to initiate the discharge (see also Chapter VIII). If $f(t)$ is the probability that a discharge will occur during the interval after a time t measured from the instant of application of an adequate voltage to the gap, then

$$f(t)\,dt = WP_0 \exp\left[-\int_0^t WP_0\,dt\right].$$

With a constant voltage, W is constant and hence, for a constant P_0,

$$f(t)\, dt = WP_0 \exp(-WP_0 t).$$

The number n of time lags which have a duration greater than t is then given by

$$n = Ne^{-WP_0 t} \tag{2.15}$$

and the mean statistical time lag is

$$t_s = \frac{1}{WP_0}. \tag{2.16}$$

The constant W in these equations has been calculated by Braunbek [53] and by Hertz [54]. In Hertz's treatment it is supposed that each of the N_0 primary electrons leaving the cathode makes exactly the same number of ionizations $(e^{\alpha d} - 1)$ in passing to the anode. The mean number of electrons liberated by the ions due to one primary electron is equal to the multiplication factor M given by $M = \gamma(e^{\alpha d} - 1)$. Clearly for any individual case M must equal $0, 1, 2$, etc., but its average value need not be a whole number. Each time that $M = 0$ an avalanche process stops. By consideration of the various possibilities the fraction S of the primary electrons which give rise to stopping avalanches can be calculated. The probability that one electron can start the discharge is given by $W = 1 - S$. It may then be shown that

$$M = \frac{\gamma \log(1 - W)}{\log(1 - \gamma W)}. \tag{2.17}$$

As γ is usually small, equation (2.17) reduces to

$$M = -\frac{1}{W} \log(1 - W). \tag{2.18}$$

The value of W as a function of M is given in Fig. 2.36, the upper scale denoting the gap voltages in an example given by Hertz [54]. In this case the time lag decreases from infinity at the threshold breakdown voltage of 350 V to 1 sec for an overvoltage of 2 V, if $N_0 = 10$ electrons per sec.

Hertz's calculations have been extended by Wijsman [167], who concludes that the probability P_0 that an avalanche initiated by one electron from the cathode leads to breakdown is given by

$$\begin{aligned}
P_0 &= 1 - 1/q \quad \text{if} \quad q > 1 \\
P_0 &= 0 \qquad\quad \text{if} \quad q < 1
\end{aligned} \right\} \tag{2.19}$$

in which $q = \gamma \exp\left\{ \int\limits_0^d [\alpha(x')\, dx'] - 1 \right\}$ is the average number of secondary electrons liberated at the cathode as a result of the first avalanche.

Wijsman shows that the probability $R(V, T)$ that a spark occurs within a time interval T after application of a voltage V between the electrodes is given by

$$R(V, T) = 1 - \exp[-nP_0(T-T_0)] \quad \text{for } T > T_0$$
$$R(V, T) = 0 \quad \text{for } T < T_0$$

$$\left. \right\} \qquad (2.20)$$

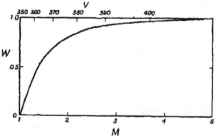

FIG. 2.36. Curve relating the probability W, that a primary electron is multiplied into a steady current, with the multiplication factor M. The upper scale gives the corresponding voltage V in a particular case (argon, $p = 10\ 6$ mm. Hg, $d = 0\ 8$ cm , $\gamma = 0\ 02$, $V_B = 350$ V).

In this expression n is the average number of electrons liberated per second at the cathode by an external irradiating source, and T_0 is the formative time lag.

The effect of preceding discharges on the statistical time lag has been studied by Paetow [55]. A typical result is shown in Fig 2 37 which corresponds to a case where a current of 1 μA passed for 1 sec. between two plane nickel electrodes. At a time of t_w sec. after extinguishing this current the value of N_0 was determined from the statistical time lag $1/(N_0 W)$, the overvoltage being so great that $W = 1$. The influence of the current of 1 μA continues to be noticeable at a time of 50 sec. after its cessation, the number of electrons liberated per sec. being then about 10. This so-called after-current is explained by Paetow as field emission caused by surface charges on an impure cathode, these surface charges possibly being produced by photons from the preceding discharge.

The formative time lag required for the growth of the discharge from the condition where the current flow in the gap is i_0 until breakdown is complete can be calculated in the approximate manner given by Druyvesteyn and Penning [11] as follows. One ionization cycle takes a time t_i which is approximately equal to the transit time of an ion from anode to cathode. A primary current I_0 corresponding to N_0 electrons per sec is assumed to change in a discontinuous manner in the time t_i into $I_0(1+M)$ and after $2t_i$ into $I_0(1+M+M^2)$. After time $t_F = nt_i$ the electron current I_t at the cathode is given by

$$I_t = I_0 \frac{M^{n+1}-1}{M-1}. \qquad (2.21)$$

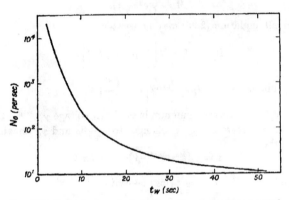

Fig 2 37. Rate of emission of electrons, N_0 per sec. from a nickel cathode as a function of time following a discharge of 1 μA for 1 sec.

Generally I_t/I_0 is large and $n \gg 1$, so that equation (2.21) may be written

$$n \log M = \log\left\{\frac{I_t}{I_0}(M-1)\right\}$$

or

$$t_F = t_i \frac{\log\{(M-1)I_t/I_0\}}{\log M}. \qquad (2.22)$$

A more exact treatment of the problem has been carried out by Schade [56] as follows. The electron current from the cathode at time t is

$$I(t) = I_0 + \gamma I_+(t).$$

The positive ion current at the cathode at time t is equal to the positive ion current at the anode at the earlier time $(t - t_i)$

$$I_+(t) = I_{a+}(t - t_i).$$

The positive ion current at the anode is

$$I_{a+}(t) = I_a(t) - I(t)$$

and the electron current at the anode is

$$I_a(t) = I(t)e^{\alpha d}.$$

From these four relations it is clear that

$$I(t) = I_0 + \gamma I_{a+}(t-t_s)$$
$$= I_0 + \gamma[I(t-t_i)e^{\alpha d} - I(t-t_i)]$$
$$= I_0 + M I(t-t_i), \tag{2.23}$$

where $\qquad\qquad M = \gamma(e^{\alpha d} - 1). \tag{2.24}$

If t_i is small, equation (2.23) may be written

$$I(t) = I_0 + M\left[I(t) - t_i \frac{\partial}{\partial t} I(t)\right]. \tag{2.25}$$

Then, by integration, $\qquad t_F = \frac{t_i}{\epsilon} \log\left[1 + \epsilon \frac{I(t)}{I_0}\right] \tag{2.26}$

where $\epsilon = M - 1$ At the minimum breakdown voltage $\gamma_0(e^{\alpha_0 d} - 1) = 1$. If, for a small overvoltage, α_0 changes to $\alpha_0 + \delta\alpha$ and γ_0 remains unchanged, then

$$\epsilon = \gamma_0[e^{(\alpha_0 + \delta\alpha)d} - 1] - 1 = \delta\alpha\, d,$$

and therefore $\qquad t_F = \frac{t_i}{\delta\alpha\, d} \log\left[1 + \delta\alpha\, d \frac{I(t)}{I_0}\right]. \tag{2 27}$

Over a limited range of values of E/p it may be assumed that

$$\alpha/p = A e^{-Bp/E}$$

and consequently for a given pressure $\alpha = ae^{-b/V}$, where a and b are constants and V is the applied voltage. Hence

$$\delta\alpha = ae^{-b/V}\left[1 - \exp\left(-\frac{b(V-V_0)}{VV_0}\right)\right] = ae^{-b/V} \frac{b(V-V_0)}{VV_0} \tag{2.28}$$

for a small overvoltage $(V-V_0)$, where V_0 is the minimum sparking voltage. From equations (2 27) and (2 28) we have, approximately,

$$t_F = \frac{a}{V - V_0} e^{-b/V}, \tag{2.29}$$

the logarithmic term in equation (2.27) being roughly constant as the term $\epsilon I(t)/I_0$ is large.

For $V = V_0$, equation (2.29) gives an infinite time for the formative time, but in this case, if equation (2.26) is considered, the correct solution as $\epsilon \to 0$ is

$$t_F = t_i \frac{I(t)}{I_0}. \tag{2.30}$$

For $t_i = 10^{-5}$ and $I(t)/I_0 = 10^{10}$, the formative time t_F is 10^5 sec.

Measurements by Schade [56] of the formative times of sparks in neon tend to support the view that the time is given by an expression of the form of equation (2.29). The times observed experimentally are plotted as points in Fig. 2.38; the full curve represents equation (2.29) when the constants a and b are determined by matching the curve to two of the observed points.

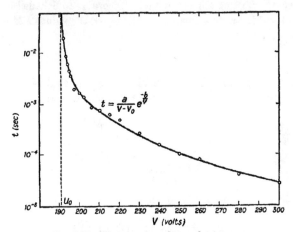

$$t = \frac{a}{V-V_0} e^{-\frac{b}{V}}$$

Fig. 2 38. Formative time for sparks in neon.

Another calculation of the formative time lag of sparks has been made by Steenbeck [57] who assumed that the current builds up according to the relation

$$i = i_0 \, e^{t/\tau}.$$

From considerations of the breakdown mechanism Steenbeck shows that τ is given by the expression

$$\frac{\alpha v' \tau \gamma}{\alpha v' \tau - 1}\left[\exp\left(\frac{\alpha v' \tau - 1}{\alpha v' \tau}\, \alpha d \right) - 1 \right] = 1, \qquad (2.31)$$

where v' is related to the velocities v_+ and v_- of the positive ions and electrons by the expression $1/v' = 1/v_+ + 1/v_- \sim 1/v_+$.

In the above calculations it has been assumed that the secondary electrons are produced by the impact of positive ions on the cathode. However, it is known that in some discharges secondary electrons are also produced by radiation from the gas The formative time lag has been calculated by Raether [58, 59], taking into account both these factors, and is given as

$$t_F = \frac{M_+ t_i + M_r t_-}{M_+ + M_r - 1}\, \log\left[1 + (M_+ + M_r - 1)\frac{I(t)}{I_0} \right], \qquad (2.32)$$

where M_+ is the part of M which is caused by positive ion bombardment, M_r is the part of M caused by radiation, and t_+ and t_- are the times of transit of positive ions and electrons respectively across the gap.

Heymann [20] has deduced an expression for the formative time lag in the case of a rising applied voltage. According to Schade [56] when the gap is subjected to a small overvoltage ΔV the value of M is given approximately by

$$M - 1 \sim \frac{\partial \alpha}{\partial E} \Delta V.$$

For small overvoltages $M-1$ is small. If the voltage rises linearly the average overvoltage will be half the final value reached:

$$(M-1)_{av} \sim \frac{1}{2} \frac{\partial \alpha}{\partial E} \Delta V,$$

or
$$(M-1)_{av} \sim \frac{1}{2} \frac{\partial \alpha}{\partial E} \frac{\partial V}{\partial t} t_F. \tag{2.33}$$

From Schade's equation (2.26) for t_F it follows that, provided $M-1$ is small,

$$t_F \sim \frac{t_i}{M-1} \log\left[1 + (M-1)\frac{I(t)}{I_0}\right] \tag{2.34}$$

and therefore, from equations (2.33) and (2.34),

$$t_F \sim \left\{\frac{2t_i}{(\partial \alpha/\partial E)(\partial V/\partial t)} \log\left[1 + (M-1)\frac{I(t)}{I_0}\right]\right\}^{\frac{1}{2}}. \tag{2.35}$$

The final overvoltage is then

$$\Delta V_m = \frac{\partial V}{\partial t} t_F = \left\{\frac{2t_i \, \partial V/\partial t}{\partial \alpha/\partial E} \log\left[1 + (M-1)\frac{I(t)}{I_0}\right]\right\}^{\frac{1}{2}}. \tag{2.36}$$

The logarithmic term causes little variation and it follows roughly that

$$t_F \propto \left(\frac{\partial V}{\partial t}\right)^{-\frac{1}{2}}; \qquad \Delta V_m \propto \left(\frac{\partial V}{\partial t}\right)^{\frac{1}{2}}. \tag{2.37}$$

Experimental curves relating ΔV_m with $(\partial V/\partial t)^{\frac{1}{2}}$ for breakdown in argon have been made by Heymann [20]. The slopes of these curves are given in Table 2.3 where they are compared with the theoretical slopes deduced from expression (2.36).

A further discussion of oscillographic studies of the formative time of sparks is given in Chapters IV and VIII.

TABLE 2.3

p	d	t_i	Experimental slope	Theoretical slope
4 56	0 104	0 17	0 023	0·017
	0 8	4 95	0 086	0 083
9 5	0 1	0 29	0 025	0 020
	0 41	2 74	0 065	0 064
8 65	0 2	0 84	0 036	0 034
	0 81	7 05	0 114	0 107

Breakdown in vacuum

When the gas pressure in a discharge chamber is reduced to such a low value that the mean free paths of electrons and positive ions moving in the residual gas is large compared with the gap between the electrodes the mechanism of breakdown becomes independent of ionization in the gas. This condition is obtained for instance with a 1 cm gap at a residual gas pressure of 10^{-6} mm Hg or less. Further reduction in gas pressure does not cause any change in the breakdown voltage unless the gas present affects the surfaces of the electrodes.

The breakdown mechanism in vacuum appears to be closely associated with the field emission of electrons from the cathode, particularly in the case of short gaps. It is generally thought that sparks take place from projections on the surface of the cathode where the electric field is appreciably higher than for the surface as a whole. High field current densities occur at those projections, with a consequent increase in local temperature which leads to vaporization of the metal and the breakdown of the gap. It is proposed therefore to review briefly some of the relevant data available on field emission before proceeding to a discussion of experiments directly concerned with breakdown. A summary of the published work on field emission up to 1943 has been given by Jenkins [60].

Following experiments by Millikan and Eyring [61], who recorded the field currents emitted from a tungsten wire cathode, Millikan and Lauritsen [62] found that the results could be plotted in the form of the logarithm of the current against the reciprocal of the field. They then proposed the following formula to express both the field and thermionic emissions:

$$i = A(T+cE)^2 \exp\left(-\frac{b}{T+cE}\right), \qquad (2.38)$$

where E is the field strength, T is the temperature of the cathode, and A, b, and c are constants. The implication in equation (2.38) is that an increase in the field is effectively equal to an increase in the temperature of the electrons in the cathode. In a later paper Millikan, Eyring, and Mackeown [63] showed that the experimental results obtained from a study of the field emission currents from points could be expressed by nearly straight line relations when plotted in the form $\log i$ against $1/E$.

In similar experiments by Gossling [64] for the field currents from wires and point electrodes the results were plotted in the form of curves relating $\log i$ with $V^{\frac{1}{2}}$, where V is the total voltage across the gap. These curves were, in general, concave towards the $V^{\frac{1}{2}}$ axis and showed that the field currents could not be explained in terms of the Schottky equation [66]. Gossling's data replotted later by Millikan, Eyring, and Mackeown [63] in the form of $\log i$ as a function of $1/E$ gave approximately straight line relationships.

Later measurements by de Bruyne [67, 68] appeared to show that field emission was independent of temperature at temperatures below that at which thermionic emission becomes important. This evidence conflicted with equation (2.38), but it seems that the increase in emission with temperature observed by Millikan and his colleagues [61, 62, 63, 69] was probably due to the presence of thoria in the wire used in their experiments, so that after heat treatment there may have been a surface layer of thorium which could give appreciable thermionic emission at 1,100° K.

In the above experiments the voltages used were relatively low and the high fields were obtained by the use of wire or point electrodes. Some results for nearly uniform fields were obtained, however, by Hayden [70] who measured a breakdown gradient of $1 \cdot 2 \times 10^6$ V/cm. for a discharge across a $0 \cdot 306$ cm. gap between molybdenum spheres. By careful outgassing of the spheres Piersol [71] subsequently recorded an increase in the breakdown gradient up to $5 \cdot 4 \times 10^6$ V/cm.

A new approach to the theory of field emission in terms of quantum and wave mechanics was made by Fowler and Nordheim [72], who deduced the following expression relating the field current and the voltage gradient:

$$I = 6 \cdot 2 \times 10^{-6} \frac{\mu^{\frac{1}{2}}}{(\phi + \mu)\phi^{\frac{1}{2}}} E^2 \exp\left[-\frac{6 \cdot 8 \times 10^7 \phi^{\frac{3}{2}}}{E}\right], \qquad (2.39)$$

where I is in A/cm.², E is in V/cm., μ is the maximum kinetic energy of an electron at absolute zero, ϕ is the normal thermionic work function. The physical interpretation of equation (2.39) is that when the field is small the escaping electrons must have sufficient thermal energy to raise them above the top of the potential barrier. With increasing field the barrier becomes progressively thinner until eventually there is a finite chance of an electron escaping through the barrier, the mechanism being referred to as the 'tunnel effect'.

The Fowler–Nordheim equation indicates that emission should be apparent at fields of about 10^7 V/cm or more. This agrees reasonably with experiment when account is taken of the fact that emission occurs mainly from surface irregularities where the actual field may be an order of magnitude different from the measured field In a paper of Stern, Gossling, and Fowler [73] a number of experimental observations have been examined in terms of the above equation. It is assumed that because of surface roughness the actual field F causing emission at localized points is equal to BF_m where F_m is the measured field. From a plot of log I against I/F_m for field emission from a fine electrolytically polished tungsten point, B is found to be 2 68 and the emitting area to be $6 \cdot 3 \times 10^{-8}$ cm.^{2}

Later measurements by Chambers [74] confirm the importance of the surface condition of the cathode, and show that by suitable heat treatment it is possible to reduce the small projections which give rise to local increased field strengths and become emitting centres. Chambers considers that the earlier results published by Del Rosario [75], who measured appreciably higher breakdown fields than those recorded by other investigators, can be explained by the heat treatment used.

Experiments on the breakdown fields for tungsten and thoriated tungsten have been made by Ahearn [76]. The results show that the breakdown field is independent of the degree of thoriation and also of the temperature, provided that the latter is insufficient to cause appreciable field emission. The highest electric field which could be applied to a cathode without breakdown was about $4 \cdot 7 \times 10^6$ V/cm This observation that the field emission currents are independent of the degree of thoriation, as deduced from thermionic emission from the whole filament, appeared to conflict with the result expected from the Fowler–Nordheim expression, which implies that the field emission should change rapidly with the work function. However, Ahearn points out that the small projections on the cathode surface which form the centre of the field emission may not be activated to the same extent as the main cathode surface. He also suggests that breakdown of the gap may be caused by the ohmic heating at these points by the field currents which in turn reduced the tensile strength of the material to a value below the electrostatic pull due to the high field.

The breakdown of gaps with mercury cathodes has been studied by

Beams [77], Quarles [78], and Moore [79]. The reasons for the choice
of mercury as a cathode material were partly because of its practical
importance in mercury arc rectifiers but also because a mercury pool
automatically assumes a smooth surface and its work function can be
varied by repeated distillation. As D.C. fields cause a distortion of the

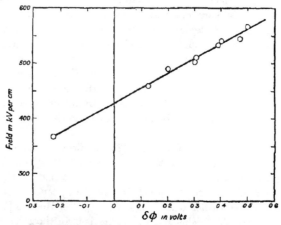

FIG. 2.39. Breakdown voltage gradient as a function of the work
function of the mercury surface.

surface it was necessary to use impulse voltages of duration 10^{-7} to 10^{-6}
sec. In investigations of the breakdown between a steel anode sphere and
a mercury cathode Beams [77] found that the breakdown field could be
increased from $3 \cdot 5 \times 10^5$ V/cm. for impure mercury up to $1 \cdot 8 \times 10^6$ V/cm.
for mercury after repeated distillation   Quarles [78] continued the
measurements with a molybdenum anode sphere of 2 cm. diameter and
a mercury cathode and related the breakdown field to the change $\delta\phi$
in the work function of the cathode with the results shown by the curve
of Fig 2.39. The observed difference in the breakdown field for a given
change in work function is appreciably greater than that predicted by
the Fowler–Nordheim formula The latter gives a 30 per cent. difference
in breakdown field for a 1 V change in work function whereas experiment
shows the difference to be 60 per cent. Later experiments by Moore [79],
who also studied the effect of the work function on the breakdown field,
confirm the results obtained by Beams [77] and Quarles [78] and show
that the measurements are in qualitative but not in quantitative agreement with theory. Oscillographic records by Moore [79] show that the

duration of the impulse voltage used is 10^{-7} sec., and it is considered unlikely that the surface of the mercury is distorted sufficiently in such a short time to account for the fact that the measured breakdown field is appreciably lower than the theoretical value.

Following the earlier work by Snoddy [80] and by Beams [81] Chiles [82] has investigated the onset of spark discharges in vacuum by means of a rotating-mirror camera in which the mirror rotates at 1,650 r p s.

TABLE 2.4

Metal	Gap in mm.	Mean time interval in 10^{-8} sec.	Mean vapour velocity in 10^5 cm /sec	Voltage across gap in kV
Aluminium	1 1	4 1	7 3	75
Copper .	0·89	5 6	5 8	80
Magnesium	1 0	5 5	7 9	75
Tin . .	1 3	0 34	5·4	80
Tungsten (pure)	0 6	2 5	5 2	120
Tungsten (commercial)	0 81	2 4	5 4	95

The luminosity at the anode is observed to precede that at the cathode by a short time-interval which varies according to the electrode materials used, as shown in Table 2.4. The velocity of the vapour cloud from the anode, as measured by Chiles, is also included in the table.

Beams [83] has also studied the breakdown of point-plane and wire-cylinder gaps in vacuum when subjected to impulse voltages of ~ 1 μsec duration. The electrode materials used were platinum, tungsten, or steel for the smaller electrode and nickel or steel for the larger electrode. With a negative voltage on the wire the discharge is initiated by a field current from the wire. With a positive voltage on the wire discharges occur for fields at the cathode much lower than that required to cause field emission, and it is concluded that the discharge must then be initiated by the emission of positive ions from the anode On the other hand, experiments by Mason [84] appear to show that positive ion emission from the anode is unimportant. Mason points out that if the breakdown of the gap is to be initiated by the emission of positive ions, at least one electron must be liberated from the cathode in order to release one or more positive ions when it hits the anode. The field required to cause the liberation of an occasional electron from the cathode may be expected to be considerably lower than that to cause appreciable field emission. Hence, if another source of initiating electrons is provided and the breakdown voltage is found to be affected thereby, it should be possible to test the possibility of the influence of positive ion emission

by electron bombardment of the cathode. Mason therefore compared
the breakdown voltage characteristics of gaps with cold and hot cathodes
for an electrode arrangement in which the cathode consists of a 0 028-in.
diameter tungsten wire bent in a semicircle of 1-in. diameter and the
anode of a 2-in. diameter copper disk with its surface perpendicular to

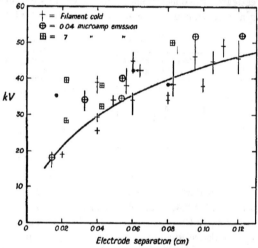

FIG. 2.40   Breakdown voltage as a function of electrode
separation with a cold cathode and also with a cathode heated
to give different electron emission currents (measured at
200 V) The curve drawn through the measured points gives
the voltage required to give a constant field of 600 kV/cm.
at the cathode.

the plane of the cathode semicircle. The breakdown voltage curves are
given in Fig 2 40. At each point the vertical line connects the highest
and lowest voltages recorded in a particular series of tests, the breakdown
voltage generally increases gradually for successive breakdowns. The
full curve in Fig 2.40 shows the voltage necessary to give a constant
field of 600 kV/cm Most of the points for a cold cathode lie along this
curve, but a slightly higher voltage appears to be required to cause breakdown for a heated cathode in contradiction to the expected result if the
emission of positive ions by electron bombardment plays a role in the
mechanism

Oscillographic studies by Hull and Burger [85] of the breakdown of a
2-mm. gap between tungsten electrodes show that the current starts
as a pure electron discharge but changes to a tungsten vapour arc within

a time of less than 10^{-7} sec. The arc voltage drop is below 1,000 V with currents as high as 20,000 A. On the termination of the current the tungsten vapour condenses and the breakdown voltage reverts to its original value. The time lag for the transition from an arc to a pure electron discharge depends on the penetration of heat into the electrodes and is between 1 and 10 μsec. for heavy discharges. The approximate voltage gradient required to cause breakdown is 500 kV/cm. for tungsten and is 100 kV/cm. for carbon.

Mason [84] has also measured the impulse breakdown characteristics of a gap between aluminium electrodes in which the normal D.C. breakdown gradient is about 300 kV/cm. The oscillographic records show that, for an impulse voltage rising to 60 kV in 1·5 μsec., the breakdown voltage is higher than that under D.C. conditions. The result is explained by the hypothesis that few electrons are emitted from the cathode until the impulse voltage wave exceeds the D.C. breakdown voltage. The impulse voltage then continues to rise until such time as metal vapour from the electrodes crosses the gap to cause final breakdown. A few measurements for gaps of 0·02 cm. to 0·07 cm. may be explained in this way if the vapour velocity is $\sim 10^5$ cm /sec.

Some results for nearly uniform fields have been obtained by Anderson [86], who found that the breakdown voltage was affected appreciably by the smoothness of the cathode surface, and that higher gradients were necessary for breakdown after the electrodes had been subjected to a low pressure glow discharge in hydrogen. Different breakdown voltages were obtained according to the material of the cathode surface as shown by the results given in Table 2 5 which refers to the breakdown of a 1-mm. gap after conditioning by a glow discharge.

TABLE 2.5

Material	Breakdown kilovolts
Steel	122
Stainless steel	120
Nickel	96
Monel metal	60
Aluminium	41
Copper	37

The breakdown of gaps between a stainless steel sphere of 1 in. diameter and a 2-in. diameter steel disk in vacuum, for voltages up to 680 kV, has been investigated by Trump and Van de Graaff [87]. The resultant curves of breakdown voltage and of breakdown voltage gradients are given in Fig. 2.41. Whereas a voltage gradient of several million V/cm. is required to break down gaps of 1 mm. or less a gradient of the order of 100 kV/cm. is all that is necessary to cause breakdown of a gap of 5 cm. This gradient is inadequate to produce breakdown by the mechanism of field emission and some other mechanism must therefore be involved.

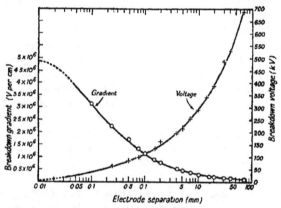

FIG. 2.41 Curves for the breakdown voltage and the breakdown voltage gradient between a steel sphere, of 1 in. diameter, and a steel disk, of 2 in. diameter, in vacuum.

Trump and Van de Graaff [87] have elaborated an earlier suggestion by Van de Graaff that the breakdown mechanism for the longer gaps may be accounted for by an interchange of charged particles and photons between cathode and anode. An electron leaving the cathode is accelerated by the voltage between the electrodes and bombards the anode with the resultant emission of a positive ion and photons. The positive ion and some of the photons then cause further electron emission from the cathode. If A denotes the average number of positive ions produced by one electron, B the average number of secondary electrons produced by one of these positive ions, C the average number of photons produced by one electron, and D the average number of secondary electrons produced by one photon, then the process becomes cumulative and breakdown ensues when

$$AB + CD > 1. \qquad (2.40)$$

The various coefficients are functions of the voltage between the electrodes and of the electrode materials. Some experiments to determine their values have been carried out by Trump and Van de Graaff [87] for a 2-cm. gap between parallel plate steel electrodes. The small value of the coefficient A, between 2×10^{-4} and 20×10^{-4} in the energy range studied, indicates that B must be large if this particle mechanism is to be important. The measured number B' of electrons per positive ion has a value ranging from about 1 to 2 as the applied voltage is increased from 100 to 200 kV. The investigators point out that the true value of B may be many times larger than B' because of the fact that in an actual gap the cathode is subjected to a high voltage gradient which has the effect of assisting the escape of the electrons. A heavy positive ion having energy of the order of 100 kV will produce brief but intense local heating at the point of impact. Under conditions of high cathode gradient space-charge limitation of current is reduced and a relatively large number of electrons may escape. No experimental data appear to be available yet for the coefficients C and D.

In measurements of the breakdown between coaxial cylinders, of area 500 cm.^{2} and spaced at 7 mm., Seifert [88] has recorded the frequency of breakdown as a function of applied voltage for voltages up to 100 kV. The breakdown voltage is found to depend on (1) the number of previous breakdowns, (2) the energy dissipated per breakdown, (3) the electrode smoothness, cleanliness, and outgassing, (4) the arrangement of the solid insulation in the system as regards high fields and surface conditions, (5) the geometry and polarity of the electrodes, with particular reference to the bombardment of insulators by electrons, and (6) the rate of pumping. Irregular background currents of the order of 1 mA at voltages above 60 kV are usually accompanied by fluorescence on the quartz supports, and it is believed that the removal of solid insulation from the high field regions should lead to an increased breakdown voltage.

Intermittent post-breakdown discharges

The characteristics of low pressure discharges for currents of ~ 1–$1,000\,\mu A$, i e for the currents flowing after breakdown has occurred but before a normal glow discharge is established, have been little studied. The region referred to is that shown in Fig. 2.42 which shows a negative current-voltage characteristic and is associated with flashing neon-lamp circuits and other relaxation phenomena [89].

Oscillographic studies of the currents flowing in this region show marked variations with the nature of the gas being studied. Craggs and Meek [50] have carried out preliminary work in several gases at pressures varying from 0·1 to about 2 cm. Hg following earlier work by other authors [90]. Valle [91] (see also Schulze [92]) has made a particularly detailed study of the macroscopic properties of discharges with negative characteristics but these early studies, neglecting space charge effects and the movements of charged particles and relying instead on lumped

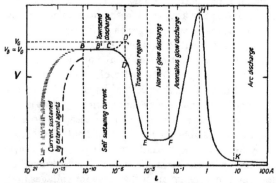

FIG 2.42   Schematic characteristic of a gas discharge between flat parallel plates   Scale for the current   logarithmic between 100 and 0 1 amp, still more contracted for lower values of ι — — · — current with photo-emission from the cathode ; · · · · · (CD') positive characteristic of the Townsend discharge, the breakdown potential V_B is lower than the starting potential of the glow discharge V_G, ———— (CD) negative characteristic of the Townsend discharge, $V_B = V_G$.

characteristics (e g inductance) of the discharges, need to be supplemented by considerations of ionization phenomena on the lines developed by Druyvesteyn and Penning [11].

It has been suggested [50] that photoionization may be an important process in argon even at pressures of a few centimetres in agreement with the results of Weissler [93] at pressures varying from an atmosphere down to ~ 10 cm. Hg.

The discharge occurring in voids in solid dielectrics [94] (see the work of Whitehead [153] and references there cited) sometimes show oscillographic characteristics similar to those found by Craggs and Meek [50]. A study of these discharges is of value since they may greatly affect the useful life of an insulator and lead to its premature failure.

Geiger counters

Geiger counters are widely used devices for the detection, registration, and measurement of ionizing atomic particles of all kinds, and their

operating characteristics have been studied by many investigators with
the result that the various mechanisms involved now appear to be
fairly well understood. Comprehensive accounts of Geiger counters
have been given in recent books [95 to 98] and review articles [99, 135].
In this section we shall give only a brief account of present knowledge
concerning Geiger counter discharge mechanisms in so far as they relate
to low pressure discharge problems in general.

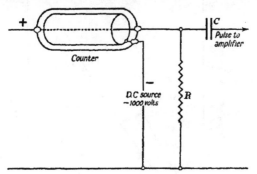

FIG 2 43. Basic circuit of Geiger counter output stage

The basic method of using a Geiger counter is shown in Fig. 2.43 where
the counter is shown in its typical but not inevitable form of an axial
wire as the anode and a cylinder as the cathode. These are sealed within
a glass tube or other suitable container which is filled with a gas generally
at a pressure of ~ 10 cm. Hg. The static characteristic of such a tube,
with a low circuit resistance, is very similar to that given by Loeb for
high pressure corona discharges in that at a certain voltage (usually
~ 1,000 V) a sudden increase in current occurs. This voltage is that at
which, with a suitably high resistance of 10^6 to 10^{12} Ω, the tube will
begin to count ionizing particles passing through it, i.e. will provide
short voltage pulses in the output part of the circuit of Fig. 2 43. The
pulses are discrete, each one corresponding to one incident particle, and
the tube recovers its operating potential between successive pulses
provided that the circuit and tube adjustments are satisfactorily made.
The pulse shape depends greatly of course on these variables [99].

When a particle passes through a counter an electron avalanche may
be initiated by ionization produced in the counter gas by the particle
and travels radially towards the wire. This localized avalanche does
not spread along the counter unless the voltage is at or above the so-
called Geiger or Geiger–Müller threshold. Typical characteristics below

the Geiger threshold are given in Fig. 2.44 where the amplification factor
is the ratio of the charge in the pulse to the initial ionization produced in
the counter. The amplification factor, over an appreciable voltage range
near the counting threshold, is independent of the initial ionization
caused by the particle, and so the pulse size depends directly upon the

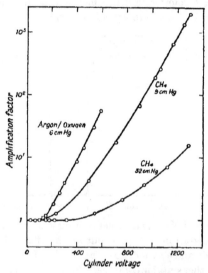

FIG 2.44   Amplification-factor variation with
voltage, for A/O$_2$ mixture (94 per cent. A), and
for commercial methane at pressures of 9 and
32 cm. Hg, near the proportional threshold.

initial ionization. This is the proportional counting region in which
pulse sizes for different kinds of atomic particles vary as shown in Fig.
2.45 for α and β particles. As the Geiger threshold is approached (at
about 3,100 V in Fig. 2 45) the constancy of the amplification factor
vanishes and instead the pulse size becomes independent of the initial
ionization. This is because the original localized Townsend avalanche
can be propagated along the counter, in a direction at right angles to
its initial radial direction, so that at a sufficiently high voltage, at the
Geiger threshold, the discharge fills the whole length of the counter. A
further increase in voltage gives no corresponding increase in pulse
size. The transition region (D of Fig. 2.45) covers a voltage range in
which the distance of propagation varies for each particle counted.

K

There are for the purpose of the present account three main phenomena
to be briefly described, viz.:

1. the mechanism of counters in the proportional region,
2. the mechanism of discharge propagation, and
3. non-proportional counting.

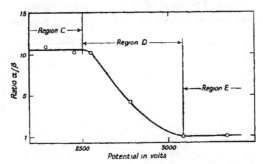

FIG. 2 45 The transition between the region of proportional
amplification of the initial ionization in a counter (region *C*)
to the Geiger–Muller region (region *E*). The ordinates are
the ratios of the sizes of the voltage pulses produced by
alpha particles shot into the counter through a thin window
to the sizes of the pulses produced by beta particles whose
ranges are equal to the length of the counter; the abscissae
are the potential differences across the counter

Proportional counting

The avalanche is produced from the initial electrons ejected from the
gas atoms or molecules in the counter by the counted particle. The
electrons in this avalanche are swept rapidly into the wire and the
positive ions are swept slowly across the counter to the cathode cylinder
from the immediate neighbourhood of the wire where most of them were
produced (see the discussion on negative and positive point coronas in
Chapter III). The movement of the positive ion space charge causes
the useful voltage pulse on the wire (see Corson and Wilson [99] for an
accurate account of pulse formation in Geiger and proportional counters
and in ionization chambers). Whilst the positive ions are moving towards
the cylinder other ionizing particles can be counted there or in other
parts of the tube since the size of the pulse voltage is small compared
with the wire voltage.

Rose and Korff [101] have given a useful account of proportional
counting on the assumptions that photons and positive ions in the
avalanche do not cause emission of electrons from the cathode, that

recombination and attachment processes are negligible, and that
fluctuations in the ionization processes are negligible. The amplification
factor A derived by Rose and Korff is then found to be

$$A = \exp\left\{2(aN_m\,Cr_1V_0)^{\frac{1}{2}}\left[\left(\frac{V_0}{V_t}\right)^{\frac{1}{2}}-1\right]\right\}, \qquad (2.41)$$

where a is the rate of increase of ionization cross-section with energy,
N_m is the number of molecules of type m per unit volume, $\frac{1}{2}C$ is the

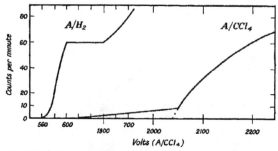

FIG. 2.46   Characteristic curves of counters   The curve on the
left is that of a non-proportional Geiger–Muller counter, using a
quenching resistance of 10^8 ohms. The threshold of this counter
is at 575 volts   The curve on the right is for the same counter
after CCl_4 has been added to a pressure of 1 cm. Hg   The voltage
scales are overlapped but different. The CCl_4 counter is exposed
to both alpha and beta rays   Below 2,035 volts it counts alpha
particles but no beta particles. The right-hand portion of the
curve and the dotted extension is the beta-particle response curve.
The quenching resistance is 10^6 ohms in this case.

capacitance per unit length of the counter, r_1 is the radius of the wire,
and V_0 and V_t are respectively the operating and proportional threshold
voltages [109]   The formula is only applicable when the above assumptions hold, e.g. for values of A between 10^4 and 10^5. Transition curves
showing the change from proportional to non-proportional counting
are shown in several papers [102, 103]. Fig. 2.46 has been obtained by
Korff [109].

Further work on the calculation of avalanche sizes in Geiger counters,
or for wire-cylinder electrode arrangements in general, has been carried
out by Brown [105], Curran et al. [106], Medicus [107], and Fetz and
Medicus [108].

Rose and Korff [101] obtained considerable experimental evidence
in favour of their formula for certain gases. The amplification factor
was determined from a knowledge of the specific ionization of the P_0

α-particles used in the experiments. Such data exist for most of the gases used, and the constant for CH_4 was taken as being the same as that for Ne, a procedure which is justified by their similarity in structure. The amplification factor is given by

$$A = cP/en, \qquad (2.42)$$

where c is the capacitance of the anode system, P the pulse size, n the initial ionization, and e the electronic charge. The pulse size in suitable units may also be expressed [110] as

$$P = (1/c)nf(V) = \sigma lpf(V)e/c, \qquad (2\,43)$$

where σ is the specific ionization of the counted particle in the counter gas at pressure p, l is the path length in the tube, and $f(V)$ is the gas amplification at a voltage V. It is clear that a conventional amplifier of finite sensitivity will enable only the upper part of the characteristic to be examined.

It was found that the gases investigated could be separated into two groups. Monatomic and diatomic gases such as A and H_2 and the mixtures A/H_2, A/O_2, and Ne/H_2, for one group, give a rapid variation of pulse size with voltage. The second group comprises the polyatomic gases and mixtures of polyatomic with other gases where the proportion of the polyatomic constituent is high (see Fig. 2.46). Examples of such mixtures are A/CH_4, A/BF_3, H_2/CH_4, and illuminating gas. These gases are used as fillings in the so-called self-quenching counters which may be operated with low external resistances. The relative performance of the two types of gas is seen in Fig. 2.46. The A/H_2 counter was necessarily used with a high external resistance, of $10^9\ \Omega$, whereas the self-quenching type may be used with $10^6\ \Omega$ or less. It is seen from Fig. 2.46 that in the proportional region below 600 V the A/H_2 counter requires a stable power supply. The A/CCl_4 counter was used with a source emitting α and β rays, the latter being responsible for the rapidly rising part of the curve above 2,035 V below which only α rays are recorded. The property of discriminating in this way is referred to more fully later and is an important property of proportional counters.

The essential differences between the two curves of Fig. 2.46 have been further emphasized by Rose and Korff. Fig. 2.47, taken from their paper, refers to a copper cathode counter 3 cm. long, 1 cm. in diameter, with a wire of diameter 0·075 mm. and filled with CH_4 and A at a pressure of 10 cm. Hg. It is seen that even in the presence of a large concentration of a self-quenching gas a monatomic gas still has a considerable effect. The curves which are extended approximately to the non-proportional threshold were derived from equation (2.41) and were fitted to one of the experimental points as the proportional threshold could not in these measurements be detected (see Fig 2.44). Stuhlinger [111] found proportionality in argon-filled counters to extend over a range of approximately 2,000 V. The correctness of Rose and Korff's theory is clearly indicated in Fig. 2 47 except for the 50 per cent. CH_4 mixture.

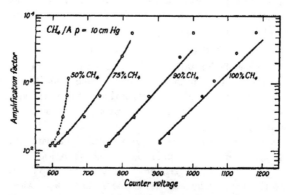

FIG 2 47   The amplification factor (logarithmic scale) plotted against counter voltage for various relative concentrations in CH_4/A mixtures at 10 cm Hg pressure. The points are experimental and the full curves are theoretical with one point adjusted The relative concentration of CH_4 in per cent. is shown for each curve. The dimensions of the counter were: wire diameter 0 075 mm.; Cu cylinder, diameter 1 cm. and length 3 cm.

Rose and Korff explained the differences between the behaviour of simple and complex gases by pointing out that the production of ultra-violet radiation by electronic excitation is highly probable only in the simpler gases. Such photons can cause photo-emission from the counter cathode and so a rapidly rising characteristic as in the curve for A/H_2 given in Fig. 2.46. Further experiments were made which supported this suggestion.

As the voltage is raised towards the Geiger threshold the discharges tend to spread along the counter [112]. Stever[113] and others[114, 115] showed that in certain cases where the counters contained complex gases, such as CH_4, C_2H_5OH/A, etc., the discharge could be localized even above the Geiger threshold by sealing glass beads on to the counter wire. This and many other experiments of a similar nature [116, 117, 118, 119, 120] indicate that in such gases the discharge is propagated along the wire probably by a step-by-step photoionization process and that no

cathode mechanism, such as photo-emission, is involved. In counters
filled with the elementary gases we have seen above that photons
produced in the primary avalanche can cross to the cathode and there
cause emission of electrons. In this case, therefore, the discharge may
spread along the counter either at the wire or by cathode emission or by
both means. Fig. 2.48 shows schematically the two mechanisms, which
are discussed further on p. 137 et seq.

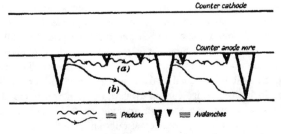

Fɪɢ. 2.48. Mechanisms of discharge spread in a Geiger counter.

Measurements of the velocity of Geiger discharge propagation have
been made for the complex gases [121, 122, 123, 124, 125] and for the
simple gases [125] by various workers. In the latter case since the
photoionization process is relatively clear it had been hoped that the
measurements would be of use in determining photoionization cross-
sections (see Chapter I), but this hope was not realized since it has proved
up to the present to be impossible to decide which of the mechanisms
of Fig. 2 48 is operative in a simple gas Geiger counter.

Non-proportional counting

It has already been shown that in the proportional part of the counter
characteristic there are two general types of gas filling, namely those
which do and those which do not give photo-emission of electrons at
the cathode. It is to be expected that these differences will persist in
some form above the Geiger–Muller threshold.

The mechanism of non-proportional counting has been studied for
some years [126, 127] but the most important contributions to the
subject appeared in 1940 [128, 129]. The Geiger–Muller threshold
voltage, i e. that at which the pulses are all of the same size, is defined
[102, 103, 129] as the starting voltage. The difference between the
latter and the operating cylinder voltage on the plateau is the over-

voltage. Ramsey's [128] empirical formula for the wire voltage V as a function of time t during a counter discharge is

$$V = V_0 - (k/c) \log(t/t_0 + 1), \tag{2 44}$$

where c is the capacity of the counter wire system, k is a constant depending only on V_0 and t_0 is also a constant. The change in V is caused by the movement of the positive ion sheath, formed by the avalanche process, away from the wire The electrons in the avalanche are swept quickly into the wire, but as they are formed very near the latter, where the field is high, their movement does not contribute appreciably to changes in V. Ramsey's formula has also been derived theoretically [129] on the simple assumption that none of the positive ions in the space-charge layer is large enough to produce ionization by collision. The discharge may thus be said to be quenched, in a way very similar to that encountered in high-pressure corona discharges from points (see Chapter III). The electrons move through an average distance less than the wire diameter so that their displacement does not contribute appreciably to the change in wire potential and are swept into the wire before the ions have moved appreciably outwards towards the cathode The space-charge layer then expands relatively slowly until the ions reach the cathode. Stever [113] has given a full account of this part of the discharge, which may occupy ~ 100 μsec , and there are many later papers on the subject [96]. At this stage the difference between slow and fast (self-quenching) counters becomes apparent. In the former case the positive ions cause emission of secondary electrons from the cathode; such secondaries may or may not be multiplied near the wire, depending on the field strength there.

In order to avoid this repeated form of discharge, which may lead to breakdown with the formation of a stable glow discharge in the counter, it is necessary to use a high series resistance, as in the basic circuit of Fig. 2.43, of perhaps 10^{10} to 10^{12} Ω which seriously limits the maximum possible rate of counting. This 'slow counter' behaviour is encountered if simple counter gas fillings (e.g H_2, A, Ne, etc.) are used. With the 'fast counters' (e.g. those containing CH_4 or C_2H_5OH/A, etc.) low resistances of $\sim 10^6$ Ω or less may be used and the permissible counting rates are higher.

The negligible effect of the discharge in a fast counter in the production of secondary electrons from the cathode has been explained by Korff and his collaborators [130, 131] It is pointed out that firstly the ionization and collision processes in, for example, an alcohol/argon counter are such that the argon ions lose their charge to alcohol molecules by the Kallmann–Rosen change of charge process [95] so that very few argon ions reach the cathode. Secondly the alcohol ions so produced, or the methane ions in a counter containing methane only, produce no appreciable secondary emission at the cathode. This is because complex molecules will tend to dissociate (a) after receiving, at $\sim 5 \times 10^{-8}$ cm. from the surface, the $(I-\phi)$ eV excitation energy (I = ionization potential of the complex molecule and ϕ = cathode work function) resulting from the neutralization of the complex ion by an electron extracted from the cathode but (b) before reaching the cathode after the ensuing $\sim 2 \times 10^{-12}$ sec. since the mean life of the excited ion will be $\sim 10^{-13}$ sec. Hence no second electron will be ejected from the cathode in such a process.

When argon is used as a counter filling no such mechanism can occur, so that many of the argon ions proceed to cover the 5×10^{-8} cm. between the point of neutralization and the cathode surface. Secondary emission can occur if the atom approaches to within $\sim 2 \times 10^{-8}$ cm. of the surface [95] and, since the mean life of an excited argon atom is $\sim 10^{-7}$ sec., many such atoms can reach the surface and will cause secondary emission if the excitation energy $(I-\phi) > \phi$ or if $I > 2\phi$. Korff [95, pp. 98–108] has given a fuller account of Geiger counter cathode mechanisms with more supporting evidence. Later papers are cited by Curran and Craggs [96]. A recent discovery [133] that a mixture of O_2, N_2, A, and Xe with Ag cathodes gives self-quenching may be explicable if the cathode work function, which may be greatly increased by oxidation [134], rises to a value greater than one-half the ionization potential of Xe, i.e. to about 6 eV.

At the present time there is appreciable interest in counters with low starting potentials of a few hundreds of volts [136, 137, 138] (see the above discussion on the Penning effect, pp. 83–4), in counters filled with NH_3 which gives self-quenching [139], in counters with electro-negative gas fillings [140–3], and in argon/CO_2 fillings [144]. The latter results [140–4] may yield important information on negative ions. Indeed approximate values of attachment probabilities have already been determined in this way [145, 146].

Spark counters [147, 148, 149] and parallel plate counters [150, 151] are also of interest in the field of discharge physics. Spark counters are normally used in gases, chiefly air, at relatively high pressures (~ 1 atm.) and the applied voltage is such as to cause steady corona currents to flow across the discharge gap. The passage of an ionizing particle, especially one giving a high specific ionization, then gives a visible and often audible spark. It is usual to connect an appreciable capacitance in parallel with the gap, when the effect can be very striking. The counters, if of the non-uniform field type, are generally more sensitive to particles passing through the discharge region in the strongly divergent field near the wire or point, and marked directional effects can be obtained.

The mechanism of operation of spark counters whether of the wire-plane or parallel plane types does not seem completely clear, since the behaviour of different gases is often inconsistent with any simple qualitative and general theory. Connor [151], whose work on this subject is the latest available, gives a discussion of the effects involved and suggests that space-charge distortion of the electric field, caused, for example, by an α-particle passing through the counter in the region of maintained corona, is sufficient to cause spark breakdown. The general variation of the counter response, with varying voltage, to electrons and more heavily ionizing particles seems largely, but not entirely, consistent with this explanation Connor's paper includes most of the principal references to earlier work on the subject.

Discharge propagation at low gas pressures

Since the various mechanisms of discharge propagation are of interest in breakdown processes we shall proceed to give a short account of a theory of the manner in which a discharge spreads along the wire of a Geiger counter. The theory is by Alder *et al.* [121] and should be applicable to certain other discharges.

A Geiger counter is essentially a coaxial wire-cylinder arrangement and a Townsend discharge started at any point within it, by means of radiation passing through it from an external source, will tend to grow near the wire since the field there is high. The discharge can be considered as starting from a single avalanche which extends along the wire over a distance x_0. If it is supposed that the high-energy photons from this primary avalanche can cause photoioniza-

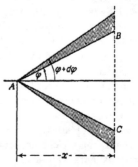

Fig 2 49 Diagram illustrative of photon induced discharge spread in Geiger counters.

tion in the gas very near to the limit of x_0 then an ion pair formed in this way may give rise to another avalanche in the radial field of the wire, and so the discharge spreads in a series of steps. Fig. 2.49 illustrates the calculation, BC is a plane perpendicular to the wire at distance x from the primary avalanche which contained N_0 photoionizing quanta. The last photon of the arbitrary primary avalanche gives an ion pair at, say, distance x_0. If the mean duration of an avalanche is ν then the rate of diffusion along the wire is v, where $v = x_0/\nu$. The number of quanta dN falling on the plane BC in a cone lying between the semi-vertical angles ϕ and $(\phi+d\phi)$ is

$$dN = N_0 e^{-\mu r} \frac{2\pi r^2 \sin \phi}{4\pi r^2} \, d\phi \qquad (2.45)$$

If x is much less than the tube radius then the total number of quanta falling on the plane BC is

$$N(x) = \int_0^{\frac{1}{2}\pi} \frac{N_0}{2} e^{-\mu x/\cos \phi} \sin \phi \, d\phi, \qquad (2.46)$$

or $\qquad\qquad N(x) = \frac{N_0}{2}[e^{-\mu x}+\mu x E_i(-\mu x)], \qquad (2.47)$

where μ is the absorption coefficient for the photoionizing radiations and

$$E_i(-\mu x) = \int_{\mu x}^{\infty} e^{-t}/t \, dt.$$

If β is the probability of production of a photoelectron, then in order that propagation may take place $\beta \cdot N(x_0) = 1$, or

$$\frac{N_0 \beta}{2}[e^{-\mu x_0}+\mu x_0 E_i(-\mu x_0)] = 1. \qquad (2.48)$$

Since $x_0 = v\nu$, v can be introduced and is determinable in terms of N_0, β, μ, and ν. Alder and his collaborators did not determine N_0 and v, but normalized the last equation at several points to experimental curves of discharge spread velocity as a function of counter tube voltage and showed that the forms of the theoretical and experimental curves agreed quite well. The above authors (also Hill and Dunworth [122] and Wantuch [123]) worked with complex gas mixtures, for example, alcohol/argon, and made allowance for the varying composition of the gas in some detail The propagation speeds in simple gases have been measured by Balakrishnan, Knowles, and Craggs [124], some of whose results are given in Fig. 2.50. These results cannot yet be used for the determination of μ since N_0 and v are not known Also the above treatment is little more than illustrative of a general possible mechanism, though most authorities on Geiger counter mechanisms seem agreed that discharge spread along the wire by photoionization is probably the operative process.

This theory of discharge propagation in Geiger counters, which has relevance to various avalanche studies, has been presented in a different form by Wilkinson [119], using the same mechanism but arriving at the result through other arguments. Wilkinson finally obtains for the velocity of propagation v

$$v = \frac{4}{\pi}\left[1 - \left(\frac{1}{2}\right)^{e^m}\right]\left(\frac{e^m - 1}{\log[(Q_0 \epsilon/e)(1 - e^{-m})^2]}\right)^{\frac{1}{2}} v_e, \qquad (2\,49)$$

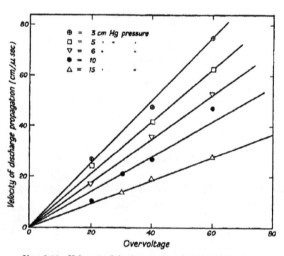

FIG 2 50   Velocity of discharge propagation in hydrogen

where ϵ is the probability per ion that an avalanche shall cause another avalanche further along the counter wire. Q_0 is the charge/unit length of the counter wire given by

$$Q_0 = \frac{V}{2\log(b/a)}, \qquad (2.50)$$

where V is the applied counter voltage and b and a are the cathode and anode radii. v_e is the mean (radial) electron velocity in the vicinity of the wire, and m is given by

$$m = q/Q_0, \qquad (2.51)$$

where q is the charge generated per unit length of counter in the discharge. The mean free path for photon absorption is, of course, involved in the treatment but in a fairly complicated manner. Reference should be made to the original paper for a detailed comparison of theoretical and experimental values of v. The determination of q, Q_0, and m present no difficulty, ϵ is not known with any accuracy but probably $\epsilon \sim 10^{-5}$. The calculation of v_e from known mobility data involves some extrapolation from low E/p values but the results agree fairly well ($\sim 3 \times 10^7$ cm./sec.) with those of Raether's work [59]. The agreement between theory and practice in the determination of v, as given by equation (2.49) is satisfactory for a photon absorption coefficient of about 600 cm.$^{-1}$ obtained in an argon-alcohol filled counter [121]. It is more difficult to obtain experimental data for v in simple gases (e g. H_2, A, etc.) and also to interpret them [124, 125] as the photons are not confined to the region near the counter wire since absorption of them is less complete. A counter discharge may propagate by photo emission from the cathode and not by photoionization in the gas near the wire [113] as is the case with complex counter gas fillings.

It is necessary perhaps to stress again that the particular value of this Geiger counter work in spark studies is twofold. Firstly information may thereby eventually be obtained on photoionization cross-sections in simple gases and, secondly, the analysis of the discharge propagation mechanisms in Geiger counters is of value in studies relating to avalanches and the work on avalanche-streamer transitions.

Propagation mechanisms have been studied for low pressure discharges of various kinds, particularly by Snoddy, Beams, and their collaborators [154, 155, 156, 157] As an example of the latest work from this school the research of Mitchell and Snoddy is of great interest [157]. It is recalled that the general mechanism of propagation of leader strokes is thought to be by virtue of electron motion in the field near the advancing leader tip [158] and the region in which ionization by collision occurs is located ahead of the tip, so that a *few* free electrons (produced, for example, by photoionization) are necessary in this region. The drift velocity v_e of electrons in a field E is [159]

$$v_e = \sqrt{\left(\frac{2Ee\lambda}{\pi m}\right)} \qquad (2.52)$$

with the usual symbols. This gives a speed of $3\cdot7 \times 10^7$ cm./sec. when $E/p = 40$ V/cm./mm. Hg in air at 1 atm. [157].

Schonland [160, 161] showed that the velocity of the pilot streamer in lightning discharges is of this order but that the velocity of stepped leaders [162, 163] which progress in regions of fairly strong ionization left from preceding streamers is much greater. The speed of propagation is then

$$v = N^{\frac{1}{2}} v_e \, d. \qquad (2.53)$$

N is the concentration of electrons per cm.^{3} in front of the advancing discharge and d is the distance over which the field of the discharge tip is effective. Schonland [161] describes conditions in which $N \sim 1,000$, and $d \sim 6$ cm. (tip radius ~ 1 cm.). The conditions of equation (2.53) differ from those of equation (2.52) since N is relatively great compared with the ionization in front of the tip of the pilot streamer.

Mitchell and Snoddy [157] studied discharges propagating in air and in hydrogen at a pressure of a few cm. Hg. They found that with low applied voltages the leader strokes had propagation speeds comparable with those given by equation (2.52). The speed was governed by E/p and, for the minimum value of E/p which gave propagation, $v = 1.7 \times 10^7$ cm /sec. (the minimum $E/p = 20$ V/cm./mm. Hg, which is about the minimum required for Townsend α-ionization in air [164]). The minimum current in the leader was about 1 A.

With higher applied voltages the impulse discharges appear more analogous to the continuous lightning leaders in that their greater speed indicates propagation in an ionized channel. Mitchell and Snoddy considered that the free electrons ahead of the tip were produced by photoionization, possibly enhanced in their case by reflection of light from the tube walls as it seems unlikely that even at a few cm. Hg pressure photoionizing radiations could travel distances of at least a tube diameter [117]. For this case Schonland's formula (equation 2.53 above) can be modified since N, if caused by photoionization due to highly absorbable radiation, will vary with the distance from the leader tip. The field at the discharge tip is approximately $E = V/2r$ V/cm. where V is the applied potential and r is the tube radius; in conditions where no tube is used, as in sparks in air at 1 atm. or where the discharge tip had a radius less than the tube radius, r would be replaced by the streamer tip radius (see Loeb and Meek [165]). Then the distance d in which a breakdown can occur, i.e. when $E/p > 40$ V/cm /mm. Hg, is given approximately by

$$d = (rV/80p)^{\frac{1}{2}}. \tag{2.54}$$

At a distance l ahead of the tip the number of photoelectrons per cm.^{3} is given by

$$N = N_0 \exp(-\mu lp/760), \tag{2.55}$$

where N_0 is the concentration just ahead of the tip and μ is the absorption coefficient of the operative radiation taken as 10 cm.$^{-1}$ [166] (see Chapter I for data on absorption coefficients from which it will be seen that this value should be taken with reserve since it appears to be too low). On the assumption that the entire region d must be filled with conducting filaments or avalanches starting at the original photoelectrons, then the minimum speed is controlled by the time required for the region of d farthest from the stem to be so filled. From equation (2.53)

$$v = N_0^{\frac{1}{3}} . 3 \times 10^7 \, d \exp(-10dp/2280)$$

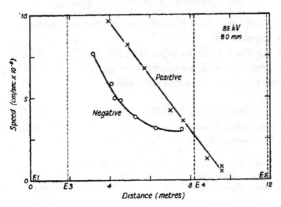

FIG 2.51   Speed-distance curves of low potential discharge in air at 8 0 mm pressure and with 85 kV applied potential   Position of fixed electrodes is indicated by vertical dotted lines

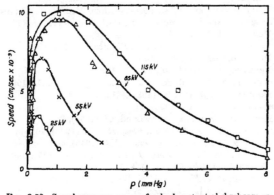

FIG. 2.52.   Speed-pressure curves for high potential discharge in air   Negative applied potential.

and inserting for d from equation (2.54) we have

$$v = 3 \times 10^7 \, N_0^{\frac{1}{3}} (rV/80p)^{\frac{1}{3}} \exp\left[-\frac{1}{228} \left(\frac{prV}{80} \right)^{\frac{1}{2}} \right].$$

By assuming N_0 constant over the range of pressures of the experiments and putting in values of v, r, p, and V for one set of conditions it was found [157] that $N_0 \sim 10^2$ ions/cm.³ so that

$$v = 1 \cdot 5 \times 10^8 (rV/80p)^{\frac{1}{2}} \exp\left[-\frac{1}{228}\left(\frac{prV}{80}\right)^{\frac{1}{2}}\right]. \qquad (2.56)$$

At low pressures the effect of diffusion, ignored in the above treatment, should be introduced.

Results for the discharges in air are given in Figs. 2.51 and 2.52. The speeds of propagation in hydrogen were generally $\sim$ 20 to 50 per cent. higher than in air for the same conditions.

REFERENCES QUOTED IN CHAPTER II

1. J. S TOWNSEND, *Electricity in Gases*, Oxford University Press, 1915.
2. J. S. TOWNSEND, *Electrons in Gases*, Hutchinson, 1948.
3. G. HOLST and E OOSTERHUIS, *Phil Mag* **46** (1923), 1117.
4. L. B. LOEB, *Proc Phys. Soc.*, Lond **60** (1948), 561.
5. D. H. HALE, *Phys. Rev.* **56** (1939), 1199.
6. L. B. LOEB, *Fundamental Processes of Electrical Discharges in Gases*, J. Wiley, 1939
7 S. FRANCK, *Arch. Elektrotech.* **21** (1928), 318.
8 H. C BOWKER, *Proc. Phys. Soc*, Lond **43** (1931), 96
9 M E BOUTY, *C R. Acad Sci*, Paris, **136** (1903), 1646; **137** (1903), 741
10 R F. EARHART, *Phys Rev* **29** (1909), 293, **31** (1910), 652
11. M. J. DRUYVESTEYN and F. M. PENNING, *Rev. Mod. Phys.* **12** (1940), 87.
12 J. S TOWNSEND and S. P. MCCALLUM, *Phil. Mag* **6** (1928), 857
13. F. M. PENNING, ibid. **11** (1931), 961.
14. F. M. PENNING and C. C J. ADDINK, *Physica* **1** (1934), 1007.
15. F. M. PENNING, ibid **1** (1934), 1028
16. F. M. PENNING, ibid. **10** (1930), 47, **12** (1932), 65
17. H. JACOBS and A P. LA ROCQUE, *J. Appl Phys.* **18** (1947), 199.
18. F. EHRENKRANZ, *Phys Rev.* **55** (1939), 219.
19. R. SCHOFER, *Z. Phys.* **110** (1938), 21.
20 F. G. HEYMANN, *Proc. Phys. Soc.*, Lond. **63** (1950), 25.
21. H. JACOBS and A. P LA ROCQUE, *Phys Rev.* **74** (1948), 163
22 J. CUEILLERON, *C. R. Acad. Sci*, Paris, **226** (1948), 400.
23. D. H. HALE and W. S. HUXFORD, *J Appl. Phys* **18** (1947), 586
24. W. E BOWLS, *Phys. Rev* **55** (1938), 293.
25. F. M PENNING, *Proc. K. Ned. Akad Wet.* **34** (1931), 1305.
26 S. P. MCCALLUM and L. KLATZOW, *Nature*, Lond. **131** (1931), 841.
27. J. S. TOWNSEND and S. P. MCCALLUM, *Phil Mag* **17** (1934), 678.
28. F. LLEWELLYN JONES and J. P. HENDERSON, ibid. **28** (1939), 185.
29. H. E. FARNSWORTH, *Phys. Rev.* **20** (1922), 258
30 F. LLEWELLYN JONES, *Phil. Mag.* **28** (1939), 192
31. H FRICKE, *Z. Phys.* **86** (1933), 464.
32. D. H. HALE, *Phys. Rev* **55** (1939), 815.
33. B. FREY, *Ann. Phys. Lpz.* **85** (1928), 381.
34. D. Q. POSIN, *Phys. Rev.* **50** (1936), 650.

35. E. Meyer, *Ann. Phys. Lpz.* **58** (1919), 297; **65** (1921), **335**.
36. F. Llewellyn Jones, *Phil. Mag* **28** (1939), 328.
37. F. Llewellyn Jones and W. R Galloway, *Proc Phys Soc.*, Lond **50** (1938), 207.
38. R Grigorovici, *Z. Phys* **111** (1939), 596.
39. R. B. Quinn, *Phys. Rev.* **55** (1939), 482.
40. J J. Thomson and G. P. Thomson, *Conduction of Electricity through Gases*, Cambridge University Press, 1933.
41. W. R. Carr, *Philos. Trans.* A, **201** (1903), 403
42. S. S. Cerwin, *Phys. Rev.* **46** (1934), 1054.
43. J. S. Townsend and S. P. McCallum, *Phil. Mag.* **6** (1928), 857.
44. S. P. McCallum and L. Klatzow, ibid. **17** (1934), 291.
45. H. Buttner, *Z. Phys* **111** (1939), 750.
46. J. Slepian and R. C. Mason, *J. Appl. Phys.* **8** (1937), 619.
47. L. G. H. Huxley, *Phil Mag.* **10** (1930), 185; **5** (1928), 721.
48. J. H. Bruce, ibid. **10** (1930), 476.
49. H F. Boulind, ibid **18** (1934), 909; **20** (1935), 68.
50. J. D. Craggs and J. M. Meek, *Proc. Phys. Soc.*, Lond. **61** (1948), 327.
51. L. G H. Huxley and J. H. Bruce, *Phil. Mag.* **23** (1936), 1096
52 M. von Laue, *Ann. Phys. Lpz* **76** (1925), 261.
53. W. Braunbek, *Z. Phys.* **39** (1926), 6; **107** (1937), 180.
54. G. Hertz, ibid. **106** (1937), 102.
55. H. Paetow, ibid. **111** (1939), 770.
56. R Schade, ibid **104** (1937), 487, **108** (1938), 353.
57. M. Steenbeck, *Wiss Veroff Siemens-Werk*, **9** (1930), 42.
58. H. Raether, *Z. Phys.* **117** (1921), 394, 524.
59. H Raether, *Ergebnisse der exakten Naturwissenschaften*, **22** (1949), 73.
60. R. O. Jenkins, *Reports on Progress in Physics* (1942–3).
61. R. A Millikan and C. F Eyring, *Phys. Rev.* **27** (1926), 51.
62. R A. Millikan and C. C. Lauritsen, *Proc. Nat. Acad. Sci Wash.* **14** (1928), 1, 45.
63. R. A. Millikan, C F. Eyring, and S S Mackeown, *Phys. Rev.* **31** (1928), 900.
64. B. S. Gossling, *Phil Mag* **1** (1926), 609.
65. N. A. de Bruyne, ibid **5** (1928), 574.
66. W. Schottky, *Z. Phys.* **14** (1923), 80.
67. N. A. de Bruyne, *Proc. Camb. Phil Soc.* **24** (1928), 518
68. N. A. de Bruyne, *Phys. Rev.* **35** (1930), 172.
69. R A. Millikan and C. C Lauritsen, ibid. **33** (1929), 598.
70. J. L. R. Hayden, *Trans. Amer. Inst. Elect. Engrs.* **41** (1922), 852.
71. R. J. Piersol, *Phys. Rev.* **25** (1925), 113.
72. R. H. Fowler and L. W. Nordheim, *Proc. Roy. Soc.* A, **119** (1928), 173.
73. T. E. Stern, B. S. Gossling, and R H. Fowler, ibid. **124** (1929), 699.
74. C. C. Chambers, *J. Franklin Inst.* **218** (1934), 463.
75. C Del Rosario, ibid. **205** (1928), 103.
76. A. J. Ahearn, *Phys. Rev.* **50** (1936), 239.
77. J. W. Beams, ibid. **44** (1933), 803.
78. L. R. Quarles, **48** (1935), 260.
79. D. H. Moore, ibid. **50** (1936), 344.

80. L. B. SNODDY, ibid. **37** (1931), 1678.
81. J. W. BEAMS, ibid **44** (1933), 803.
82. J. A. CHILES, *J. Appl. Phys.* **8** (1937), 622.
83. J. W. BEAMS, *Phys. Rev.* **43** (1933), 382.
84. R C. MASON, ibid **52** (1937), 126.
85. A. W. HULL and E. E. BURGER, ibid. **31** (1928), 1121
86. H. W. ANDERSON, *Elect. Eng. N Y.* **54** (1935), 1315.
87. J. G TRUMP and R. J. VAN DE GRAAFF, *J. Appl. Phys.* **18** (1947), 327.
88. H. S. SEIFERT, *Phys Rev* **62** (1942), 300.
89. J. D. COBINE, *Gaseous Conductors*, McGraw-Hill, New York, 1941.
90. J. T. TYKOCINER, R. E. TARPLEY, and E. R. PAINE, *Bull. 278*, Illinois Univ. Eng. Exptl. Station.
91. G. VALLE, *Phys Z.* **27** (1926), 473
92. W. SCHULZE, *Z. Phys.* **78** (1932), 92.
93. G L. WEISSLER, *Phys. Rev.* **63** (1943), 96.
94 A. E. W. AUSTEN and W. HACKETT, *J. Inst. Elect Engrs.* **94**, Part I (1944), 298.
95 S. A. KORFF, *Electron and Nuclear Counters*, van Nostrand, New York, 1946.
96. S. C. CURRAN and J. D CRAGGS, *Counting Tubes*, Butterworth, London, 1950.
97. B. ROSSI and H. H. STAUB, *Ionization Chambers and Counters· Experimental Techniques*, McGraw-Hill, New York, 1949.
98. D. H. WILKINSON, *Ionization Chambers and Counters*, Cambridge, 1950.
99. D. R. CORSON and R R. WILSON, *Rev Sci. Instrum.* **19** (1938), 207.
100. S. C. CURRAN, J. ANGUS, and A. L COCKROFT, *Phil. Mag.* **40** (1949), 36
101. M. E ROSE and S A. KORFF, *Phys. Rev.* **59** (1941), 850.
102. C. G. MONTGOMERY and D. D. MONTGOMERY, *J. Franklin Inst.* **231** (1941), 447.
103. C. G. MONTGOMERY and D. D MONTGOMERY, ibid. 509.
104. W F. G SWANN, ibid. **230** (1940), 281.
105. S. C. BROWN, *Phys. Rev.* **62** (1942), 244
106. S. C. CURRAN, A. L. COCKROFT, and J ANGUS, *Phil. Mag.* **40** (1949), 929.
107. G. MEDICUS, *Z. Phys* **74** (1932), 350; *Z. angew. Phys.* **1** (1949), 316.
108. H. FETZ and G. MEDICUS, ibid **1** (1948), 19.
109. S. A. KORFF, *Rev. Mod. Phys.* **14** (1942), 1.
110. S. A. KORFF, ibid. **11** (1939), 211.
111. E. STUHLINGER, *Z. Phys.* **108** (1938), 444.
112. B. COLLINGE, *Proc. Phys. Soc* **13** (1950), 15, 63.
113. H. G. STEVER, *Phys Rev.* **61** (1942), 38.
114. M. H. WILKENING and W R KANNE, ibid. **62** (1942), 354.
115. A. NAWIJN, *Het. Gasontladings Mechanisme van den Geiger-Muller Teller*, Hoogland, Delft, 1943.
116. J. D. CRAGGS and A. A. JAFFE, *Phys Rev.* **72** (1947), 784
117. A. A. JAFFE, J D. CRAGGS, and C. BALAKRISHNAN, *Proc. Phys. Soc B*, **62** (1949), 39.
118. F. ALDER, E. BALDINGER, P. HUBER, and F. METZGER, *Helv. Phys. Acta*, **20** (1947), 73.
119. D. H. WILKINSON, *Phys. Rev.* **74** (1948), 1417.
120 S. H. LIEBSON, ibid. **72** (1947), 602.

121. P. Huber, F. Alder, and E. Baldinger, *Helv. Phys. Acta*, **19** (1946), 204.

122. J. M. Hill and J. V. Dunworth, *Nature*, Lond. **158** (1946), 833.

123. E. Wantuch, *Phys. Rev.* **71** (1947), 646.

124. A J. Knowles, C Balakrishnan, and J. D Craggs, ibid. **74** (1948), 627.

125. C. Balakrishnan and J. D. Craggs, *Proc. Phys. Soc.* A, **63** (1950), **358**.

126 H Greiner, *Z. Phys.* **81** (1933), 543.

127. A. von Hippel, ibid. **97** (1936), 455.

128 W. E. Ramsey, *Phys. Rev.* **58** (1940), 476.

129. C G Montgomery and D. C. Montgomery, ibid. **57** (1940), 1030

130. S. A. Korff and R. D. Present, ibid **65** (1944), 274.

131. M Rose and S. A. Korff, ibid **59** (1941), 850.

132. D. R Young, *J. Appl Phys* **21** (1950), 222

133. L. G. Shore, *Rev. Sci. Instrum.* **20** (1949), 956. See also O. Riedel, *Z. Naturforsch.* **5a** (1950), 331.

134. J H. de Boer, *Electron Emission and Adsorption Phenomena*, Cambridge, 1935

135. H. Friedman, *Proc. Inst. Radio Engrs.* **37** (1949), 791.

136 P. B. Weisz, *Phys. Rev* **74** (1948), 1807.

137. J. A. Simpson, *Rev. Sci Instrum.* **21** (1950), 558.

138 F. M Penning, *Z Phys* **46** (1927), 335.

139. S. A. Korff and A. C. Krumbein, *Phys Rev.* **76** (1949), 1412

140. S S. Friedland, ibid **71** (1947), 377.

141. R. D. Present, ibid **72** (1947), 243.

142. S. H. Liebson and H Friedman, *Rev. Sci. Instrum.* **19** (1948), 303.

143. E. Franklin and W. R Loosemore, *J. Inst. Electr. Engrs.* **98** Part II, (1951), 237.

144 S. C. Brown and W. W. Miller, *Rev. Sci. Instrum.* **18** (1947), 496.

145. M. E. Rose and W E Ramsey, *Phys. Rev.* **59** (1941), 616.

146. C. G. Montgomery and D. D. Montgomery, *Rev. Sci. Instrum.* **18** (1947), 411.

147. W. Y. Chang and S. Rosenblum, *Phys. Rev.* **67** (1945), 222.

148. A. Rytz, *Helv Phys Acta*, **22** (1949), 1.

149 R. M. Payne, *J. Sci. Instrum* **26** (1949), 321.

150. R. W Pidd and L. Madansky, *Phys. Rev.* **75** (1949), 1175.

151. R D. Connor, *Proc. Phys. Soc.* B, **64** (1951), 30.

152. H. Raether, *Z. Phys.* **110** (1938), 611.

153. S. Whitehead, *Dielectric Breakdown of Solids*, Oxford, 1951.

154. J. W. Beams, *Phys Rev.* **36** (1930), 997.

155. L. B. Snoddy, J. R. Dietrich, and J. W. Beams, ibid. **50** (1936), 469.

156. L. B. Snoddy, J. R. Dietrich, and J. W. Beams, ibid. **52** (1937), 739.

157. F. H. Mitchell and L. B. Snoddy, ibid **72** (1947), 1202.

158. C. E. R. Bruce, *Proc. Roy. Soc.* A, **183** (1944), 228.

159. H. Jehle, *Z. Phys.* **82** (1933), 785.

160. B. F J. Schonland, *Proc Roy. Soc.* A, **164** (1938), 132.

161. B. F. J. Schonland, *Phil. Mag.* **23** (1937), 503.

162. B. F. J. Schonland and H. Collens, *Proc. Roy. Soc.* A, **143** (1934), 654.

163 B. F. J. SCHONLAND and H. COLLENS, ibid **152** (1935), 595.

164. F. H. SANDERS, *Phys Rev.* **41** (1932), 667.

165. L. B. LOEB and J. M MEEK, *The Mechanism of the Electric Spark*, Stanford, 1941.

166. A. M. CRAVATH, *Phys Rev.* **47** (1935), 254.

167. R. A. WIJSMAN, ibid. **75** (1949), 833.

CORONA DISCHARGES

THE purpose of this chapter is to discuss the more recent work on the fundamentals of high-pressure corona discharges, which are the transitory, faintly luminous and audible, glows discerned in a discharge gap at voltages below, though frequently near to, the sparking value. There have, of course, been many experiments carried out on corona and described in the literature (an excellent review of the earlier work up to 1926 is given by Whitehead [1]), but it is proposed to make an arbitrary selection of the published papers and to omit much work that is of appreciable technical importance in favour of a discussion of the various physical mechanisms encountered in corona discharges.

The electrostatic field in point-plane gaps

Much of the best recent work on high pressure corona has been carried out by Loeb and his school. With the exception of Weissler's [2] and Pollock and Cooper's [3] recent work most of the studies have been made in air and with point-plane gaps, the reason for the adoption of this apparently unsuitable geometry is that space charge localization is then more effective Accurate field calculations can be made for confocal paraboloid electrodes, and Kip [4] concluded from tests with such electrodes and with hemispherically-ended point electrodes that the use of the latter, assuming the geometry to be that for the former, was sufficiently accurate with most long gaps. Fitzsimmons [5], for example, used the following paraboloid electrodes: brass points of focal lengths 0·009 and 0·019 cm. with hollow copper electrodes of focal lengths 1, 2, 3, and 5 cm. The focal lengths of these hollow electrodes give the effective gap lengths, and the field E in V/cm. at a distance x cm. from the end of the point electrode along the common axis of the two electrodes is given by

$$E = \frac{V}{\log_e(f/F)} \frac{1}{x+f},\tag{3.1}$$

where V is the point potential in volts, f and F are the focal lengths of the point and the hollow electrode respectively. Since $V/\log_e(f/F)$ is a constant for any particular value of V with a given pair of electrodes, the (E, x) curve for a given V will be a hyperbola.

For point-plane gaps, using 0·03- and 4·7-mm. diameter points, the data given by Loeb [6, p. 515] for E/p plotted as a function of the distance

x from the point are shown in Fig. 3 1. The values of E/p are given for the starting potentials of positive-point corona in air at 1 atm. The inset curves of Fig. 3.1 give α as a function of x using the data of Sanders [7] and Masch [8]. If the inhibiting effect of the space charge may be neglected, and this is reasonable for a positive point only, then the

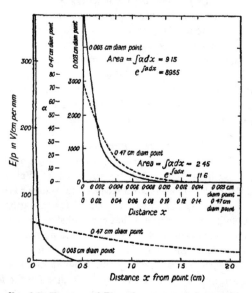

FIG. 3.1. Variation of E/p and α in a point-plane corona gap

number of electrons in a single avalanche advancing towards the positive point may be found from $n = e^{\int \alpha \, dx}$ by integration of the area beneath the α-curves of Fig. 3.1. It is found that for a 0·003-cm. diameter point the avalanche contains some 9,000 electrons.

For a gap between a hyperboloidal point and a plane, with R being the radius of curvature of the tip, the field intensity E_p at the point is given by

$$E_p = \frac{2V}{R\log(4x/R)}, \qquad (3.2)$$

where x is the point-plane distance and V is the point potential [9]. This formula was used by Eyring, Mackeown, and Millikan [10] in their field current studies. The field near a hyperboloid of revolution, radius R, varies with distance x from the point according to the expression

$$E = \frac{R}{R+x} E_p, \qquad (3.3)$$

where E_p is defined as above.

Negative-point corona in air

In his experiments on negative corona Trichel [11] used the circuit of Fig 3.2. The gap current and voltage were measured with a resistance

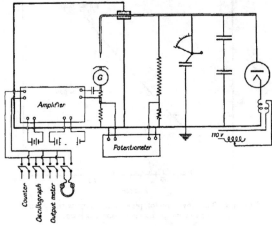

FIG 3.2. Schematic diagram of the electrical circuits employed for negative-point corona measurements

in series with the corona gap of 1,000 Ω. Another resistance of 10,000 Ω was connected in the gap lead to supply a signal, derived from the corona current, to an amplifier and cathode-ray oscillograph. Resistances of this order were found to have an inappreciable effect on the frequency of the corona pulses. Four points were used, viz. platinum wires with hemispherical ends, of radii 0·5 and 1·5 mm. respectively, a brass wire similarly shaped but with radius 4·73 mm., and finally a steel needle with a sharp point (30° angle). Trichel's basic discovery was that in air the corona current consists of a series of regular pulses, at fairly low currents, the frequency of which increases with mean current (Fig. 3.3) and is ∼ 50 kc./s. at 10 μA. The frequency is largely independent of gap length for a given point but increases with point sharpness.

Apart from the current flowing in the corona gap due to the drift of charged particles from one electrode to the other [18] the corona pulse, measured usually as the voltage across a resistance placed in series with the corona gap, can be caused by the displacement current The elementary explanation of this by Corson and Wilson [12] has been developed in detail by other writers on Geiger counters, notably by Wilkinson [13], and is as follows

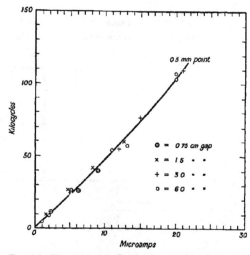

FIG 3 3   The variation of pulse frequency with current
in a negative-point air corona discharge

The electrode system, of capacitance C, charged to voltage V_0, has energy $\frac{1}{2}CV_0^2$. If a particle of charge e moves from x_0 to x in a field E the energy of the system will change to $\frac{1}{2}CV^2$, where

$$\tfrac{1}{2}CV^2 = \tfrac{1}{2}CV_0^2 - \int_{x_0}^{x} eE\, dx.$$

The pulse voltage (V_0-V) is then given by

$$(V_0-V) = \frac{1}{V_0 C}\int_{x_0}^{x} eE\, dx$$

if $(V_0-V) \ll V_0$.

The variation of the frequency of the negative corona pulses in air, as a function of pressure is shown in Fig. 3.4. At voltages near onset the pulse recurrence rate becomes irregular.

Trichel put forward the following mechanism to explain these phenomena. He supposed (see below for the results of later work) that a

random positive ion approaches the point at a time when the field is high
enough to accelerate it to the point and there produces at least one
secondary electron, which then gives by the Townsend mechanism, as
it leaves the region of the point, a dense cloud of positive ions. This
space charge cloud further reduces the field at its outer periphery, where

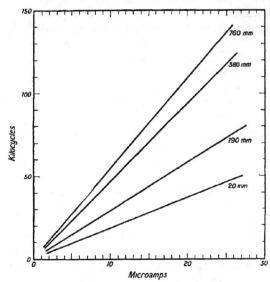

FIG. 3.4. The variation of pulse frequency with corona current,
for different pressures, in Trichel's air corona.

the field is already weak because of the appreciable distance from the
point, and the electrons lose energy rapidly and form negative ions by
attachment. The positive ions then move into the point without
producing further ionization as the field is then too low and the pulse
is formed. The field near the point may have risen again, by the
time the last few positive ions are approaching the point, to a value
sufficient to accelerate one of them to such an extent that another
triggering electron is produced by Townsend's γ-mechanism [6], and
the cycle repeats itself The cathode spot, visible as a glow and shown in
Fig. 3.5, is larger than the size of a single avalanche, possibly because
of the spread of the discharge by photoionization in the gas near the
point.

The discharge has visual characteristics similar to those of a glow, although it is necessary to use a telemicroscope to discern the discharge structure. The visual characteristics of corona have been discussed in detail by Loeb [6]　Fig. 3.6 shows some of the diagrammatic results; the minute dark space next to the point is not apparent in the sketch. At higher fields, where the frequency of pulses is irregular, a second glow appears, then a third, and so on. These glows are separated in space and are located at fixed points. The discharge appears to pass from one cathode spot to another and this accounts for the irregular pulse frequency. It seems that the faint glow, 'far' from the electrode in Fig. 3 6, is the region of attachment and the negative space charge field here may be great enough, Trichel suggests, to cause excitation and so luminosity. The high field between

FIG. 3 5

the cathode and positive space charge will give rise to sharp lines in the negative glow. Schematic representations of ion density, potential, pulse shape, and visual discharge appearance are given in Fig. 3 7 [11]. The space charge conditions are shown for two points A and B in the corona pulse.

This pioneer work of Trichel was extended by Loeb, Kip, Hudson, and Bennett [9], since Bennett and Hudson had both found difficulty in producing the negative pulses in dust-free air with very sharp points. Bennett found regular pulses for all points with radius greater than about 0·0025 cm. but irregular corona with sharper points　He used hyperboloidal points and calculated from equation (3.3) the negative corona onset fields as given in Table 3.1.

TABLE 3.1

Bennett's Data of Corona Onset Fields in 10^5 V/cm. for Negative Points of Different Radii R cm. and Various Gap Lengths x cm.

x cm	R cm					
	0 00058	0 0012	0 0025	0 0050	0 0115	0 047
8	8 0	3 7	2 22	1 73	0 64	0 62
4	8 0	3 8	2 21	1 73	0 68	0 62
2		3 9	2·21		0 73	
1	7·7	3 8	2 19	1 75	0 74	0 63
0 5	7·6	3 7	2 22	1 78	0 70	0 72
0 1	..	..	..	(1 86)	..	(0 79)

The fields for the sharpest points are roughly of the same order as those found by Millikan and Lauritsen [14] for current-conditioned wires giving cold emission. As the onset fields are roughly constant for

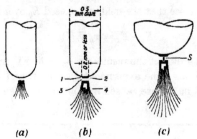

Fig. 3.6. Drawings of typical pre onset and active spot negative corona phenomena. (a) shows pre-onset glow, (b) shows the structure of the typical active spot accompanying Trichel pulses with Crookes dark space (1), negative glow (2), Faraday dark space (3), and positive column (4), and (c) shows the incipient retrograde streamers with large points and high current densities

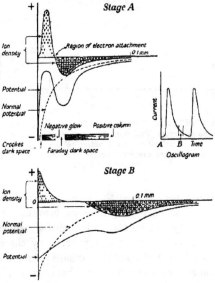

Fig. 3.7. Schematic diagram of space charge distributions in negative point corona

the larger points but increase for the sharper points, it was supposed that the onset condition might be that a certain critical field E_0 should be exceeded throughout a certain distance x_0 from the point. From equation (3.3) we see that the field will exceed E_0 over the distance x_0 if

$$E_p - E_0 = x_0 E_0 / R. \tag{3.4}$$

Fig. 3 8 gives a plot of the results for $E_0 = 50$ kV/cm. It is apparent that this model has some experimental support, and it seems possible that the Trichel pulses are caused by some mechanism, such as attach-

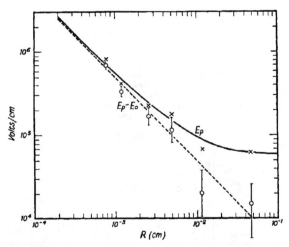

Fig. 3 8   Field plots, illustrative of the use of equation (3.4), in negative point corona Full line with crosses represents the logarithm of onset field in volts/cm. as a function of the logarithm of point radius in cm   Dashed line with circles represents $\log(E_p - E_0)$ as a function of $\log R$.

ment (see below) in the air near the point, rather than at the point, provided that sufficient trigger electrons are present. This implies that the γ-mechanism may not be the main process responsible for pulse repetition. The latter argument is supported by the experiments of Bennett and others [9, 15] who find that Trichel pulses can be obtained in room air with points of various metals, and that no measurable differences in onset potential or current attributable to different cathode materials exist, provided that the points are properly conditioned   Conditioning generally results when the points are used for some time in room air. Regular Trichel pulses cannot be obtained in carefully filtered or dried air and the corona then consists of bursts of irregular Trichel-type pulses.

Hudson [9] found with a microscope that conditioned points collect small dust particles, and it was then shown that regular Trichel pulses can be obtained with artificially dusted, clean but unconditioned points and that they persist longer in moist air, apparently because the dust particles adhere better to the point in such conditions. The active spots of the discharge generally occur adjacent to a dust speck and the function of the latter appears to be similar to the Paetow effect [16] (see the discussion of glow-arc transitions in Chapter XII). Kip [9] also found that the Trichel pulses are preceded by irregular corona with very small currents ($< 10^{-9}$ A). Irradiation of negative point gaps with ultraviolet radiation has two opposing effects (a) it produces more free electrons and (b) it tends to decrease the corona current by producing low-mobility ions in the gap. However, it does not seem clear [9] which effect predominates in any given circumstances.

Trichel's explanation of the pulse formation in negative corona as being due primarily to the positive ion cloud needs to be qualified, as Bennett [15] found no Trichel pulses in H_2 and Weissler [2] found none in H_2, N_2, or A. Further, the duration of Trichel pulses suggests that negative ion transit across the gap is involved. A Townsend γ-mechanism may not be responsible by itself because of the corona's insensitivity to cathode metal. Since the positive ion cloud has largely moved into the point before the negative ions become ineffective in reducing the field there, γ will be small and the resulting pulses irregular. If fresh electrons are available from a Paetow mechanism or ultraviolet irradiation, the displacement current pulses due to movement of the negative ions will be regular although they will tend to be smaller than the random pulses as the latter may be formed after the negative ion cloud has moved a greater distance from the point.

With larger points and higher currents, a luminous streak may be seen extending towards the point from the near edge of the glow column shown in Fig. 3.6. Loeb *et al.* [6, 9] attribute this to the formation of retrograde streamers (bright filamentary discharges) which lead eventually to breakdown with larger points. The streamers cannot advance quite to the cathode because the positive space charge there chokes off the discharge, although with larger points such streamers may reach the point and cause breakdown. Loeb [6, p. 520] suggests that spark breakdown might otherwise occur for negative points when the negative ion space charge reaches such a value at the positive plane electrode that a positive streamer could start here and propagate back to the point. This possibility is again referred to later in this chapter.

Positive-point corona in air

Positive-point corona studies, following the visual observation by Loeb and Leigh [17] of streamers near the point in N_2, showed that the discontinuities in current were small compared with the Trichel pulses. Kip [18] made studies of gap voltage and current for positive discharges in dry air for hemispherical point-plane gaps, with a 50 kV D.C. source

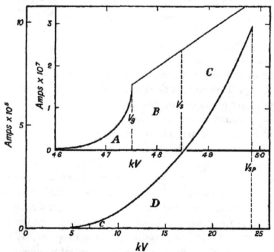

FIG 3.9   Typical voltage/current characteristic for positive point corona.

stabilized to better than 1 per cent. Currents varying from 10^{-10} to 10^{-3} A were measured, and Kip [18] has given curves showing current - voltage characteristics for various conditions

The positive-point currents vary with voltage, in air, as shown schematically in Fig. 3.9. It is necessary, for corona to be observed, that E/p should be sufficiently high to avoid attachment over a distance of about 0·1 mm. from the point [19, 20] (see Chapter I). Free electrons can then give Townsend avalanches directed towards the point. The current increases steadily above i_0, which is the initial current due to external ionization, in region A of Fig. 3.9. At about V_g there is a sudden increase in current and oscillograms show the presence of pulses of current, $\sim 10^{-3}$ sec. in duration and composed of small bursts of charge ($\sim 10^7$ ions), each burst lasting $\sim 10^{-5}$ sec. (region B). This form of corona is referred to by Loeb et al. as burst corona or burst pulses [6].

For fine points, the current increases until the voltage reaches V_s when the corona becomes self-sustaining and external irradiation then becomes unnecessary. Region B is the Geiger counter region and occurs with all but the finest points. Large pulses, referred to by Loeb as inductive kicks, appear, at voltages just above V_g, on the oscillograph screen (circuit of Fig. 3.10). They appear either by themselves or before the

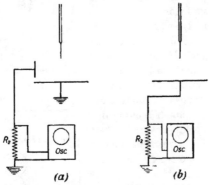

Fig 3 10. Kip's arrangement for the measurement of drift and displacement currents in corona discharges, using a cathode-ray oscillograph.

burst corona. Near or above V_s the burst corona is always initiated by an inductive kick. A high circuit resistance tends to give kicks without burst corona, but at higher voltages the recurrence rate of the kicks increases and finally gives more or less continuous burst corona. Kip found that the insertion of a gauze near the plane electrode (Fig. 3.10) caused the plate current to take the form of a smooth pulse, which is the ion current from a burst pulse or kick. The kicks are inductive effects and are observed with an unscreened plate. Fig. 3.10 (a) gives the inductive effects only, recorded when an insulating screen is arranged to prevent ion collection by the side electrode, and Fig. 3.10 (b) gives both inductive and current pulses. From an oscillogram taken with a screened plate the number of ions (10^{-9}–10^{10}) in a pulse and their transit time across the gap (the time between the kick and the current pulse on the C.R.O. time base) may be found. Later work of English [21] (Fig. 3.11) with improved apparatus showed the true corona wave forms to be somewhat different from those found by Kip.

Visual examination of the corona in air shows that the burst corona is associated with a brilliant blue glow near the point and the kicks are due to streamers of 0·3–2 cm. in length which often progress in a straight line to the plate. The streamers disappear as the voltage is increased beyond V_s (Fig. 3 9) and only the burst corona is observed. For short gaps, spark breakdown occurs before stable corona is established as the field is never low enough to prevent streamer propagation across the whole gap As the voltage is raised still further streamers reappear and spark breakdown follows.

English found later [22] that, contrary to the findings of Kip, streamers could be distinguished from burst pulses in air for the finest positive points. Below 0·01 mm. radius, if the point was dusted with MgO powder, it was found that streamers and bursts could be distinguished with points down to 0·0001 mm. radius. The streamers have a duration of about 0 4 μsec , independently of point size, but the burst pulses have durations varying between 100 and 300 μsec. or more. These experiments were made with 6-cm. gap length. A further interesting result is that each centimetre length of visible streamer contained 10^{10} ions, the pulses being measured with an improved oscillographic technique in which the overall amplification is 6,000 and a resistance of 1,000 Ω is connected in series with the corona gap.

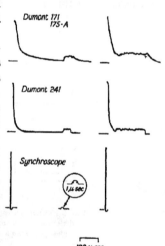

FIG 3.11 Effect of amplifier characteristics. A streamer pulse, a burst pulse, and a streamer triggering a burst, as seen on different oscilloscopes. Positive point.

In a 6-cm. gap with a 0 05-mm. point, the streamers gradually lengthen, as the voltage is raised, to 1·1 cm. and then breakdown occurs [Loeb 6, p. 526]. Successive streamers follow down the same paths, and if on striking the cathode an efficient cathode spot is produced the rearrangements of space charges leads to breakdown. Air blasts cause movements of the old streamer channels, as can be seen by the fact that the new streamers apparently follow the air movements. V_{sp} (Fig 3.9) is the sparking potential.

The corona mechanisms in air may be described as follows. The current-voltage characteristics for point-plane gaps in general consist of three ranges [23]. First, for 10^{-14}–10^{-8} A, there is the dark current region (narrower for sharper points) in which Townsend's α is the predominant controlling factor. Secondly, for 10^{-8}–10^{-6} A, there is the visible intermittent corona range for which cathode mechanisms, as well as α, are important. This corona is not self-sustaining and externally produced electrons are necessary to initiate it. The third region is that of self-sustained 'continuous corona' in which the current is 10^{-5} A or more up to spark or arc breakdown.

In the investigations of positive corona in air at the lowest voltages only avalanches are discerned, produced in the region near the point where detachment is possible, i.e where $E/p > 90$ V/cm./mm. Hg. As the voltage is raised the avalanches increase in size and, in spreading laterally round the point by a photoionization mechanism, yield either burst pulses which are visible as a light glow covering the point in a thin layer, or streamers, which are filamentary discharges extending several millimetres or even centimetres into the gap.

Loeb has pointed out [24] that the onset voltage V_f for field intensified currents is such that E_f/p at the point exceeds 90 V/cm./mm. Hg when the O_2 ions lose their electrons in collisions with neutral molecules (see Chapter I). Below E_f the negative ions do not give a Townsend process to any appreciable extent since their mobility is too low. At a higher potential, V_g, the threshold of pre-onset burst pulse corona is reached, characterized by the appearance on a suitably arranged oscillograph of long (50–2,000 μsec) pulses of corona and an abrupt increase in the mean current flowing in the gap (see Fig. 3.9). Loeb suggests [24] that Fitzsimmons [5] confused V_f with V_g.

At perhaps 50–150 V above V_g the pre-onset streamers appear, to be followed, some 100–500 V higher, by the end of the Geiger counter characteristic and the appearance of steady, self-sustaining, burst pulse corona.

Loeb [25] puts the threshold for burst pulse onset at a voltage V_g for which the field surrounding the point is such that

$$f\beta\exp\left[\int\limits_{r}^{a+r} \alpha\,dx\right] = 1. \qquad (3.5)$$

In this expression r is the point radius and β is the chance that a photo-electron is produced, by absorption of a photoionizing photon from an avalanche, at a distance, from the point surface, greater than a. a is the distance from the point at which an electron either becomes free or at which Townsend's first coefficient α in the decreasing field becomes negligible. (The range of a is generally assumed to extend to where α falls to ~ 1.) f is the ratio of the numbers of photoionizing photons and electrons in the avalanche. As an approximation

$$\beta = 0 \cdot 5 \exp(-\bar{a}\mu), \tag{3.6}$$

where the factor 0 5 is included to allow for the fact that about 50 per cent. of the photons produced near the point will be absorbed in it. μ is the relevant photon absorption coefficient $\bar{a}$ is an average path for the photons in the hemispherical shell between r and $(r+a)$. Roughly

$$\bar{a} = a/\cos\bar{\theta}, \tag{3.7}$$

where $\bar{\theta}$ is the average angle of photo-emission with the normal to the point surface·

$$\bar{\theta} = \int_{0}^{\frac{1}{2}\pi} \theta \sin\theta \, d\theta \Big/ \int_{0}^{\frac{1}{2}\pi} \sin\theta \, d\theta = 1 \text{ radian.}$$

Thus, finally, $$\beta = 0 \cdot 5 \exp(-1 \cdot 86\mu a). \tag{3.8}$$

Loeb proceeds to put $a = 0 \cdot 04$ cm. and, for assumed values of $\mu = 10$ cm.$^{-1}$ and 100 cm.$^{-1}$, deduces the corresponding values of β as $0 \cdot 238$ and $3 \cdot 05 \times 10^{-4}$, and the values of f as 7×10^{-5} and $5 \cdot 5 \times 10^{-2}$. Cravath [26] has estimated $f \sim 10^{-4}$. Additional experimental data on these various quantities are needed before further advances in the theory of corona threshold can be made [27].

Discussion of these new formulae involving f, etc., is continued in · Chapter VI where the recent papers of Loeb [25], Loeb and Wijsman [28], and Hopwood [27] are treated further.

The resemblance between equation (3.5) for the positive burst pulse onset and the criterion for negative corona onset, which is

$$\gamma\left(\exp \int_{r}^{a+r} \alpha \, dx\right) = 1, \tag{3 9}$$

is clear. The latter expression, discussed in Chapter II, introduces the Townsend γ-mechanism, since for a negative point the impact of positive ions will cause secondary emission of electrons. It should be stressed, however, that the values of α used in rapidly divergent fields, such as those in the point-plane corona gaps discussed in this chapter, may need correction, according to the work of Morton and others [29, 30] (see p. 65).

At or near V_g (Fig. 3.9) the streamers may be observed to initiate a burst pulse or to occur alone, but they do not occur during a burst pulse since the lowering of the field near the point by the space charge caused by the burst pulse inhibits the unilateral growth of the streamers.

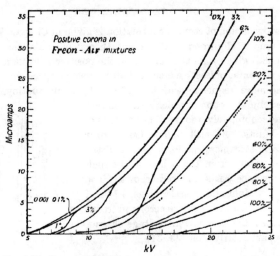

FIG 3 12. Current v potential curves for positive corona in freon-air mixtures The percentage number indicates the amount of freon by volume in dry, clean air. Total gas pressure 745 mm. of Hg The dotted curves are representative of the values obtained in a second run, about 15 minutes after the first one was finished

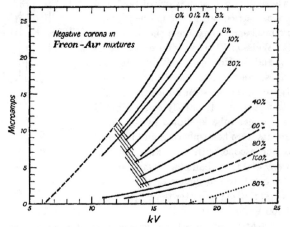

FIG. 3.13. Current v. potential curves for negative corona in freon-air mixtures. The percentage number indicates the amount of freon by volume in dry, clean air. Total gas pressure· 745 mm of Hg To the left of the shaded area and below the dotted continuation of the pure air curve (0 per cent), the current values for all mixtures fluctuate violently. The dotted 80 per cent curve below the 100 per cent one is due to the formation of several secondary corona spots 1 to 2 cm. from the tip of the point electrode on the stem

(Further discussion of streamer formation is given in Chapter VI where
an account of Hopwood's [27] recent theory of avalanche-streamer
transitions is given) As the voltage on the positive point is raised still
further, the burst pulses, each choked off by its own space charge, occur
more and more frequently with (at V_s, Fig. 3.9) the fluctuating, but self-
sustained, burst corona in which space charges
are never completely removed between successive bursts.   At still higher fields, near
breakdown, streamers can form again and the
extension, away from the point, of the high
field region encourages unilateral growth. The
diagrams of Mohr and Weissler (Figs. 3.12 and
3.13) for positive and negative corona in freon-
air mixtures show the current-voltage characteristics particularly well [23, 31]

Positive and negative point coronas in various gases

The above general account of corona mechanisms has been drawn from the work of Loeb and
his school as carried out in the pre-war years.
It is now necessary to examine the later publications Weissler [2] has made experiments
in H_2, N_2, and A. He used carefully outgassed
pyrex tubes, metal parts baked *in vacuo*, and
purified gases, although some interesting results
were obtained with deliberately contaminated
gases. A typical tube is shown in Fig 3.14.
The Pt points were hemispherical and polished
under a microscope with tin oxide, the Pt
plates were 4 cm. in diameter. Gap lengths
ranged from 3 1 to 4·6 cm , and the voltage
supply was stabilized to better than 1 per cent
above 5,000 V.

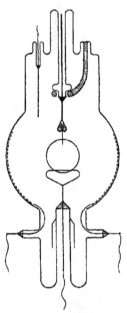

FIG. 3.14. Weissler's experimental tube   The Pt wire
and Pt plane electrodes, the
circular observation window,
and (upper left hand of
diagram) the W electrode
which can be heated to remove oxygen from, for ex-
ample, impure argon.

In H_2 at 1 atm. with a 3·1 cm. gap between a positive point and a
plane, corona onset occurred at 3·5 kV. Avalanches, but no streamers,
began to form at 9 to 10 kV and led at higher potentials to sparkover.
The addition of O_2 ($\sim 0\cdot1$–1 per cent.) immediately resulted in the
formation of pre-corona onset streamers which increased in strength as
the voltage was raised to breakdown. An increase in the amount of O_2

led to pre-onset streamers followed by burst pulse corona and finally breakdown streamers. Weissler interprets these results by suggesting that photoionization, which is necessary for streamer formation, is easier in O_2/H_2 mixtures to such an extent that 1 per cent O_2 in H_2 gave copious pre-onset streamers. The space charges so produced inhibit further streamer formation and lead to lateral spreading of the corona over the point, i.e burst pulses. When the field becomes high enough, as the voltage is raised, to clear the space charges, streamers may again be formed and will lead to breakdown. In pure N_2 at atmospheric pressure the effects are generally similar for the same gap The main features of the phenomena are summarized in Table 3 2.

TABLE 3.2

Essential Data of the Positive Point-plane Corona

Gas	Pre-onset streamers	Onset potential (kV)	Visual character	Burst pulses	Current†† (µA)	Onset of breakdown streamers (kV)	Approx. breakdown potential (kV)
H_2 pure§	none	3 5	localized†	none	4	10	18
$H_2 + 0.1\% O_2$	yes	5	intermittent streamer-corona	none		8	
$H_2 + 0.5\% O_2$	yes	4 5	streamer-corona	none	3	9	
$H_2 + 1\% O_2$	yes	4	uniform‡	small	3	15	18
N_2 pure	none	4 8	localized	none	1	11	20
$N_2 + 0.2\% O_2$	incipient streamers	4 5	streamer-corona	none	1	9 3	
$N_2 + 1\% O_2$	incipient streamers	4 8	localized, for higher volts uniform glow	yes	1	12 5	20
Room air§	yes	5 5	uniform	yes		15	
A pure‖	none	none	none	none			3 5
$A + 0.1\% H_2$	yes	3 3	localized	none	0 5		
$A + 0.5\% H_2$	yes	3·5	localized	small	0 5		
$A + 1\% H_2$	yes	3 45	localized	small	0 5		
$A + 0.3\% N_2$	none	none	none	none	.	3 5	3 5
$A + 0.5\% N_2$	yes	3 6	streamer-corona	small		4 6	4 6
$A + 1\% N_2$	incipient streamers	3 5	localized	small	0 5	6 5	6 5
$A + 0.1\% O_2$	yes	3 8	streamer-corona	none	.	4 1	4 1
$A + 0.4\% O_2$	yes	4 5	streamer-corona	small	0 6		7
$A + 1\% O_2$	yes	4 0	uniform	small	0 6	..	15

† Localized glow, due to electron avalanches approaching the point in the highest field region.
‡ Uniform glow, caused by the burst pulse corona
§ Gap length 3 1 cm
‖ Gap length 4 6 cm
†† Current at a potential which is about 20 per cent above the corona onset potential

In argon, with conditions as given above, a spark crossed a 4·6 cm. gap at approximately 3·5 kV and no corona pulses of any kind were observed. With 0·3 per cent. N_2, a few strong streamers were noticed at 3·5 kV and breakdown followed at once; with 0 5 per cent N_2 onset streamers were observed up to 4·1 kV at which localized corona, with occasional burst

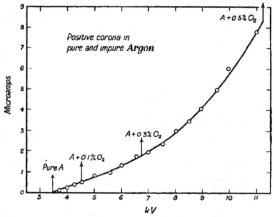

FIG 3.15 Current-voltage characteristics for positive corona in pure and impure argon. The arrows indicate the upper limit of the current-voltage curve and the sparking threshold for a specific mixture of O_2 in A.

pulses, was observed. A few strong breakdown streamers were noticed immediately before spark breakdown at 4·6 kV. The only difference noticed with oxygen added to argon was that above onset streamers still existed in the presence of burst pulse corona (see Fig 3.15).

Weissler suggested that the argon discharges are explicable on the grounds of ease of photoionization, because of the presence of many metastable argon atoms, and also because the small Ramsauer cross-section in argon (see p. 8) means that the Townsend α reaches effective values at low E/p. The addition of H_2 or N_2 to argon quenches the metastable states and so allows streamers to form before breakdown. The destructive effect of O_2 on the metastable states was so great that streamers are weak at the low potentials and cannot choke themselves

Positive corona in H_2, N_2, and A was studied [2] at low pressures, i e. 100–300 mm. Hg. The corona effects were similar to those at 760 mm. Hg, but the corona naturally tended to be more diffuse. A point of great interest was that streamers could still form and be propagated due,

Weissler suggests, to the localized field near the point since streamers could hardly be expected in plane parallel gaps in these conditions.

Weissler pointed out, with regard to negative point corona (see Table 3.3), that negative Trichel pulses [11] should be expected to occur in all

TABLE 3 3

Essential Data of the Negative Point-plane Corona

Gas	Localized corona			Trichel pulses	Approx breakdown potential (kV)	Peculiarities
	Onset potential (kV)	Onset current (μA)	Visual character			
H₂ pure	2 6†	140	unsteady glow	none		alternation of uniform and localized glow
H₂+0 1% O₂	4 2	70	unsteady glow	yes	.	as above
H₂+0 6% O₂	5 8	2	unsteady glow	yes	.	as above
N₂ pure§	3 7	90	steady glow	none	14	
N₂+0 1% O₂	4 3	50	steady glow	yes		
N₂+1% O₂	5 5	0 1	intermittent‡	yes		5 to 7 kV intermittent corona
Room air	8	0 1	intermittent	yes		8 to 11 kV intermittent corona
A pure‖	.	. .			3 5	no corona
A+0 3% N₂	2 5	200	steady glow	none	4 1	
A+1% N₂	3 9	250	steady glow	none	.	
A+0 1% H₂	3 2	400	unsteady glow	none	4 1	. .
A+1% H₂	3 3	300	unsteady glow	none	.	.
A+0 1% O₂	3	1	steady glow	yes	.	
A+0 4% O₂	3	0 1	intermittent	yes	24	3 to 6 kV intermittent corona
A+1% O₂	3 3	0 1	steady glow	yes	.	

† The figure 2·8 occurs in Weissler's table, this seems to be a misprint
‡ Due to lack of triggering electrons
§ Gap length 3 1 cm
‖ Gap length 4 6 cm

gases forming negative ions, since the low mobilities of the latter, as compared with electrons, allow negative space charges to build up round the point. Thus, in pure H_2 corona onset was at 2·6 kV, appreciably lower than for positive corona, with a continuous glow spread over the point. This glow was unstable and sooner or later contracted to a concentrated discharge in the region of highest field strength. No Trichel pulses were observed, but the introduction of small amounts of O_2 immediately produced them, for example with 0·6 per cent. O_2 they were noticed at 5·8 kV whilst the steady corona began at 7·5 kV.

The fact that the onset potentials for positive and negative corona in air differ by a greater amount than the corresponding values of hydrogen may be due to photoionization governing both mechanisms in air (where the process is efficient) and so giving similar onset potentials, whilst in pure hydrogen the negative corona currents are governed by secondary emission from the point due to positive ion bombardment. This may be so, as with a negative point in hydrogen the positive ions are not hindered near the point by negative ion space charges since attachment is small. With a positive point in pure hydrogen the main source of electrons is the distant plate, since photoionization is small, and as the field near the plate is low, one cannot expect appreciable photo-currents to be produced Corona currents in this case are thus small, the onset voltages being higher than with a negative point.

The results for negative corona in pure N_2 were similar to those for pure H_2 and Trichel pulses were again only observed if O_2 was introduced. The negative corona in pure argon was difficult to observe, and breakdown occurred without preceding discharges, although the addition of H_2 or N_2 to the argon allowed corona to be observed. If O_2 was the adulterant, Trichel pulses were obtained, even if the impurity was only 0·1 per cent. The equality of spark breakdown voltages for positive and negative points in pure argon gives [2] further qualitative support for the suggestion that strong photoionization is in fact operative in both cases, since the persistent metastable atoms should be plentiful in argon and should tend to accentuate photoionization

Weissler draws attention also to the peculiar phenomenon, also noticed by Bennett [9], of the pitting of a negative point. Since the positive ions in Weissler's experiments strike the cathode with energies not much greater than 1 eV, it seems remarkable from a knowledge of cathode sputtering that etching, i.e the formation of crater-like pits in the cathode, should be noticeable. The effect was marked with points of W, Pt, Cu, and Pb, but not Al, in H_2 and N_2 Weissler suggests that though experimental data are scanty a cathode-sputtering mechanism [32] seems the most likely explanation. However, it seems also possible that atomic recombination of the products of dissociation in the above molecular gases at the cathode might, since it is extremely exothermic, give volatilization and hence pit formation. This would be disproved if pitting should be obtained in argon, but no data appear to be available. Since the effect is noticed at the cathode it seems necessary to add to the above tentative explanation the suggestion that, perhaps, an absorbed layer of H or N atoms (neutralized positive ions) forms on the cathode as an intermediate but constantly occurring stage in the process.

English [33] has studied corona from positive and negative points in the form of water-drops, following earlier work by Macky [34] and Zeleny [35]. The results are of interest in connexion with atmospheric electricity, and the electrical effects associated with rain [36].

The onset of discharge at atmospheric pressure occurs at the same potential for positive and negative points and is ascribed to electrostatic disruption of the drops accompanied by pre-onset streamers in the former case. This effect cannot therefore be connected directly with corona phenomena, but the electrical effects associated with the formation of the resulting droplets are of interest and are described in detail by English. At higher voltages the space charges produced round the drop reduce the field at the drop, as with metal point coronas, and the drops become stable This occurs at 7·5 kV for drops of 0·13 cm radius, but if the voltage is increased to 10 5 kV the field distortion by space charges is overcome by the increased field, particularly at the outer surface of the space charge zone at the point, and the water drops again become unstable

With a large negative point of 0·13 cm radius, pulses were observed with the oscillograph. The first kind resembled Trichel pulses and the second kind, of a more complicated form, was associated with the droplet formation For water-drops the negative corona onset is at least 3 kV higher for the point described than for positive onset, as would be expected from the low value of γ (see Chapter II) but the various transition voltages (spray-to-drop, etc.) complicate the issue.

English [21] has recently repeated, with an improved oscillograph of band-width approximately 6 mc./sec , the earlier work of Kip, Trichel, and others on negative and positive point-plane corona in air. The measurements covered the pressure range 756–210 mm. Hg. The duration of positive breakdown streamers, with a 0·3-mm. diameter Pt point was about 0·8 μsec. The intermittent corona onsets were approximately equal but showed a small pressure-dependent difference. Hudson's work on the effect of MgO coating on negative points, for the purpose of stabilizing the Trichel pulses, was confirmed, and it was found that the dusting of fine positive points ($<$ 0·01 mm. radius) gives bursts or streamers, or both, where previously neither had been observed. The mechanism for the dusted positive point is not clear.

Electrical measurements [22] showed that in the particular conditions obtaining, positive streamers contained 10^{10} ions per centimetre length. For a polished 0·38-mm diameter Pt point, the duration of positive pre-onset streamers and negative Trichel pulses near onset was about 0·4

μsec., whilst positive breakdown streamers lasted 0 8 μsec. Even with
the latest techniques, however, English [22] does not consider that the
rise times of positive streamers and burst pulses have been accurately
measured, although Trichel pulses can probably be accurately resolved
as their build-up depends on relatively slow ion movements.

English [21] has also measured the current-voltage characteristics
for negative and positive point coronas in air and has discussed their
shapes (see Loeb [6] and the above discussion on pp 157-8) in terms
of the various pulse formation mechanisms. Further work on these
characteristics was carried out earlier by many workers [37, 38, 39,
40, 41].

Gaunt and Craggs [42] have developed a technique for studying corona
in which the visible and ultra-violet radiation from the corona pulses
is detected by a photo-multiplier tube and displayed with a cathode ray
oscillograph. The circuit differs from that used by Saxe and Meek [43]
who had earlier developed a multiplier technique for the study of impulse
corona. English [44] independently also observed the visible radiation
from corona discharges in a similar fashion. There are at present unexplained differences between the results of Gaunt and Craggs and those
of English as regards the duration of the light from streamers, the former
measuring this as ∼ 0·5 μsec and the latter about ∼ 0·1 μsec Gaunt
and Craggs [51] following other authors [45, 46] have reported observations on the emission spectra of corona discharges which usually (in
contrast with spark channels) show band spectra

Another recent development in corona research is the study of the
luminous cloud of discharge accompanying the growth of the filamentary
streamer discharges described above. The recent work of Komelkov
[47, 48] is described in Chapter IV together with the work, by Saxe and
Meek [43]. Saxe, in studies of impulse corona, stressed the possible
importance of this phenomena to a greater degree than had previous
workers since, although [49, 50] it had been previously recorded on
photographs of leader strokes and D.C. corona, the significance of
it had been largely overlooked. Saxe found that the radius of the
luminous cloud [43] was 10-20 cm. The radius of the final channel of
the leader stroke is ∼ 0·2 cm. and the transition from the cloud to
leader stroke is fairly sharp Further confirmation of the nature of the
diffuse pre-streamer discharge is given by the Lichtenberg figures published by Allibone [52], and by the D.C. corona photographs of Loeb
[50] and Gaunt and Craggs [42]. The latter (Fig. 3.16, Pl. 1) show that
the luminous clouds consist partly, if not entirely, of thin luminous

channels, the mechanism of which is not clear. Several possible theories are briefly reviewed by Craggs and Meek [53] and further discussion of the phenomenon is given in Chapters IV and VI.

Corona discharges in freon-air mixtures

It is clearly of interest, in order to decide the importance of negative-ion formation, to study corona formation in strongly electronegative gases   This has been done by Weissler and Mohr [23, 31] for negative and positive corona, using freon (CCl_2F_2)-air mixtures at a total pressure of 745 mm. Hg in a 3·1 cm. gap between a point and a plane, both electrodes being formed of platinum. The point had a diameter of 0·5 mm., and the plane was 4 cm. in diameter. The corona pulses were observed by measuring the voltage developed across a $3 \times 10^4 \, \Omega$ resistor in series with the plate.

For a freon content up to 0·1 per cent., the phenomena were similar to those found with dry air, i e. Trichel pulses were seen at about 6 kV ($i \sim 10^{-8}$ A). The pulses were irregular, in the form of bursts each containing a few pulses, similar in these respects to those found by Hudson [9] in enclosed gaps containing dust-free dry air (the importance of dust, etc , in producing triggering electrons at the point is discussed on p. 156). At 12·5 kV ($i \sim 12 \, \mu$A) continuous corona developed, Trichel pulses were seen in this region, decreasing in size as the current increased   With 0·1–1 per cent. freon in dry air, otherwise as above, the only notable difference was the appearance of a hysteresis effect, i.e. a change in threshold value from 6 kV to 9–10 kV for intermittent corona (Trichel pulse) onset. With 1–20 per cent. of freon, the bursts of Trichel pulses changed as shown in Fig. 3.17 and the current-voltage curves showed gradually decreasing currents at any given voltage as the amount of freon was increased. With 60 per cent or more of freon present the Trichel bursts could be distinguished only with difficulty because of their small size. The onset potential for Trichel pulses in pure freon was 12·6 kV.

The visual appearance of the corona in these mixtures gradually changed from the glow discharges seen in air, with the usual dark and glowing regions, to a diffuse general glow characteristic of high freon contents. At intermediate freon concentrations (40–80 per cent) two discharge forms were observed alternately, at the same potential (similar effects in low-pressure corona were found by Craggs and Meek [54]).

In discussing the results for negative corona, Weissler and Mohr assume that the Trichel pulses in air are caused by the formation of a negative ion cloud at such a distance from the point that $E/p < 90$ (see Chapter I). When successive avalanches have together contributed sufficient negative ions to reduce the field near the point to such a value that further avalanches cannot develop, then the discharge ceases and the negative ion cloud diffuses away (in $\sim 10^{-4}$–10^{-3} sec.) until the field near the

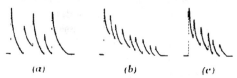

(a)   (b)   (c)

Fig. 3 17   (a) Irregular Trichel-pulse burst in dry, clean air near onset of negative corona (7,000 volts) (b) The same in 6 per cent freon in air, $p = 745$ mm. of Hg   (c) The same in 10 per cent freon in air, $p = 745$ mm of Hg It is possible that the individual pulses in (b) and (c) return to the zero line. Because of the high speed of pulse formation (less than one μsec) the details of this oscillograph trace must be instrumental, but the general decay of the pulses seems to be real

point becomes sufficiently great to give avalanche formation again Now the electron affinities of free Cl (3·8 eV) and F (3·5–4 eV), which are products of dissociation of CCl_2F_2, are much greater than for O_2 (0·07–0·19 eV) or indeed for O (2 2 eV), although the latter is considered unlikely to exist for any appreciable time in air. The negative ion space charges will therefore be formed closer to the point in freon than in air Thus the corona currents or size of avalanches will be lower, for a given voltage, in freon than in air. Negative ions in these gases may be stable up to $E/p = 200$, as shown by rough calculations of the corona-point fields [23, 31]. It is also suggested that the production of free Cl and F is responsible for the peculiar hysteresis effects observed in these experiments (see above) An alternative explanation has been given by Warren et al. [55, 56, 57].

The positive corona in freon-air mixtures showed the characteristics of the discharges in air in which various foreign products, for example oxides of nitrogen, had been produced if the freon content was 10 ^{4} to 10^{-1} per cent Using the gap described above, the normal onset streamers, continuous corona, and breakdown streamers were observed. With 1 per cent. freon the onset for intermittent corona was 5 kV, and strong streamers were observed and persisted above the continuous corona onset at 9·15 kV, burst pulses were negligible in number. The results of several runs at varying freon concentrations, represented as visual discharges, are shown in Fig. 3.18 The addition of more than 1 per cent of freon gave more burst pulses, with a decrease in size and frequency of the streamers.

The case of 1 per cent. freon represents the mixed gas effects studied previously by Weissler [2] who attributed the easy formation of streamers in such conditions to the increased probability of photoionization

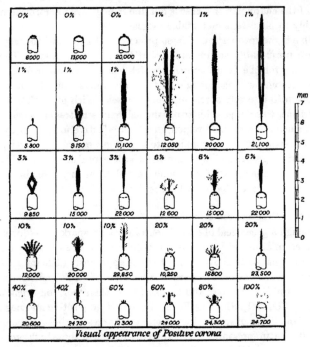

Fig. 3 18. Visual appearance of positive corona in freon-air mixtures
The percentage number indicates the amount of freon by volume in dry,
clean air, the other number the point potential in volts

compared with that for a pure gas. The unusually abundant formation of streamers in the continuous corona region in freon-air mixtures may be explained by the presence of negative ions which neutralize the positive space charges in the streamer channels and facilitate the formation of succeeding streamers. The narrow, spindle-like discharges shown in Fig. 3.18, for a mixture containing 1 per cent. freon, have current densities varying from 1.7×10^{-2} A/cm.^{2} at 10 kV, near the onset of continuous corona, to 1 A/cm.^{2} at 21 kV. The formation of these discharges can be attributed to the peculiar case of streamer formation in this gas mixture. Mohr and Weissler[31] suggest that this may be because (a) lateral branching is unlikely with strong streamers and (b) the existence of stable negative ions gives an increased lifetime to the channel and successive streamers may follow the same track. Such spindles have been observed in pure N_2 near breakdown, but the persistence of the channel in this case is not due to negative ion formation but probably to persistent metastable atoms.

With a higher proportion of freon present ($>$ 1 per cent) the negative ions are presumably so numerous that the positive ion space charge, responsible for the field distortion which leads to streamer formation, cannot become strong enough to form streamers and the discharge develops into laterally spreading burst pulses. With more than 40 per cent. freon it was not possible to distinguish between streamers and burst pulses, since the high field region, in which effective ionization by free electrons is produced, is confined to a region very close to the point because of the great stability of the negative ions in F and Cl. Kip[6] also observed a similar effect for very sharp points in air, probably again because of the restricted high field region.

High-pressure coronas

Although a great deal of experimental work on spark breakdown and the determination of corona onset voltages has been carried out at pressures equal to or above 1 atm. there is apparently little information on corona mechanisms, comparable with that of Loeb and his school, below 1 atm. [2] Pollock and Cooper [3], in their work on breakdown in N_2, O_2, CO_2, SO_2, SF_6, CCl_2F_2, A, He, and H_2 in positive and negative point-plane gaps, observed the corona discharges in these gases up to some 20–30 atm. pressure. The corona discharges at these high pressures are not, apparently, very different from those at $\sim$ 1 atm. The corona onset curves (see Figs 7 26 and 7.27, Chapter VII) give the potentials at which onset streamers appear on an oscillograph. At sufficiently high pressures in a positive point-plane gap corona does not precede sparkover in most attaching gases and the $+C$ curves merge with the $+S$ curves (Fig. 7.26, oxygen mixtures, Chapter VII); ($C \equiv$ corona onset, $S \equiv$ spark breakdown). At lower pressures in a positive point-plane gap there is, again in the gases which readily form negative ions, a fairly wide corona region in which, as the voltage is gradually raised, burst corona with a D.C. component follows the onset streamers and is in turn followed by breakdown streamers just before sparkover. Negative ion formation tends to choke the onset streamers (see p. 173) but they may be increased in number by irradiation, for example by radium, and this effect is clearly shown for low pressures of CCl_2F_2 in the oscillograms of Craggs and Meek [54] Pollock and Cooper's curves show in some cases the influence of irradiation. At the higher voltages, with more photoelectrons being produced near the point, burst pulse corona is produced and the whole region round the point may glow, in contrast with the unilateral propagation of the onset streamers Finally, breakdown streamers form at the higher fields which are strong enough to overcome the choking effect of space charge and to permit streamer propagation across the gap to the cathode. Positive points in H_2, He, A, and N_2 also give streamers but no bursts, although in hydrogen a rapid and regular succession of small streamer-like pulses was observed In the nonattaching gases, however, the individual streamers are able to propagate entirely across a short gap before a sufficient number of streamers form simultaneously to give burst corona, i.e. the region of intermittent corona consisting of onset streamers extends to the breakdown voltage. Thus the formation of negative ions in a gas appears to be able, in some gases, to stabilize burst pulse corona and to choke off intensive individual streamers.

Considerable attention has been paid to corona problems by Russian workers since 1945. Popkov [58] has developed probe techniques for studying space charge conditions in corona discharges. Hey and Zayentz [59, 60] have studied streamer formation with Wilson chambers and have also made detailed studies of time lags in impulse corona Wilson chamber work of a similar nature has also been carried out by Gorrill [6] and others [61, 62] (see Chapter IV for a discussion of the work of Raether and his collaborators) The conclusions of Hey and Zayentz are in general agreement with those of Loeb's school in that pulses pass through a sequence of forms as described above

The American work, to March 1948, is reviewed in a detailed report by Loeb [50], which should be studied carefully by interested workers in this field, and which deals with both negative and positive coronas.

REFERENCES QUOTED IN CHAPTER III

1 S WHITEHEAD, Electrical Discharges in Gases, Benn, London, 1928.
2. G L. WEISSLER, Phys. Rev 63 (1943), 96
3. H C POLLOCK and F. S COOPER, ibid 56 (1939), 170.
4 A. F. KIP, ibid. 54 (1938), 139.

5 K E. FITZSIMMONS, *Phys Rev* **61** (1942), 175
6. L. B LOEB, *Fundamental Processes of Electrical Discharge in Gases*, Wiley, New York, 1939
7 F. H SANDERS, *Phys Rev* **41** (1932), 667, **44** (1933), 1020.
8. K. MASCH, *Arch Elektrotech* **26** (1932), 589.
9 L. B LOEB, A F KIP, G G HUDSON, and W. H BENNETT, *Phys Rev.* **60** (1941), 714
10 C. F EYRING, S. S MACKEOWN, and R A MILLIKAN, ibid **31** (1928), 900
11 G W. TRICHEI, ibid **54** (1938), 1078
12 D. R CORSON and R R WILSON, *Rev Sci Instrum.* **19** (1948), 207.
13 D H WILKINSON, *Ionization Chambers and Counters*, Cambridge, 1950
14 R A MILLIKAN and C.C LAURITSEN, *Proc Nat Acad Sci Wash* **13** (1928), 45.
15 W H BENNETT, *Phys. Rev.* **58** (1940), 992
16. H. PAETOW, *Z Phys* **111** (1939), 770.
17. L. B. LOEB and W. LEIGH, *Phys. Rev* **51** (1937), 149.
18 A F. KIP, ibid **55** (1939), 549.
19 L B LOEB, ibid **48** (1935), 684
20 W DE GROOT and F M PENNING, *Ned. T. Natuurkde* **11** (1945), 156.
21. W. N. ENGLISH, *Phys Rev* **74** (1948), 170
22 W. N ENGLISH, ibid **71** (1947), 638
23 G. L. WEISSLER and E I MOHR, ibid. **72** (1947), 289.
24. L. B LOEB, ibid **73** (1948), 798.
25 L. B LOEB, *Rev Mod Phys* **20** (1948), 151.
26 A M CRAVATH, *Phys Rev.* **8** (1935), 267
27. W. HOPWOOD, *Proc. Phys Soc* **62** (1949), 657.
28. L B LOEB and R A WIJSMAN, *J. Appl Phys.* **19** (1948), 797
29. P. L. MORTON, *Phys Rev.* **70** (1946), 358
30 L. H. FISHER and G. L WEISSLER, ibid **66** (1944), 95.
31 E I MOHR and G L WEISSLER, ibid **72** (1947), 294
32 A. VON ENGEL and M STEENBECK, *Elektrische Gasentladungen*, Springer, Berlin, 1934.
33 W N. ENGLISH, *Phys Rev.* **74** (1948), 179
34 W. A MACKY, *Proc Roy Soc* A, **133** (1931), 565.
35. J. ZELENY, *Phys. Rev.* **16** (1920), 102.
36 J A CHALMERS, *Atmospheric Electricity*, Oxford, 1949
37 S P. FARWELL, *Trans Amer Inst Elect Engrs* **33** (1914), 1631.
38. H PRINZ, *Arch Elektrotech.* **31** (1937), 756.
39 J ZELENY, *Phys Rev* **25** (1907), 305
40 E BENNETT, *Trans. Amer Inst Elect Engrs* **33** (1914), 571.
41. J. D. COBINE, *Gaseous Conductors*, McGraw-Hill, New York, 1941.
42. H. M. GAUNT and J. D CRAGGS, *Nature*, London, **167** (1951), 647.
43 R F SAXE and J M MEEK, ibid. **162** (1948), 263.
44 W. N. ENGLISH, *Phys Rev.* **77** (1950), 850
45 A. C YOUNG and A G CREELMAN, *Trans. Roy Soc. Can* **26** (1932), 39.
46 Y MIYAMOTO, *Arch. Elektrotech.* **31** (1937), 371
47. V. KOMELKOV, *Bull. Acad Sci. U S.S.R.* (Tech. Sci.), No. 8, (1947), 955.
48. V. KOMELKOV, *Doklady Akad. Nauk, U.S.S.R.* **68** (1947), 57.
49 T E. ALLIBONE and J. M. MEEK, *Proc Roy Soc* A, **166** (1938), 97; **169** (1938), 246.

50 L. B LOEB, *J. Appl. Phys.* **19** (1948), 882.
51. H M GAUNT and J. D. CRAGGS, in course of preparation.
52. T. E. ALLIBONE, *Nature*, London, **161** (1948), 970.
53. J. D. CRAGGS and J. M. MEEK, *Research*, **4** (1951), 4.
54 J D. CRAGGS and J. M. MEEK, *Proc. Phys Soc.* **61** (1948), 327.
55. J W WARREN, W. HOPWOOD, and J. D CRAGGS, ibid. B **63** (1950), 180.
56. J. W. WARREN and J D. CRAGGS, in course of preparation.
57 J W WARREN and J. D CRAGGS, in course of preparation.
58. V. I. POPKOV, *Elektrichestvo*, No. 1 (1949), 33.
59. V. HEY and S ZAYENTZ, *J Phys U S S.R.* **9** (1945), 413
60 V. HEY and S ZAYENTZ, *J Exp. Theor. Phys. U.S S.R* **17** (1947), 437, 450.
61. H. RAETHER, *Ergebnisse d. exakten Naturwiss.* **22** (1949), 73.
62. L. B LOEB and J. M MEEK, *The Mechanism of the Electric Spark*, Stanford, 1941.

PLATE 1

FIG 3.16 Corona discharge positive point,
in argon (containing about 0.5 per cent
nitrogen) at a voltage slightly lower than
that for spark breakdown

Fig. 4.1 Chopped spark discharges across a gap of 8 cm in air between parallel plates (Time interval of 5.3 × 10⁻⁸ sec between first and last photographs)

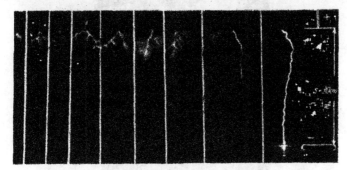

Fig. 4.2 Sparks at different stages of their growth across a 30 cm gap in air between a positive point and a plane (Time interval of 2.5 × 10⁻⁸ sec between first and last photographs)

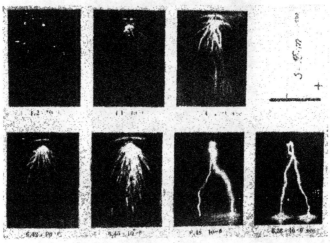

Fig. 4.3 Sparks at different stages of their growth across a 16 cm gap in air between a negative point and a plane

PLATE

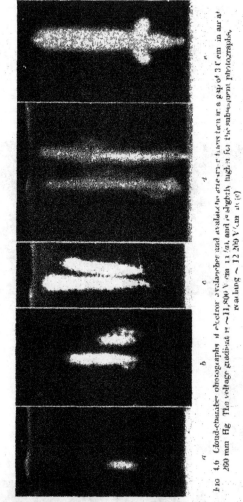

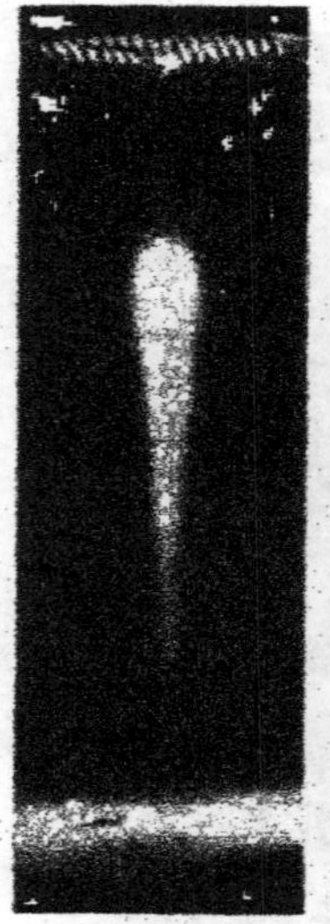

Fig. 4.5. Cloud-chamber photograph of an electron avalanche

a b c d e

Fig. 4.6. Cloud-chamber photographs of electron avalanches and avalanche-streamer transition in a gap of 3.6 cm in air at 260 mm Hg. The voltage gradient is ~11,800 V/cm in (a), and is slightly higher for the subsequent photographs, reaching ~12,200 V/cm in (e).

PLATE 4

FIG. 4.8. Streamer formation in a point-plane gap in air as recorded in a cloud chamber. (a) Positive point. (b) negative point

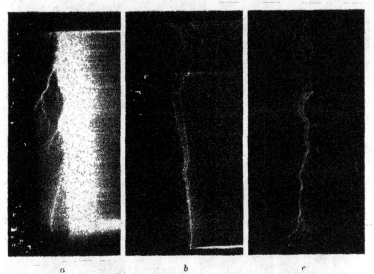

           a   *b*   *c*

FIG. 4.10 Typical photographs of the leader-main stroke sequence for sparks between a positive point and a plane in air (see Fig. 4.12 for time intervals)

     (a) Gap – 100 cm Series resistance   12,000 Ω
     (b) ,,   130 cm   ,,   ,,   370,000 Ω
     (c) ,,   190 cm   ,,   ,,   1,000,000 Ω

PLATE 5

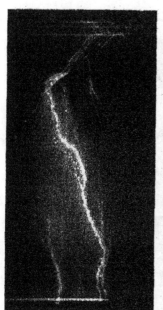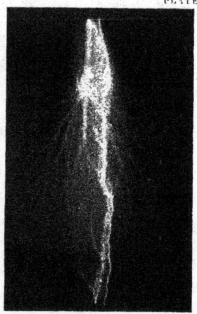

FIG. 4.15. Rotating camera photograph
of a negative point-plane discharge
across a 100 cm. gap in air. Series
resistance = 100,000 Ω (a) Figs. 4.18
and 4.19 for time intervals)

FIG. 4.17. Rotating camera photograph of a
negative point-plane discharge across a 76 cm.
gap in air. A point of length 1.25 cm. projects
from the plane

Series resistance = 1.9 MΩ

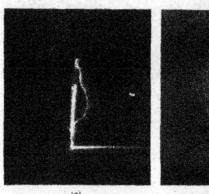

(a)   (b)

FIG. 4.16. Rotating camera photographs of negative point-plane discharges across
a 25 cm. gap in air

(a) Series resistance = 0.6 MΩ   (b) Series resistance = 1.0 MΩ

PLATE 6

Fig 4 45 Negative Lichtenberg figure in air at 1 atm containing CCl₄ Electrode diameter = 0 2 cm Applied voltage = 4 5 kV

Fig 4 42 Lichtenberg figures in compressed air (a) Positive discharge, 30 3 atm, 30 kV (b) negative discharge, 31 atm, 50 kV

Fig 4 44 Change from primary figure (a) to single positive streamer (b) by increase of pressure in air Electrode diameter = 0 25 cm for both (a) and (b)

Fig 4 46 Positive primary figure with radial spike structure recorded in air at 1 atm containing CCl₄ Electrode diameter = 0 25 in Applied voltage = 30 kV

Fig 4 47 (a)

Fig 4 47 (b)

Lichtenberg figures for point plane electrode arrangement
(a) Positive point   (b) Negative point

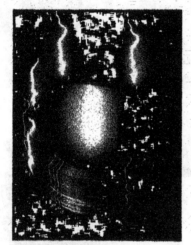

Fig 3 3 Photograph of stepped-
leader from

Fig 3 1. Photograph of multiple-stroke
lightning discharge

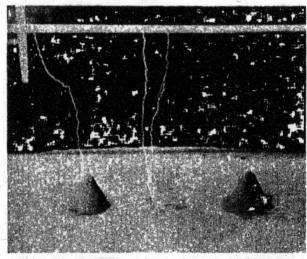

Fig 3.8. Photograph of sparks between a metal rod and a sand electrode

PLATE 8

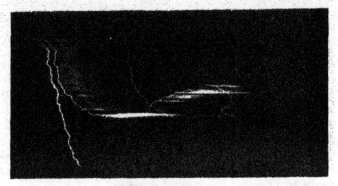

Fig. 7.14.
Moving film photograph showing a change in the path of a discharge containing three strokes

Fig. 7.15. Still photograph of lightning discharge to ground and within the cloud

EXPERIMENTAL STUDIES OF THE GROWTH OF
SPARK DISCHARGES

A DESCRIPTIVE account of the more recent experimental studies of the growth of spark discharges is given in this chapter, which is subdivided according to the techniques used in the various investigations. Some, but by no means all, of the phenomena described can be explained in terms of electron avalanches and streamer mechanisms, according to the theoretical considerations discussed in Chapter VI. But because of the incomplete nature of the theory of the spark mechanism, and the various conflicting opinions on the subject, the various experimental results are presented here without comment.

Suppressed discharges

If an impulse voltage is applied to a discharge gap for a limited time only, the growth of a spark in the gap can be arrested before the spark channel has fully formed across the gap. This method, involving the use of a second gap connected in parallel with the discharge gap being studied, has been used by several investigators. By adjustment of the length of the second gap, the duration of the voltage on the first gap can be varied, a spark being formed across the second gap and causing the voltage to collapse before the spark formation in the first gap is complete.

Investigations of sparks in air by means of this technique have been made by Holzer [1]. A series of photographs of the growth of sparks across an 8-cm. gap are given in Fig. 4.1, Pl. 2, where it is seen that the spark channel is completed after a time of $5 \cdot 28 \times 10^{-8}$ sec. In the earliest visible stages of the growth a bluish glow is observed at the cathode accompanied by a long reddish brush discharge at the anode. A dark space near the cathode separates the cathode glow from the anode brush. With increasing time of application of the voltage to the gap a second stage in the growth commences when a brightly luminous filamentary discharge, which is described in this book as a streamer, develops from the anode and completes the breakdown of the gap. The speed of this streamer has been measured by Holzer for several gap lengths and is found to vary from $1 \cdot 4 \times 10^8$ cm./sec. for a gap of 2 cm. to $6 \cdot 6 \times 10^8$ cm./ sec. for a gap of 12 cm. in a nearly uniform field.

Holzer [1] records anomalies in the development of some of the

N

sparks observed in that one or more mid-gap streamers may be formed, the streamers growing together and to the electrodes to produce the final spark channel. When only one mid-gap streamer is present it is generally in the cathode half of the gap The spark channels formed by the joining together of several mid-gap streamers are found to be of uneven cross-section and brightness, and Holzer draws attention to the possible bearing of this observation on the phenomenon of bead lightning (see p 242).

Other studies of suppressed discharges in uniform and nearly uniform fields have been made by Torok [2] using an adjustable sphere-gap in parallel with the gap under observation, and discharges across gaps ranging up to 75 cm. between spheres of several diameters have been photographed In the longer gaps bright filamentary streamers emanating from the high-voltage sphere are recorded, for both polarities, the remainder of the gap being bridged by a light purple haze. In short gaps, up to 2 cm. approximately, several discharges were detected in which bright streamers appear in the mid-gap region with a faint glow extending in both directions to the electrodes The currents flowing in the gap during the growth of the streamers have been measured by Slepian and Torok [3] for gaps up to 75 cm. between spheres of 50 cm. diameter, and for gaps up to 110 cm. between a similar sphere and a plane. The values of the currents recorded vary over a wide range up to as high as 4,000 A.

The growth of sparks in non-uniform fields between a point and a plane, for positive and negative points, has been investigated by Holzer [1]. Sequences of photographs recorded for positive and negative discharges are given in Figs. 4.2 and 4 3, Pl. 2. Fig. 4.2 shows that the breakdown between a positive point and a negative plate is initiated at the point, from which streamer discharges develop until eventually, if the voltage is maintained for a sufficient period, a streamer crosses the gap to form a spark channel and breakdown is complete. Each of the photographs necessarily refers to a separate discharge, and because of the irregular nature of streamer development in non-uniform fields there is little similarity between them. However, from an analysis of a number of photographs of discharges under similar conditions it is possible to deduce the average speed with which the positive streamer crosses the gap. The speed is found to increase rapidly and, in the case of a 20-cm. gap, whereas the streamer discharge had travelled a distance of only about 3 cm. across the gap from the anode in 3 μsec., it travelled the remaining 17 cm. in an additional time of only 0·9 μsec. Final streamer velocities of from $3\,1\times10^7$ to $6\,6\times10^8$ cm./sec. are given by Holzer [1] for gaps of 10 cm. to 40 cm respectively.

The breakdown process between a negative point and a positive plane is initiated by the growth of streamers from the point, as shown in Fig. 4.3. After these streamers have grown part-way across the gap many mid-gap streamers develop throughout the gap Finally a positive streamer develops from the plane to meet a negative streamer from the point and the spark channel is thereby completed. The initial speed of the negative streamer, as determined from the analysis of a number of photographs, is given by Holzer [1] as $1\,4\times10^6$ cm./sec. The speed increases to about $4{\cdot}2\times10^7$ cm./sec when the negative streamer reaches the centre of the gap. In the 16-cm gap studied, the positive streamer is initiated when the negative streamer has travelled about 6 cm. and grows at a speed which decreases from an initial value of about 10^7 cm./sec to about 10^6 cm./sec when the streamers meet.

Other studies of suppressed discharges have been made by Matthias [4] and by Zinerman [5].

Cloud-chamber investigations

The cloud chamber forms a convenient method for the study of certain electrical discharge phenomena, in particular for the weak, and otherwise invisible, processes of ionization such as the electron avalanches which precede the formation of the visible corona streamers and spark channels. The technique has been developed extensively for this purpose by Raether and his colleagues [6 to 24] in a series of experiments which have yielded much useful information concerning the spark mechanism.

The discharge gap is enclosed in the cloud chamber, which is filled with the particular gas being studied together with a certain amount of water, or other, vapour. The presence of ionized gas in the discharge gap is revealed by the preferential condensation of the vapour on the ions which takes place when the gas in the chamber is caused to expand. The instant of occurrence of the gas expansion is controlled electrically in relation to the time of application of a voltage pulse between the electrodes of the discharge gap.

In the arrangement generally used by Raether [15] the discharge gap of length 3·6 cm. is formed by parallel plates, with openings in the cathode to allow for the expansion of the gas in the gap, as shown in Fig. 4.4. A voltage pulse lasting for a time of the order of 0·1 μsec. is applied to the plates and simultaneously photo-electrons are released from the cathode by illumination through a thin window in the anode discharge external to the chamber. By variation of the magnitude and duration of the voltage pulse the development of the electron avalanches and subsequent processes can be arrested at chosen stages in their development

In the experiments the expansion ratio is chosen so that the vapour condenses on the positive ions. An electron avalanche appears therefore

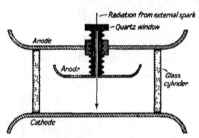

Fig. 4.4   Sketch of cloud-chamber arrangement for studying spark growth

as a cone-shaped cloud, with apex towards the cathode and axis in the direction of the field   Photographs of an electron avalanche recorded in this manner in nitrogen are shown in Fig. 4.5, Pl. 3.

The speeds with which the avalanches cross the gap depend on the voltage gradient and on the pressure and nature of the gas, but are generally of the order of 10^7 cm./sec., in general agreement with the value to be expected for electrons accelerated in a field near to the breakdown value. Some measured values of the speeds u are given in Table 4 1, which includes also the final breadths b and lengths l of the avalanches recorded [23]. It is doubtful whether too great reliance should be placed on the exact values of the breadths given in Table 4.1, partly because of the fact that the photographed outlines of the avalanches depends to some extent on the cloud-chamber expansion ratio, and also because of the diffusion of the avalanche track during the brief time elapsing between the formation of the avalanche and the subsequent expansion. Further, the presence of liquid drops, and the necessity for having a condensable vapour in the chamber, may also influence the diffusion processes and render exact interpretation difficult.

With increasing voltage the amplification in the avalanches is enhanced, and at a critical value a transition from an avalanche into a streamer takes place. The transition is illustrated by the series of photographs given in Fig. 4 6, Pl. 3, and is sketched diagrammatically in Fig. 4.7. This transition occurs in air when the gradient in the gap is such that the multiplication in the avalanche $e^{\alpha x}$ is of the order of e^{20}. The avalanche then becomes unstable in its growth, and whereas its

TABLE 4.1

Gas	Pressure in mm Hg at 0°C	E/p V/cm $/mm$ Hg	$b = 2\bar{r}$ cm	l cm	u cm /sec
CO_2	285	36	0 17	1 87	$1\ 72 \times 10^7$
O_2	290	34	0 19	1 82	1 63
Air	285	37	0 14	2 10	1 47
N_2	280	38	0 15	2 10	1 20
			0 12	1 75	
			0 09	1 40	
	143	39	0 22	2 57	1 24
	94	42	0 30	2 57	1 29
			0 29	2 34	
.			0 18	1 62	..
			0 17	1 40	
H_2	467	22	0 11	1 75	0 68
	305	26	0 15	2 34	0 80
.			0 14	2 10	
	121	31	0 28	2 10	0 92
			0 14	1 52	
A	528	12	0 23	1 87	0 43
.	290	16	0 35	2 33	0 53
			0 23	1 87	

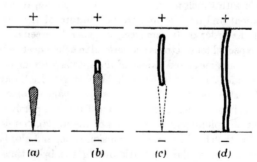

Fig 4 7. Sketch illustrating transition from an avalanche into a mid-gap streamer which develops to form a conducting path across the gap.

initial velocity is about $1\ 2 \times 10^7$ cm /sec., in air at 270 mm. Hg for E/p ~ 40, its velocity suddenly increases to about 8×10^7 cm./sec. and it proceeds towards the anode as a brightly luminous negative streamer, in contrast to the previously invisible avalanche growth. At the same time a positive streamer develops to the cathode, as shown in Fig 4.7 (c), at a speed of from 1 to 2×10^8 cm./sec. For voltages applied to the gap in excess of the minimum required to cause breakdown the transition from an avalanche to a streamer occurs earlier in its growth, the higher the voltage the shorter being the distance travelled by the avalanche before the transition occurs.

The crossing of the gap by the streamers means that a highly ionized channel has been formed joining the two electrodes. The creation of this channel constitutes the spark breakdown of the gap, and the discharge of the external circuit then takes place through this channel. The characteristics of the spark channel are discussed again in Chapter X.

Raether's cloud-chamber measurements show that the avalanche–streamer transition is characteristic of the breakdown of longer gaps at higher gas pressures. Raether [13, 15] concludes that this mechanism is applicable to discharges in air when pd is about 1,000 mm. Hg $\times$ cm. or more. For gaps in which pd is less than 1,000, approximately, it is considered that the classical theory of the spark, dependent on a cathode γ-mechanism still applies.

From considerations of the measured diameters of electron avalanches and of the value of the Townsend ionization coefficient α corresponding to the applied voltage in the gap, Raether [13] has estimated the approximate ion density at the head of the avalanche. This enables him to determine approximately the space-charge field surrounding the avalanche and he concludes that the avalanche–streamer transition occurs when the space-charge field is roughly equal to the external applied field. A similar criterion for the onset of a spark was put forward independently by Meek [25] . A detailed discussion of the theory of avalanche–streamer transitions is given in Chapter VI.

Fig. 4.8, Pl 4, shows some photographs obtained by Raether [15] for point-plane discharges in air. The results demonstrate the characteristic effects of positive and negative discharges as recorded in corona studies by other techniques (see Chapter II). The positive discharge consists of filamentary streamers, the negative discharge being formed of streamers of a more diffuse nature. The velocities of these streamers in air for point-plane and sphere-plane gaps have been measured by Raether [15] and are given in Table 4.2, where they are compared with those for a uniform field. The ratio E_{max}/E_{min}, of the maximum field to the minimum field in the gap, is that given by Raether [15] for the particular electrode arrangements used.

Other studies have been made by Nakaya and Yamasaki [28], who observed the growth of sparks in a point-plane gap of length 2·5 cm.

TABLE 4.2

Electrode arrangement	Point-plane	Sphere-plane	Uniform field
E_{max}/E_{min}	100	13 5	1
Positive streamer velocity in cm./sec	$2\ 5 \times 10^7$	10^8	$1\ 5 \times 10^8$
Negative streamer velocity in cm /sec	2×10^6	$2\ 5 \times 10^7$	$7\ 9 \times 10^7$

placed in a cloud chamber. The duration of the voltage impulse applied to the gap was controlled by variation of the length of a sphere gap connected in parallel with the point-plane gap. The records for discharges in air are generally similar to those obtained by Raether. In nitrogen the branching of the positive discharge is more pronounced than in air, the streamers being more sharply defined and about 25 per cent. longer. The negative streamers in nitrogen are also less diffuse than in air. The number of streamers observed in the discharge in oxygen is much reduced in the positive discharge, but the difference from air or nitrogen is even more marked in the negative discharge where the streamers consist of short needle-like tracks surrounding the point and extending only a short distance into the gap. The positive discharge in hydrogen takes the form of a number of thick smooth bands with diffuse boundaries, the number of these bands increasing with increasing voltage. A single broad band is recorded in hydrogen for the negative discharge Discharges in carbon dioxide are of interest in that there is a greater resemblance between the appearance of the records obtained for positive and negative points than in the other gases, and the negative discharge is larger in size than the positive discharge

The effect of admixtures of various vapours containing iodine or chlorine on the appearance of the discharges in air was also studied by Nakaya and Yamasaki [28]. With the addition of 1 per cent. of chloroform the discharge becomes reduced in length and smoother in form, and the branching is less complex than in air. The influence of the chloroform on the negative discharge is even more pronounced.

Cloud-chamber investigations of the growth of sparks in air have also been carried out by Kroemer [26], Snoddy and Bradley [27], and Gorrill [29], with results in general agreement with those already described.

The Kerr cell electro-optical shutter

The rapid changes which take place in the formation of a spark discharge are examined in the cloud chamber by the sudden removal of the voltage applied to the gap, so that the spark is arrested at a chosen stage in its development. The need for this suppression of the growth of the spark is obviated by the use of an electro-optical shutter which enables the investigator to observe the visual appearance of the formation of a simple spark at selected time intervals, which may be as short as 10^{-8} sec. or less [30, 43, 44]

In its usual form the electro-optical shutter consists of two Nicol prisms or sheets of polaroid between which is placed a Kerr cell. The latter contains a liquid such as nitrobenzene, which can be subjected to electrical stress by the application of a voltage to two metal plates immersed in the liquid. Light passing through one of the Nicol prisms becomes plane-polarized, and in the absence of the Kerr cell the transmission of this light through the second Nicol prism is governed by the relative position of the planes of polarization of the two prisms. When the Kerr cell is in place the polarization of the light is affected by the liquid and can be controlled by the electrical stress applied. The various components can be arranged so that no light is transmitted through the apparatus when the cell is unstressed. Then on the application of a voltage pulse to the cell, light is transmitted for the duration of the pulse. In a detailed account given by Dunnington [30] of the operation of the shutter it is shown that closing times of the order of 4×10^{-9} sec. may be attained.

The electro-optical shutter has been used successfully in the study of sparks, across gaps up to 1 cm. in length, by several investigators [31 to 37, 60] The development of the visible streamer processes is observed, but clearly the invisible electron avalanches cannot be recorded by this technique

A diagram illustrating the observed growth of the spark in air across a 0·5-cm. gap, as recorded by von Hámos [34], is given in Fig. 4.9. A positive streamer is found to grow across the gap from the anode to meet a small discharge at the cathode. The speed of growth of the positive streamer is given by von Hámos as about 5×10^7 cm./sec. Similar results were obtained for sparks in air across gaps ranging from 0·4 to 0·7 cm.

The growth of sparks across gaps of 0·1 to 1·0 cm. in air at pressures between 760 and 200 mm Hg has been investigated by Dunnington [31], who finds that changes occur in the breakdown mechanism with variation in gap length and gas pressure With sufficiently short gaps and low gas pressures the luminous breakdown region, or streamer process, starts at the cathode and proceeds across the gap to the anode As the gap is lengthened, or the pressure increased, there are two initial breakdown regions, one at the cathode and one in mid-gap Further increase in gap or pressure tends to subdue the cathode region. In all cases, during the later stages of development, there appears a third breakdown region at the anode. The mid-gap streamer connects with the cathode first and with the anode last. The appearance of the mid-gap streamer is recorded by Dunnington for gaps in which the relation between the gas pressure p in cm. Hg and the gap length in millimetres is such that $p^{\frac{1}{2}}d > 11\cdot9$.

The Kerr cell studies of the spark mechanism have been extended by White [35] to various gases including nitrogen, hydrogen, oxygen, carbon

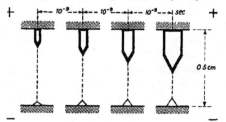

Fig 4 9 Kerr cell photographs, at 10^{-9} sec intervals, of the growth of a spark across a 0 5-cm gap in air.

dioxide, helium, and argon, for gaps up to 1 cm. In hydrogen the first visible signs of breakdown begin at the cathode, from which a streamer grows across the gap with approximately uniform velocity. When it has reached about half-way across the gap a streamer appears at the anode and the two streamers then grow together to meet at a point about three-quarters the distance across the gap from the cathode. In nitrogen the breakdown may be one of two forms depending on the gap length and gas pressure. For atmospheric pressure and gaps $< 3\cdot5$ mm the breakdown is similar to that for hydrogen. For gaps $> 3\cdot5$ mm. the breakdown starts in the mid-gap region as well as at the cathode, the two streamers appearing simultaneously as near as can be determined The streamers grow together and the breakdown is completed in a manner similar to that for short gaps. The cathode streamer velocity in an 8-mm. gap is $3\ 2\times10^{7}$ cm./sec in hydrogen and is $1\cdot4\times10^{7}$ cm./sec. in nitrogen. The time required for a complete luminous filament to develop across the gap is $1\cdot8\times10^{-8}$ sec. in hydrogen and $2\ 7\times10^{-8}$ sec. in nitrogen.

The earliest observable feature of the breakdown in helium is a diffuse glow of about 1 mm. diameter extending the length of the gap. An isolated streamer then appears and grows towards both cathode and anode. Later a spot forms at the surface of the cathode and a streamer appears at the anode. The several breakdown regions then grow together. A similar form of mechanism is recorded in argon except that no pre-breakdown glow is observed and no anode streamer occurs. The time required for the completion of a bright continuous filament joining the two electrodes is several times longer in helium and argon than in hydrogen or nitrogen. The breakdown in carbon dioxide is characterized by greater branching and tortuousness of the streamer than in the other gases.

In all the above measurements with the Kerr cell shutter the method of triggering the spark may have affected the spark mechanism. This is a complication which has been generally realized by the investigators One method of triggering the spark is to apply an overvoltage to the gap, which is already illuminated by a steady source of ultraviolet light. In this case the space charges built up in the gap by the steady photo-current preceding breakdown may influence the spark mechanism. In another method used the gap is illuminated by an intense flash of ultra-violet light from a nearby spark, and the resultant photo-current pro-duced may again be sufficient to affect the growth of the spark (see p. 368).

Besides its use in the study of the spark mechanism the Kerr cell shutter has been employed by White [35, 37] and Wilson [36] to record the variation of the time lag of spark breakdown as a function of gap length, amount of overvoltage, and the intensity of illumination These measurements are discussed again on p. 207 A Kerr cell using a square voltage pulse to define the exposure time has also been developed by Holtham and Prime [38] for the study of spark channels and their rate of expansion (see Chapter X). Other Kerr cell shutters using square voltage pulses to give brief exposure times are described elsewhere [39 to 42]

Rotating camera studies of the spark mechanism

Photographic studies of sparks and of lightning discharges have been carried out by a number of investigators using cameras in which a move-ment of the image in a direction transverse to its growth is accomplished by means of a rotating lens, rotating mirror, or rotating film [51]. The rotating-lens type of camera, first suggested by Boys [45], has been applied successfully by Schonland [46, 47] to record the growth of lightning in South Africa (see Chapter V), but the mechanism of spark development has generally been studied by the rotating-film camera. In all these forms of cameras, because of the relative movement between the lens and the film a distinction can be made between any portion of the dis-charge which becomes luminous before another portion. Knowledge of the relative speed of the lens and the film then enables the speed of development of the discharge to be calculated.

Following the results obtained in the studies of the growth of lightning discharges an attempt was made by Allibone and Schonland [48] using the rotating-camera technique to photograph the development of sparks in the laboratory. The resultant records showed that the spark is preceded by a streamer which develops from the high-voltage point electrode, the streamer being referred to as a 'leader stroke', a term originally proposed by Schonland to denote the analogous discharge process observed in the lightning mechanism. The discharge occurring after the leader stroke or leader strokes have bridged the gap is referred to as the main stroke.

Leader strokes were photographed only over a small fraction of the interelectrode spacing in Allibone and Schonland's measurements and the results were greatly extended in subsequent investigations by Allibone and Meek [49, 50]. The latter used a camera of the rotating-film type, with a time resolution of 24 μsec. per mm. of film and a quartz ultraviolet flint-glass lens which was later replaced by an $f/4.5$ quartz-rocksalt lens. By adjustment of the circuit conditions Allibone and Meek found that the speed of growth of the spark could be controlled, an increase in the series resistance in the circuit causing a decrease in the speed of the leader stroke. In this way, by variation of the series resistance R from 9,000 Ω to 1 9 MΩ, it was possible to record leader strokes developing completely across the gap. The high voltage necessary to cause breakdown of the long gaps, up to 200 cm., investigated in these experiments was produced by an impulse generator of 0·01 μF capacitance capable of developing 2,000 kV. The results refer almost entirely to sparks in air in non-uniform fields, as between a point and a plane, for which the speed of growth of the spark is appreciably slower than in uniform fields.

Typical photographs of discharges in air at atmospheric pressure between a high-voltage positive point and an earthed plane are shown in Fig. 4 10, Pl. 4, for different values of the series resistance R in the circuit. The leader stroke starts from the anode point and traverses the whole gap, branching in a downward direction. The main stroke follows in detail the tortuous path traced out by the leader stroke. The leader stroke is recorded as an intense filamentary channel accompanied by a less bright, but more diffuse, voluminous shower of discharge which is particularly noticeable in Fig. 4.10 (b). The importance of this shower of discharge has been stressed in recent papers by Komelkov [52 to 56], Saxe and Meek [57, 58] and Allibone [101]. The duration of high luminosity at any point of the leader stroke is stated to be no more than 0·5 μsec. After the passage of the leader stroke the luminosity falls to a low value, except in the cases of gaps subjected to voltages greatly in excess of the minimum breakdown value.

The leader stroke is sometimes of a composite character, similar to

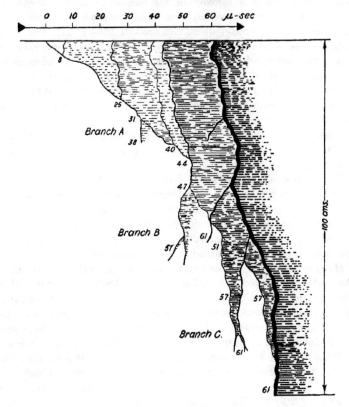

FIG. 4.11. Analysis of a rotating camera photograph of a positive point-plane discharge across a 100-cm. gap in air   Series resistance = 100,000 Ω.

that observed in lightning discharges (see Chapter V). An analysis of a typical record [49] of a discharge across a 100-cm. gap, for $R = 0.1$ MΩ and a capacitance $C = 100$ $\mu\mu$F connected directly across the gap, is given in Fig. 4.11. The first leader exhibits branching and travels relatively slowly. The subsequent leader strokes travel rapidly without branching over the previously ionized track and then more slowly with branching as they extend the track. Where the stepped leaders extend the track their velocity is comparable with that of the initial leader but there is a gradual increase as the discharge process grows across the gap. The total time between the initiation of the first leader stroke and the occurrence of the main stroke, and the number of stepped streamers recorded, both vary appreciably even under the same conditions For example, in two successive discharges across a 135-cm. gap, one has a time of 94 μsec between the first leader and the main stroke, and exhibits no stepped leaders, while the second has a time of 180 μsec. and exhibits four stepped leaders. In a few discharges, as in Fig. 4 10 (c), the positive leader does not bridge the gap completely, but is met by a short negative leader propagated from the earthed plane, in this particular discharge four well-defined stepped leaders occur from the positive point at intervals of 14, 22, and 41 μsec., the last developing with a final velocity of $2 \cdot 4 \times 10^6$ cm /sec. to meet the ascending negative leader 7 cm. above the plane.

The main stroke of the spark discharge, like that of the lightning discharge, does not branch as frequently as the leader stroke but only follows the more prominent branches, as shown in Fig. 4 10. The intensity of the main stroke along branches is less than those of the leader stroke along the same branches. The speed with which the main stroke grows is too high to be recorded by the rotating camera used.

If a point electrode is placed on the negative plane the mechanism of the discharge is altered in that after the positive leader stroke developing from the anode point has travelled part-way across the gap an upward-directed negative leader stroke starts from the earthed point and the two leaders grow towards each other. In one record, for an 89-cm. gap where the earthed point projects 68 cm. above the plane, with $R = 0 \cdot 4$ MΩ, the negative leader stroke is initiated at a time 26 μsec after the start of the positive leader stroke, when the latter has grown about one-third of the gap length The main stroke follows after a further time interval of 20 μsec The velocity of the negative leader is less than one-third of that of the positive leader over the distances where the two leaders are developing simultaneously. Between the two leader strokes a voluminous shower of discharge is observed. Branching of the discharge is enhanced by the presence of a number of earthed points.

In the case of breakdown of a 75-cm. gap between a high-voltage positive sphere, of 12·5 cm. diameter, and a plane, an upward-growing negative leader occurs to meet the descending positive leader about 10 per cent. of the gap above the plane. With a high-voltage positive sphere, of 25 cm. diameter, and an earthed point the growth of the negative leader is encouraged and the junction-point of the leader strokes is about 25 cm. above the cathode in a 75-cm. gap.

The time interval between the initiation of the leader stroke and the occurrence of the main stroke, for a 76-cm. gap between a positive point and an earthed plane, is plotted as a function of R in Fig. 4.12 [50].

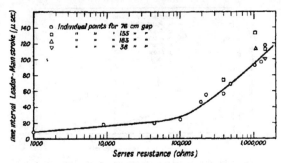

Fig. 4.12   Curve relating the time interval between leader and main strokes with series resistance for a gap length of 76 cm. between a positive point and an earthed plane in air.

Isolated points for other gaps are also shown. The values at $R = 1,000\,\Omega$ are derived from oscillographic studies of the time lags between the application of the impulse voltage and the breakdown of the gap.

The velocities of the positive leader strokes can be measured with considerable accuracy as the leaders are nearly always continuous, and reasonable consistency is obtained for the same gap and circuit conditions, particularly for the lower series resistances. In general the velocity increases with increasing distance from the anode. The initial velocity of the leader stroke and the final velocity of the leader stroke (measured at a point three-quarters of the gap length measured from the anode) are plotted in Fig. 4.13 as a function of R for a gap length of 76 cm. [50]. Some isolated points referring to other gap lengths are also shown. The initial velocity is difficult to measure at the higher resistances because of the increased number of short-stepped leaders, which become weak in intensity and ill defined. The final velocity is more or less independent of gap length at high values of R, for instance at 1 45 MΩ it varies only from $2 \cdot 6 \times 10^6$ cm./sec. for a 183-cm. gap to $2 \cdot 1 \times 10^6$ cm./sec. for a 38-cm. gap. For $R < 10,000\,\Omega$ it was not possible to measure the final velocity with the camera used because of the small time interval between the leader stroke and the main stroke and the confusion caused by the increased luminosity of the main stroke. The growth of the positive leader stroke is not affected appreciably when the plane electrode is replaced by a point. In a gap of 68 cm , where the earthed point projected 22 cm. above the plane, the final velocity of the positive leader is $2\cdot2\times10^6$ cm./sec. when the series resistance is $1\cdot3$ MΩ. For the same discharge the velocity of the upward-growing negative leader is only 10^5 cm./sec.

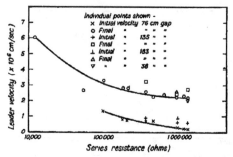

FIG 4 13 Curves relating the initial and final velocities of the leader with series resistance for a gap length of 76 cm between a positive point and an earthed plane in air

The effect of applying voltages in excess of the minimum required to cause breakdown is to decrease the time to breakdown, to decrease the number of steps in the leader stroke, to increase the luminosity of the leader stroke and the duration of this luminosity, to increase the amount of branching, and to increase the velocity of the leader stroke. The principal change in velocity with increased voltage occurs during the earlier stages of growth of the leader stroke. In a 76-cm. gap between a positive point and a plane, for $R = 0\cdot5$ MΩ, the initial velocity increases from $0\cdot4\times10^6$ to $1\,4\times10^6$ cm./sec. when the peak voltage applied to the gap is altered to twice the minimum breakdown value; the corresponding change in the final velocity is from $2\cdot0\times10^6$ to $2\cdot5\times10^6$ cm./sec.

The effect of a reduction in gas pressure on the growth of the spark has been studied for discharges across a gap of about 50 cm. in air between a high-voltage positive point and an earthed plane on which is placed a small point of height 1 cm. As the gas pressure is decreased the leader stroke grows more slowly and stepping becomes more pronounced. Leader strokes can also be detected in discharges for much lower values of R than those for which records have been obtained in air at atmospheric pressure. A drawing based on an original photograph of a discharge in air at 100 mm. Hg, for $R = 0\,1$ MΩ, is given in Fig. 4.14 [49].

The principal characteristics of discharges between a negative point and a positive plane is that, in addition to the negative leader starting from a cathode, a positive leader stroke rises from the plane to meet the

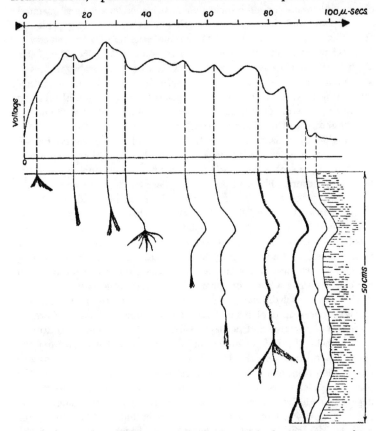

FIG. 4.14. Drawing based on a rotating-camera photograph of a positive point-plane discharge across a 50-cm. gap in air at 100 mm Hg. A point of length 1 cm projects from the plane. The voltage curve gives the variation of voltage across the gap Series resistance = 100,000 Ω.

descending negative leader A typical photograph showing the growth of a discharge across a 100-cm gap between a negative high-voltage point and an earthed plane, for $R = 0.6$ MΩ, is given in Fig. 4 15, Pl. 5 [49]. The discharge starts from the cathode point as a series of sharply defined stepped leader strokes, each of which extends the path traced out by the preceding step. When the process has developed a short distance across the gap, usually about one-fifth of the gap length, the character of the growth changes and the leader stroke proceeds in a continuous manner. Simultaneously a positive leader stroke is initiated at the plane and the two leader strokes meet in the mid-gap region In a 72-cm. gap, for $R = 0.4$ MΩ, the final negative leader velocity is 1.5×10^6 cm./sec. and the positive leader velocity is 2.6×10^6 cm./sec. Both leader strokes branch in the direction of their propagation and a voluminous shower of discharge occurs between them. A further interesting feature of negative discharges is the appearance of mid-gap streamers, several centimetres in length and branched at each end, as clearly shown to the left of the spark channel in Fig. 4 15.

Straight brush discharges from the positive plane are observed when the value of R is increased sufficiently. The brush discharge is intense for about 15 cm. of the gap, and from this intense portion considerable branching occurs. The first branches appear suddenly at about 4 cm. above the plane and are curved upwards in well-defined 'parabolic' paths. The main stroke follows the path of the brush discharge for about 6 cm. from the plane Up to this point the speed of the brush discharge was too fast to be recorded, but subsequently a normal positive leader stroke develops at a speed of $3\ 2 \times 10^6$ cm /sec The rapidly developing brush discharge from a positive electrode has been recorded only when the electrode has a plane or spherical surface of large radius.

The above brush discharge is more prominent in shorter gaps, of 25 cm. For resistances up to about 0.5 MΩ the development of the discharge is normal, i.e. negative and positive leaders meet in mid-gap. However, with higher values of resistance [49] a small glow discharge from the cathode point is accompanied by a long brush discharge from the positive plane while from out of the latter discharge, at about 8 cm above the plane, a leader stroke of the normal positive type develops to meet a negative leader stroke near the cathode after which the main stroke occurs. This form of discharge is shown in Fig. 4.16(a), Pl. 5, for $R = 2\ 0$ MΩ, the time interval between the brush discharge and the main stroke being about 2 μsec. In some of the discharges the normal positive leader stroke disappears and the main stroke takes place along the axis of the original brush discharge, as shown in Fig. 4.16 (b), Pl. 5. A similar form of discharge has been observed to occur from the spherical terminal of a Van de Graaff electrostatic generator [61].

The height of the junction-point of the two leader strokes above the plane, when expressed as a percentage of the gap length, decreases with increase in gap length for otherwise identical conditions. The height varies from 56 per cent. of the gap length for a 25-cm. gap to 32 per cent. for a 150-cm. gap, with $R = 1$ MΩ. The height decreases also with reduction in the series resistance and for a 50-cm. gap changes from 53 per cent. to 44 per cent. when R is altered from 0·25 to 0·013 MΩ. If a point is placed on the earthed positive plane, the development of the upward-growing positive leader takes place more readily and the meeting-point of the negative and positive leaders occurs nearer to the high-voltage negative point than in the point-plane gap. In some of the records of such discharges, as in Fig. 4.17, Pl. 5, which applies to a 76-cm. gap for $R = 1·9$ MΩ, a straight brush discharge develops from the negative point for about 10 cm. and is terminated by an intense elongated discharge 3 to 8 cm. in length from which a shower of well-defined negative streamers proceed toward the opposite plane electrode on which is mounted a point 1·3 cm. long. A strange feature of this discharge is that its track is not followed by the negative leader stroke which develops from the cathode at the same time as the brush discharge is initiated.

Experiments were carried out with gaps between a high-voltage negative sphere and an earthed point [50]. With a sphere of 100 cm. diameter and a gap of 100 cm. the negative leader stroke from the sphere is practically non-existent, the positive leader from the point bridging at least 95 per cent of the gap. The positive leader stroke is stepped and oscillographic measurements show corresponding peaks in the current flowing in the discharge. The velocity of the positive leader is nearly uniform, at $3·1 \times 10^6$ cm./sec. A shower of discharge is evident in the region of the sphere.

The time intervals between the initiation of the negative leader stroke and the occurrence of the main stroke, in a negative point-plane gap, for different series resistances are plotted as a function of gap length in Fig. 4.18. Considerable variations, as great as 50 per cent., may occur between records of discharges under identical conditions. Fig. 4.19 shows the variation with gap length of the time intervals between the initiation of the positive leader and the main stroke. As previously stated, the upward-growing positive leader is generally initiated at the same instant as the final step of the negative leader.

The velocity of the negative leader stroke varies over a considerable range for the same series of records. For a 57-cm. gap between a negative point and an earthed point projecting 1·25 cm. out of an earthed plane, with $R = 1·9$ MΩ, the velocities of the negative leader ranged from

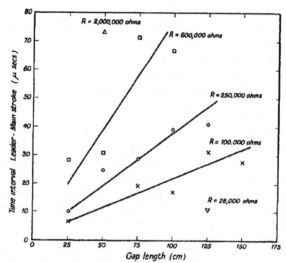

Fig. 4 18   Curves relating the time interval between the initial
negative leader and main stroke with gap length for negative
point plane discharges with different series resistances.

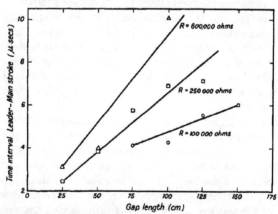

Fig. 4.19   Curves relating the time interval between the initiation
of the positive upward-developing leader and the main stroke with
gap length for negative point-plane discharges with different series
resistances.

0.8×10^6 to 0.3×10^6 cm./sec. Velocities varying from 3.2×10^6 to 1.1×10^6 cm./sec. were obtained with $R = 0.4$ MΩ for a 92-cm. gap to an earthed point projecting 11 cm above the plane.

With a reduction in gas pressure the leader strokes grow more slowly and can be recorded for lower values of R than at atmospheric pressure [49]. Stepping is more pronounced and the height of the meeting point

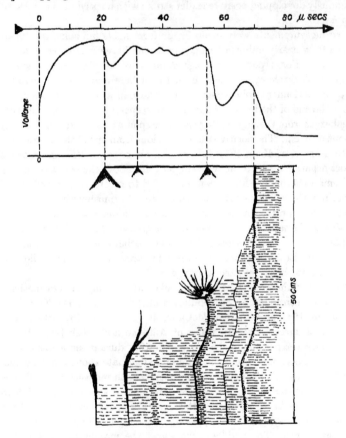

FIG. 4 20   Drawing based on a rotating-camera photograph of a negative point-plane discharge across a 50-cm gap in air at 100 mm Hg   A point of length 1 0 cm. projects from the plane   The voltage curve gives the variation of voltage across the gap.   Series resistance = 100,000 Ω

of the two leader strokes above the plane is increased. A drawing based on a photograph of a discharge at a pressure of 100 mm. Hg is shown in Fig. 4.20. The ascending positive leader stroke develops in a discontinuous manner, each step appreciably extending the ionized track, and being accompanied by a short brush discharge from the cathode point. At still lower pressures of 10 mm. Hg the discharge is diffuse but appears to be initiated by a glow which fills the gap and is followed by a continuously developing positive leader stroke with a speed of about 2×10^5 cm /sec.

No measurements were made for uniform fields but some results are given for a nearly uniform field gap of 25 cm. between spheres of 25 cm. diameter. For a positive voltage applied to the high-voltage sphere long brush discharges develop from this sphere and nearly bridge the gap but the leader stroke which initiates the main stroke seldom develops from the end of the brush. At an increased spacing, of 75 cm., the brush discharges from the high-voltage positive sphere branch after proceeding about 5–10 cm. The positive leader develops from one of these branches, with a speed of about $2 \cdot 5 \times 10^6$ cm /sec. for $R = 0 \cdot 3$ MΩ, and meets a short negative leader at about 5–10 per cent. above the negative sphere.

Some experiments have been made [62] to record the breakdown of certain of the constituent gases forming air, at atmospheric and reduced pressures. In nitrogen the leader strokes are more clearly defined than in air and the separation between steps is greater. The addition of water vapour to the nitrogen causes the leader mechanism to approximate to that in air. Leader strokes are observed in argon and in carbon dioxide, but have not been detected in oxygen

Studies of the growth of leader strokes by rotating-camera methods have also been made by Stekolnikov and Behakov [63, 64], Komelkov [52 to 56, 104], and Wagner, McCann, and MacLane [65], with results in general agreement with those of Allibone and Meek [49, 50]. In Komelkov's experiments the current flowing during the leader-stroke growth has been measured, and is found to fluctuate about a mean value, the frequency of the fluctuation decreasing with increase in R Values obtained for the leader currents [56] are plotted as a function of R in Fig 4.21. The lower curve gives the mean leader current, which is of the same value irrespective of the polarity of the high-voltage electrode and the gap length. The middle curve shows the maximum value of the leader current recorded in the final stage of its growth, and the upper curve the spark-channel current when the main stroke occurs. The current density in the leader channel is estimated to be within the range of 400 to 1,200 A/cm ^{2} for a positive leader with $R = 1,000$ Ω.

In the later measurements by Komelkov [54, 56] the leader-stroke growth has been recorded by a rotating film camera, with a resolving time of $7 \cdot 1$ μsec. per mm. of film, which is arranged in such a way that the

direction of growth of the image of the leader stroke is in the same
direction as the film movement. A slot 2 mm. wide in a cardboard sheet
is placed between the camera and the spark gap so that various parts
of the gap can be viewed. The resultant photographs show that the

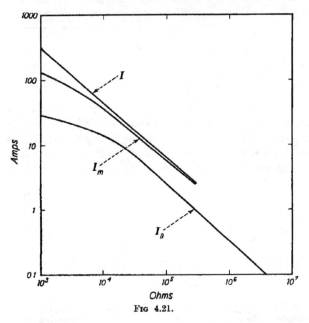

FIG 4.21.

leader stroke consists of two portions· (1) the actual leader channel,
of relatively small diameter and intense illumination, and (2) a zone of
ionization of much larger diameter and lower intensity. The latter
corresponds to the voluminous shower of discharge visible in the photographs of Allibone and Meek [49, 50] Komelkov states that the ionization zone consists of streamers emerging from the tip of the leader
channel [54, 56]. In the case of positive discharges the streamers are
filamentary and tortuous, but in negative discharges they are less
clearly defined. The radius of the ionization zone for various circuit
conditions is given in Table 4.3. The current density is estimated to be
about 50 to 200 A/cm.^{2} in the streamers and about 1,000 to 2,000 A/cm.^{2}
in the leader channel The voltage gradient in the streamers is given as
6 to 10 kV/cm. and that in the leader channel as 60 V/cm. The influence
of an insulating barrier placed between a positive point and a plane

on the mechanism of growth of the leader stroke has also been examined by Komelkov and Lifshits (105).

TABLE 4 3

Series resistance $R \Omega$	Discharge polarity	Gap length in cm	Radius of ionization zone in cm.
1,000	+-	80	13 6
3,000	+-	80	15 4
10,700	-+	80	18 5
140,000	+	80	19 2
1,000	—	25	20·4

Rotating-camera measurements have also been made by Strigel [66] for gaps of much greater length than those studied by the other investigators   The leader-stroke growth is not clearly discernible, possibly because of low values of series resistance, but it can be seen in some of the photographs of discharges between a high voltage point and an

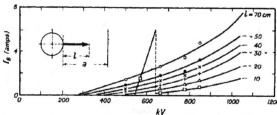

Fig. 4 22. Curves giving the variation of the final current in the leader stroke with the applied voltage for discharges across a point-plane gap. The point projects a distance of l cm from a sphere of 2 5 m diameter   The distance between the sphere and the plane is
$$a = 3\ 7\ \text{m.}$$

earthed plane. In one photograph, for a 500-cm. gap, the leader stroke is initiated at a time 0·84 μsec. before the main stroke occurs.

The magnitudes of the currents flowing in leader strokes have been measured by Schneider [103], who has also photographed the development of the discharge. The discharge gap in some of the experiments consists of a point, projecting from a sphere of 2·5 m. diameter, and a plane. Fig. 4.22 shows the variation with applied voltage of the final current in the leader stroke for several lengths of the point. The wave form of the impulse wave used is 0·5/40 μsec. Higher currents are recorded if the point is replaced by a sphere of 10 cm. diameter mounted on a rod projecting from the sphere, as shown in Fig. 4.23,

The growth of the leader stroke preceding the formation of a spark over the surface of a copper sulphate solution has been recorded by Snoddy and Beams [102] using a rotating-mirror camera. In one particular example the leader stroke grows at a speed of 5×10^5 cm./sec. but it is pointed out that this speed depends on the external circuit conditions.

Photo-electric recording

In the studies of spark growth by means of the rotating-camera technique there is difficulty in some cases, particularly at low values of series resistance, in distinguishing the leader stroke from the main stroke on account of the greater intensity of the latter. The spatial distribution of the discharge can also confuse the determination of the time relationship and introduce inaccuracies even in cases where the camera speed is high enough to separate clearly the leader stroke from the main stroke Some improvement in these respects can be made by the use of a technique suggested by Meek and Craggs [67], in which the spark is viewed by a photomultiplier connected through an amplifier to an oscillograph, and which enables the time interval between the leader stroke and the main stroke to be recorded at different points in the gap

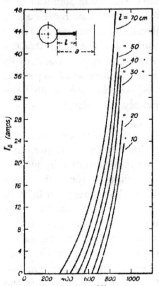

FIG. 4 23 Curves giving the variation of the final current in the leader stroke with applied voltage for discharges between a sphere of 10 cm and a plane The sphere is mounted on a rod projecting a distance of l cm from a sphere of 2 5 m diameter The distance between the latter sphere and the plane is $a = 3\,7$ m

with a higher degree of resolution than is possible with the rotating camera.

Experiments using the photomultiplier technique with a time resolution of less than 10^{-7} sec have been carried out by Saxe and Meek [57, 58] for discharges in air and in other gases between a positive point and a negative plane. Some measurements were made in which the light emission from the gap as a whole falls on the cathode of a photomultiplier, the output from which is amplified and is reproduced on an oscillograph screen By this method corona discharges were studied for conditions in which the impulse voltage applied to the gap is less than that required to cause breakdown. The light produced by the corona discharge consists of a brief pulse, which is of the same shape as the accompanying pulse of current in the gap. Both pulses rise to their peak amplitude in a time of 0·05 μsec. and decline to a small value in 0·4 μsec. By the use of slits placed between the gap and the photomultiplier the light distribution in the corona discharge has been examined and it is found that the light is emitted from a roughly hemispherical volume, with the positive point at the centre of the plane face of the hemisphere which is parallel to the negative plane and with the curved face directed towards the plane. The light intensity decays continuously with distance from the point

When the impulse voltage applied to the gap is raised to the breakdown value, or to higher values, a leader stroke develops from the point and initiates a spark. The growth of the leader stroke is studied by placing between the gap and the photomultiplier an opaque screen containing a slit, the position of which can be altered, so that changes in light in different parts of the gap can be recorded. As the leader stroke passes the slit it produces an increase in the light reaching the photomultiplier through the slit and consequently a deflexion is produced on the oscillograph screen. A deflexion is also caused by the passage of the main stroke, and from measurements of the amplitudes of the various deflexions and the time intervals, both the distribution of light in the leader stroke and its speed of growth can be determined A typical oscillogram is reproduced in Fig. 4 24. The discharge commences as an impulse corona, the light emission from which causes a detectable deflexion on the oscillogram shown in Fig. 4 24 Subsequently the light intensity passing through the slit increases gradually to a peak which corresponds to the passage of the 'core' of the leader stroke across the slit. The light then decreases rapidly to a small value and increases again suddenly when the main stroke occurs. From this evidence it is clear that the head of the leader stroke consists of a large zone of discharge which increases in intensity to a central core. The distribution of light in the head of the leader stroke is generally similar to that in the impulse corona. Emission of light can be detected up to a distance of the order of 10 cm. ahead of the central core. The results are therefore in general agreement with the allied measurements by Komelkov [54, 56], using rotating-camera methods, and with the work of Loeb et al. [68] and Gaunt and Craggs [69] on D C. corona (see Chapter III).

Curves relating the distance that the channel of the leader stroke has travelled with the time that has elapsed since the application of the impulse voltage to the gap are shown in Fig. 4.25 for several values of the

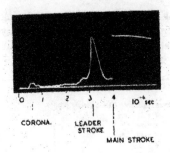

Fig 4 24

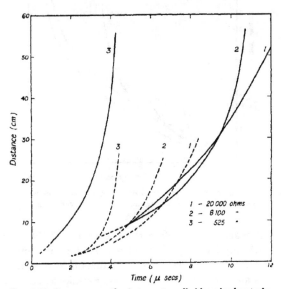

Fig 4 25 Curves giving the distance travelled by a leader stroke as a function of time for different series resistances connected to a positive point plane gap The terminating point of each curve gives the gap length data for two gap lengths are given for each resistance.

series resistance R in the circuit. The distance at which each curve stops corresponds to the gap length In general the time taken for the leader stroke to grow across the gap decreases with the value of R and

with the gap length. The velocity of the leader stroke, which is given by
the slope of the curves, gradually increases as the leader stroke approaches the plane.

The currents flowing through the earthed point during the leader-stroke
growth have also been measured oscillographically by Saxe and Meek
[57, 58] and curves showing these currents as a function of time for
several gap lengths, with $R = 20,000\ \Omega$, are given in Fig. 4.26. The

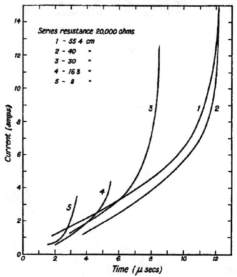

FIG 4 26. Leader-stroke currents, in a positive point-
plane gap in air, as a function of time for various gap
lengths.

final current reached just before the occurrence of the main stroke
increases with gap length and is about 3 5 A for the 8-cm. gap and
about 15·5 A for the 55·4-cm. gap. A reduction in R from 20,000 to
525 Ω causes an increase in the final leader-stroke current from 16 to
90 A, the time interval between the application of the voltage to the
gap and the main stroke being reduced from about 12·5 to 5 μsec.

From the above curves it is possible to derive further curves between
current flowing in the earthed point with the distance that the leader
stroke has travelled. Such curves for $R = 20,000\,\Omega$ are given in Fig. 4.27,
and show a certain amount of uniformity. Again, if the charge which
has flowed through the point is plotted as a function of the leader-stroke

length, the curves for the various gaps at a given R are closely the same and have a roughly constant slope of 0 88 microcoulomb/cm

In nitrogen [70] the initial corona pulse has a duration appreciably longer than in air, though its amplitude has fallen to a small value in 20 μsec. The leader-stroke growth, while generally similar to that in air,

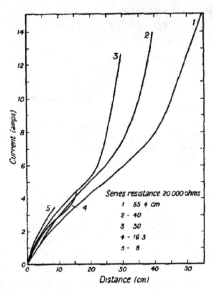

Fig 4 27. Leader-stroke currents, in a positive point-plane gap in air, as a function of the distance travelled by the leader stroke, for various gap lengths

shows much greater variations in the relation between distance travelled as a function of time for separate sparks under the same defined conditions. The initial corona pulse in hydrogen is similar to that in nitrogen, but no leader stroke in the accepted sense could be detected by the photomultiplier technique, the light from the gap decaying continually after the corona pulse until the intense light produced by the main stroke occurs In oxygen the corona pulse lasts only for about 0·03 μsec., and again no normal leader-stroke development could be recorded.

The possibilities of other photo-electric methods, involving the use of a pulsed iconoscope or image converter tube, have been examined at Liverpool by several investigators [71, 72, 73]. Results have been obtained for the growth of spark channels after breakdown of the gap, but so far the development of the leader stroke has not been photographed.

Oscillographic studies

The voltage across a discharge gap during the breakdown process has been studied oscillographically in a series of investigations by Rogowski and his colleagues. In the measurements by Rogowski,

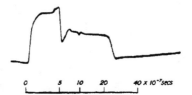

Fig 4 28   Oscillogram showing the
voltage collapse across a 0 38 cm gap in
air at 185 mm. Hg.

Flegler, and Tamm [74, 75, 76] oscillograms are shown of the impulse breakdown of short gaps between spheres and between plates in air, hydrogen, and carbon dioxide, at pressures varying from atmospheric down to a few mm. Hg. For discharges at reduced pressures the collapse of voltage across the gap is found to occur in a stepped manner, the duration of the step increasing with decreasing pressure   A typical record given by Tamm [76] for a 0·38-cm. gap between brass plates in air at 185 mm Hg is shown in Fig. 4.28. In a study of the visual growth of the discharge in air, for a gap of 0·3 cm. at gas pressures of 13 mm. Hg and 42 mm Hg, Rogowski and Tamm [75] have identified the first stage of the voltage collapse with the onset of a diffuse glow discharge, and have confirmed that the final collapse of voltage at the end of the step occurs as the result of the formation of a concentrated discharge channel which develops from the anode of the gap.

The magnitude of the voltage during the period of the step, in Rogowski and Tamm's measurements [75] for a 0·3-cm gap in air, varied from about 500 V to 1,450 V as the gas pressure is increased from 5 mm. Hg to 80 mm. Hg. As pointed out by Mayr [77] the ratio of the step voltage to the D.C. breakdown voltage of the gap is roughly constant, between 0·6 and 0·7, for the range of conditions studied. The duration of the step varies over wide limits for given conditions of gap and gas pressure and values between 2 and 12 μsec are quoted

Lower values for the step voltage are recorded by Mayr [77] for gaps of between 2 and 4 mm. in air for a hot cathode, the ratio of the step voltage to the D.C. breakdown voltage falling to about 0·2 in some cases.

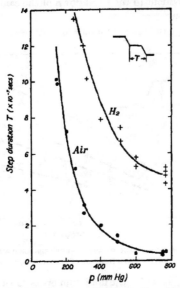

FIG. 4.29. Duration of the step in the voltage collapse in air and hydrogen as a function of gas pressure. The gap length ranges from 1 to 6 mm. between parallel plates.

In the analysis of his results Mayr [77] shows that the duration of the step varies inversely as the magnitude of the current during the step and that the product of the step current and the step duration is about 10^{-4} coulomb.

In subsequent investigations by Buss [78] for the D.C. breakdown of gaps between plate electrodes in air and in hydrogen the stepped break-down is studied for gaps up to 0·6 cm. and pressures up to 1,500 mm. Hg. The duration of the step in both gases is shown in Fig. 4.29 as a function of gas pressure. Fig. 4.30 gives the curves obtained for the D C. break-down voltage and the step voltage. The time required for the voltage to collapse from the D.C. breakdown value to the step value has also been recorded and shows a decrease with increasing gas pressure; for a change in gas pressure from 200 to 760 mm. Hg the time falls from 10×10^{-8} to

2×10^{-8} sec. for discharges in hydrogen and from $3\cdot5 \times 10^{-8}$ to $1\cdot5 \times 10^{-8}$ sec. for discharges in air. While no step is recorded in the breakdown of a 0·12-cm. gap in air at a pressure of 1,500 mm. Hg, a step of duration $2\cdot3 \times 10^{-8}$ sec. is evident in the breakdown of the same gap in hydrogen at the same pressure.

Measurements of time lags in a gap of 0·068 cm. between spheres of

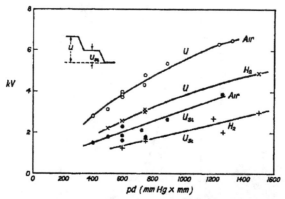

FIG. 4.30. Breakdown voltage and step voltage for discharges in air and in hydrogen as a function of pressure × gap-length

0·95 cm. diameter, for which the minimum breakdown voltage is 3,820 V, have been made by Tilles [79] using a ballistic galvanometer method which enables time lags down to 10^{-5} sec. to be recorded. On the application of a voltage in excess of the breakdown value to the gap a current is produced by a suitable circuit and is caused to flow through a ballistic galvanometer, the current being maintained at a constant value. The deflexion of the galvanometer can then be used to indicate the time for which the current has flowed, i e. the time lag to breakdown of the gap. Tilles has recorded statistical time lags up to 1 sec. for the particular gap studied, but finds that with increasing irradiation the statistical time lag decreases until, with a photo-current of about 2×10^{-12} A/cm.², the time lags are all closely constant at about 7×10^{-5} sec , the amount of overvoltage being roughly 5 per cent.

The time lags in spark breakdown have also been measured by two methods using the Kerr cell shutter. In White's experiments [37] a constant voltage is applied to an unirradiated gap and a spark is initiated by the sudden illumination of the gap by an intense flash of ultraviolet light from a nearby spark. The time lag is measured from the start of the flash until breakdown occurs in the gap. In the second method, used by Wilson [36], an overvoltage is suddenly applied to the gap, which is already illuminated from a constant source of ultraviolet light, and the time lag is measured between the instant of application of the voltage and the breakdown of the gap.

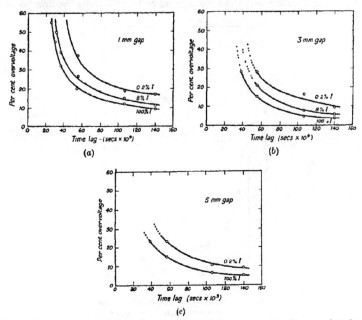

Fig. 4.31. Curves giving the time lag as a function of overvoltage for three gap lengths in air between small spheres with differing amounts of illumination I.

Using the former method White [37] has measured the variation of time lag with overvoltage and with intensity of illumination and has recorded times down to 25×10^{-9} sec. Curves for three gaps in air are given in Fig 4.31. It is pointed out by White that the rapid increase in the overvoltage required to cause breakdown in times of less than about 50×10^{-9} sec. may be attributable to the time required for the light from the illuminating spark to build up to a sufficiently intense value. This difficulty is not present in Wilson's experiments [36], where a quartz-mercury arc lamp was used to illuminate the gap, and Wilson records time lags down to 5×10^{-9} sec. for the formation of sparks in a 0·1-cm. gap.

Short time lags, down to 10^{-8} sec. with an accuracy of 10^{-9} sec., have been measured by Newman [80] and later by Bryant and Newman [81], using a method involving a number of electronic relays connected at intervals along a transmission line. At one end of the line a voltage wave is produced at the same instant that the impulse voltage is applied to the gap being investigated. The voltage wave travels along the line, to be

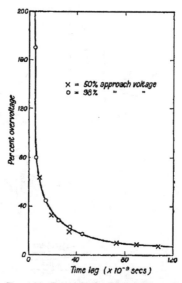

FIG. 4.32. Time lag of spark breakdown as a function of overvoltage for a short gap between spheres with intense ultraviolet illumination of the cathode.

met by a second voltage wave which starts from the other end of the line and is initiated by the breakdown of the gap. The meeting-point of the two waves is determined by the number of relays tripped and the time lag between the application of the voltage and the subsequent breakdown of the gap can then be estimated. The technique has been applied to the measurement of time lags in a gap in air between small spheres, the spacing being such that the D.C. breakdown voltage is 10 kV. Some of the values obtained by Bryant and Newman [81], for the gap when irradiated by a quartz-mercury lamp, are shown in Fig. 4.32.

The formative times of sparks in air at atmospheric pressure for uniform field gaps of length up to 0·6 cm. have been measured in experiments by Fletcher [82], using a micro-oscillograph and specially developed ultra-high-speed circuits by means of which times of less than 10^{-9} sec. can be recorded. In these investigations the effect on the time lag of varying the ultraviolet illumination of the gap was tested for a

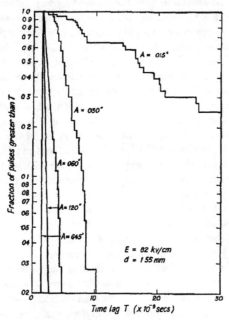

FIG. 4 33   Distribution of time lags for a 0 155-cm. gap between spheres in air with different diameter (A) aperture stops of ultraviolet illuminating system.

fixed gap of 0·155 cm. and an applied field of 82 kV/cm. For the greatest intensity used there was no detectable scatter of time lag, but as the intensity was decreased the scatter increased. This is shown in Fig. 4.33, in which is plotted the logarithm of the fraction of time lags greater than T as a function of T for different intensities of illumination. The distribution is roughly exponential for the various intensities, as has been found previously for longer time lags (see Chapter VIII). The voltage applied to the gap is about 95 per cent. in excess of the minimum breakdown voltage, and for such a high overvoltage Strigel [83] has shown that every electron released from the cathode is successful in initiating the breakdown. Therefore, on the assumption that the statistical fluctuations are due entirely to the fluctuations in the supply of initiating electrons from the cathode, the observed distributions can be used to calibrate the ultraviolet source in terms of the photo-current. If the distribution is written as $e^{-T/\sigma}$, $1/\sigma$ is the rate at which electrons are released from the cathode. σ can be measured from the curves to give a value of $0.0037/A^2$ m.μsec where A is the aperture diameter in inches of the illuminating system This corresponds to a photo-current of 4.34×10^{-8} A/cm ^{2}

The non-statistical lag obtained for the higher intensities of ultraviolet illumination is considered to be the formative lag. This formative lag was determined for various voltages and gap widths for times between 0.5 and 50 m.μsec. It was convenient to make measurements at constant voltage rather than at constant gap width, and results for several different voltages applied to gaps of different lengths are given in Fig. 4 34 which shows the formative time as a function of the applied field strength. For fields above 50 kV/cm. the formative time depends on the field alone and is independent of gap length, but for fields below 49 kV/cm. the constant voltage plot for an applied voltage of 7.5 kV begins to give longer formative times than those corresponding to the higher voltages. Similarly at fields less than 42 kV/cm. the curve for an applied voltage of 12.7 kV rises above the curve for 18 kV.

In Fig. 4.34 the values obtained by Newman [80] for the breakdown of a 2.57-mm. gap are also plotted. The broken curve in Fig. 4.34 is that interpolated by Fletcher [82] from his measurements for the same gap. Newman's measured time lags are therefore seen to be about 10^{-8} sec. longer than the micro-oscillograph values, and it is suggested by Fletcher that this may be explained on the assumption that Newman measured his time lags from the bottom instead of from the top of the rising impulse, which may have had a rise time of the order of 10^{-8} sec.

A theoretical analysis has been made by Fletcher [82] of the formative time lags in spark development, in terms of avalanche and streamer growth, and is given on p. 275 Other oscillographic results obtained by Fisher and Bederson [59] are discussed on p. 279.

Oscillographic studies have been made by Flowers [84] of the breakdown of air and other gases in short gaps between spheres, and between a coaxial wire and a cylinder, when the gaps are subjected to rapidly rising impulse voltages of peak values in excess of the minimum breakdown value, the rate of rise of voltage being approximately 25 kV in 0.05 μsec. The maximum value reached by the voltage applied to the gap, on the rising front of the voltage wave, is such that the voltage gradient at the cathode is about 500 kV/cm., and it is suggested that the breakdown is initiated in such cases by the field emission of electrons from the cathode. The measurements also show that the impulse breakdown of a 0·2-mm. gap in air remains constant at about 10 kV as the gas pressure is reduced from 1 atm down to 8 mm. Hg.

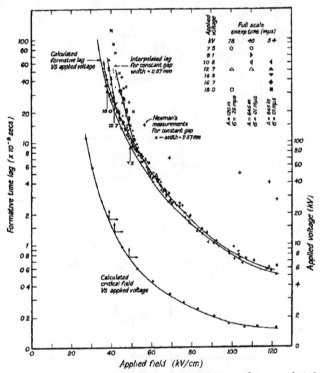

FIG. 4.34. Measurements by Fletcher [82] of the formative lag in air plotted as a function of applied field for different gap lengths (or applied voltages), and compared with a calculated formative lag and with the measurements of Newman [80] The curve is also given for the calculated critical fields below which, for a given applied voltage, multiple avalanches (and thus longer formative lags) should be required for breakdown.

The formative times of sparks across a uniform field gap of 2 cm. in air has been studied oscillographically by Messner [85], who gives the curve shown in Fig. 4.35 for the time lag as a function of overvoltage. Similar results have been obtained by Bellaschi and Teague [86] and by Strigel [90]. Curves for much longer gaps, up to 64 cm. between spheres of 200 cm. diameter, as given by Bellaschi and Teague [86], are shown

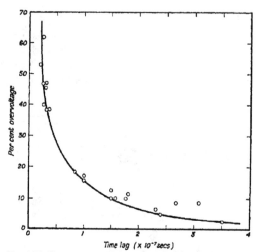

Fig. 4 35 Formative time lag as a function of overvoltage for a gap of ~ 2 cm. between spheres in air (D C. breakdown voltage = 60 kV).

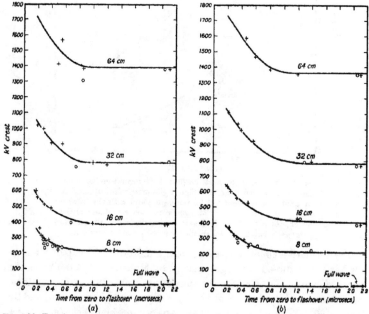

Fig. 4 36. Time lag as a function of applied voltage for various gap lengths in air between spheres of 200 cm. diameter. (a) Positive impulse. (b) Negative impulse

in Fig. 4.36. There is a tendency for the time lag at a given voltage to be slightly longer for a negative high-voltage sphere than for a positive high-voltage sphere. The difference can be accounted for by the fact that the field in the gap is not strictly uniform, and polarity effects may therefore be expected to influence the spark mechanism. The influence of overvoltage on the breakdown of gaps between spheres and between

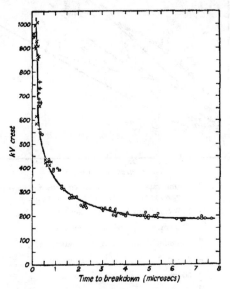

FIG. 4.37. Time lag as a function of applied voltage for a 10-in. gap between 0 5 in square rods in air.

rod electrodes has been investigated by Bellaschi and Teague [87] and by Hagenguth [88]. Fig. 4.37 shows the results obtained by Hagenguth for the impulse breakdown of a 10-in. rod-gap   The variation with gap length of the sparking voltage of a point-plane gap subjected to the minimum impulse breakdown voltage as measured by Allibone, Hawley, and Perry [89] is shown in Fig. 4.38.

Oscillographic measurements of the statistical and the formative time lags in gaps in air at pressures varying from atmospheric down to 25 mm. Hg have been made by Gänger [91]. The gaps studied range in length up to 8 cm. and include sphere-sphere and point-sphere arrangements, the peak voltage used being 50 kV. A set of curves obtained for the formative time lags as a function of the impulse ratio (the ratio of the peak voltage applied to the gap to the A.C. breakdown value), for gaps between spheres of 5 cm. diameter, is given in Fig. 4.39. Each curve corresponds to a given gap at a given gas pressure Similar curves for the time lags measured for the breakdown between a positive point and a

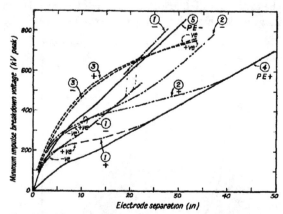

Fig 4.38. Minimum impulse breakdown voltages for gaps in air between various electrodes and an earthed plane.

$\overset{①}{+}$	12 5 cm.	diameter sphere, positive impulse.
$\overset{①}{-}$	12 5 cm.	„ „ negative „
$\overset{②}{+}$	25 cm.	„ „ positive „
$\overset{②}{-}$	25 cm	„ negative „
$\overset{③}{+}$	50 cm.	„ „ positive „
$\overset{③}{-}$	50 cm	„ „ negative „
$\overset{④}{+}$	Point electrode, positive impulse	
$\overset{④}{-}$	„ „ negative „	

negative sphere are given in Fig 4.40, and those for a negative point and a positive sphere in Fig 4.41. A continuous change in formative time lag as a function of overvoltage applied to the gap is recorded in all the curves (see also p. 279).

Lichtenberg figures

The characteristics of streamers developing from both positive and negative electrodes can be studied by means of the patterns which are produced on photographic plates by the application of impulse voltages to electrodes in contact with the plates In many of the investigations the photographic plate is placed with its back resting on a metal plate

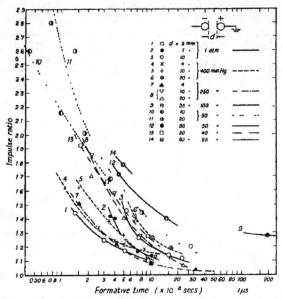

FIG 4 39 Relation between formative time and impulse ratio for various gaps and gas pressures in air between spheres of 5 cm diameter

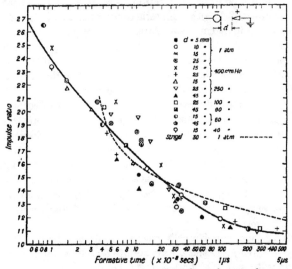

FIG. 4.40. Relation between formative time and impulse ratio for various gap lengths and gas pressures in a positive point-sphere gap in air.

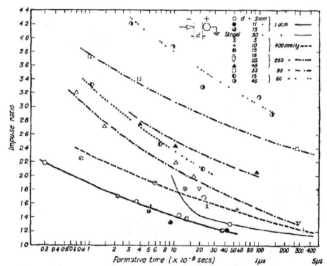

Fig 4.41. Relation between formative time and impulse ratio for various gap lengths and gas pressures in a negative point-sphere gap in air.

electrode, and a point or rod electrode is arranged to touch the photographic emulsion. In other investigations both electrodes are placed on the photographic emulsion. After a voltage pulse has been applied between the electrodes, intricate patterns appear on the photographic plate when this is developed. The character and size of these patterns, which are known as Lichtenberg figures, depend on the magnitude, the wave form, and the polarity of the applied voltage.

While Lichtenberg figures record the development of surface corona discharges the various ionization processes occurring are essentially similar to those observed in normal corona and spark discharges. An extensive study of Lichtenberg figures has been made by Merrill and von Hippel [92], who have observed the formation of the figures on photographic plates in various gases at different pressures. The results confirm that the figure is produced by the light emission from the discharge, and that the nature of the figure is characteristic of the surrounding gas

In Merrill and von Hippel's measurements a voltage wave with a wavefront of 0·5 μsec. and a wavetail of 28 μsec. is applied to a point electrode resting on the photo-sensitive emulsion. At low gas pressures a general haze is produced for both polarities and is thought to be recorded by the successive overlapping of electron avalanches With increasing gas pressure, above about 50 mm. Hg in air, the diffuse pattern splits up into a discontinuous figure, the shape of which depends on the polarity of the point electrode Typical records obtained for both a positive voltage and a negative voltage applied to the point are given in Fig. 4.42, Pl. 6, which refers to discharges occurring when the plate and associated electrodes are enclosed in air at a pressure of about 30 atm.

Fig 4.43 illustrates the mechanism of growth of the positive and the

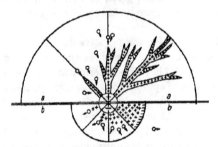

FIG 4 43. Mechanism of generation of Lichtenberg figures   (a) positive; (b) negative. O→
denotes electron.

negative discharges as explained by Merrill and von Hippel. Primary electrons near the positive electrode are accelerated into a field of increasing intensity, and positive space charges are left in the tracks of the resultant electron avalanches These positive space charges concentrate the field and cause new avalanches which lengthen and branch the path. With a negative electrode electrons are projected outwards into a field of decreasing intensity. The positive space charge left behind weakens the field in the radial direction and creates a tangential one which tends to spread the ionization. The negative figure attains its final size when the field strength at the boundary has decayed below the value necessary for effective ionization.

As the voltage is increased, for a given gas pressure, the size of the pattern increases until above a certain voltage a narrow intense streamer channel develops from the pattern. The initiation of this streamer denotes the onset of spark growth. A similar change occurs in the patterns produced at a given voltage when the pressure exceeds a certain value, as illustrated in Fig 4.44, Pl. 6, which shows the patterns produced by positive discharges in air The latter change may be related to the drop which occurs in the breakdown voltage of a point-plane gap in air when the pressure is increased above a particular value (see p. 323).

The size of the Lichtenberg figures is greatly reduced in the presence of electro-negative gases such as freon (CCl_2F_2) and carbon tetrachloride, and their characteristics are also altered. In the case of a negative figure, as shown in Fig 4.45, Pl 6, for air at atmospheric pressure containing carbon tetrachloride, the stepping in the growth of the discharge becomes much more pronounced and the speed of propagation is reduced. The corresponding positive figure, given in Fig. 4.46, Pl. 6, has a more symmetrical shape and is surrounded by intense streamers of enhanced ionization.

Similar studies of positive and negative discharge patterns formed in air at reduced pressures, and at pressures up to 20 atm., have been made by Praetorius [95], who has also determined the mean radius of the patterns as a function of applied voltage and gas pressure. Other results have been obtained by Rogowski, Martin, and Thielen [96], who have investigated the patterns produced by short impulse voltages, of 10^{-7} to 10^{-8} sec. duration, in air, hydrogen, nitrogen, and oxygen. The paths of individual electron avalanches developing from the negative electrode are recorded

In studies by Pleasants [93] the speeds of formation of positive and negative figures have been measured and values ranging between 2×10^6 and 3×10^7 cm /sec. are recorded, the speeds for positive figures being greater than those for negative figures. The influence of high magnetic fields of the order of 10,000 gauss on the shape of the figures has been investigated by Magnusson [94], who finds that with the magnetic field perpendicular to the plate the pattern is deformed in a spiral manner. The bending of the pattern occurs more readily in the case of negative figures, which show an increase in curvature with increasing distance from the centre

Investigations of the patterns produced when the surface of the photographic plate lies along the line joining two point electrodes, or along the perpendicular from a point electrode to a plane electrode, have been made by several workers [97 to 101] A typical result for a point-plane gap, as recorded by Allibone [62], is given in Fig. 4.47, Pl. 6, and shows long branched streamers from the positive point meeting shorter, less-branched streamers from the negative point. In Marx's experiments [97] the duration of the impulse voltage applied to the gap was as long as 1 m.sec., but in the later measurements by Rosenlocher [100] durations of less than 1 μsec. have been used, and the resultant records enable the growth of streamers to be studied. Honda's measurements [99] refer principally to discharges between point gaps when one point is in air and the other in oil, the photographic plate being immersed half-way in the oil.

REFERENCES QUOTED IN CHAPTER IV

1. W. Holzer, *Z. Phys.* **77** (1932), 676
2. J J. Torok, *Trans. A I.E.E* **47** (1928), 177.
3. J. Slepian and J. J. Torok, *Elect. Journal*, **26** (1929), 107.
4. A. Matthias, *Elektrotech. Z.* **58** (1937), 881, 928, 973
5. A. Zinerman and N. Nikolaevkaia, *J Exp. Theor. Phys. U S S R.* **16** (1946), 449
6. H. Raether, *Z Phys* **94** (1935), 567.
7. H. Raether, *Phys Z.* **15** (1936), 560.
8. H. Raether, *Z. tech. Phys* **18** (1937), 564.
9. H. Raether, *Z. Phys.* **107** (1937), 91.
10. H. Raether, ibid. **110** (1938), 611.
11. H. Raether, ibid. **112** (1939), 464.
12. H. Raether, *Arch. Elektrotech.* **34** (1940), 49.
13. H. Raether, *Z. Phys* **117** (1941), 375.
14. H. Raether, ibid. 524
15. H. Raether, *Elektrotech. Z* **63** (1942), 301.
16. H. Raether, *Ergeb. d. exakt. Naturwiss.* **22** (1949), 73
17. E. Flegler and H. Raether, *Phys. Z.* **36** (1935), 829.
18. E Flegler and H. Raether, *Z. tech. Phys* **16** (1935), 435.
19. E. Flegler and H. Raether, *Z. Phys.* **99** (1936), 635
20. E Flegler and H. Raether, ibid. **103** (1936), 315.
21. E Flegler and H Raether, ibid. **104** (1936), 219.
22. W. Riemann, ibid. **120** (1942), 16
23. W. Riemann, ibid. **122** (1944), 216.
24. W. Riemann, ibid. 262.
25. J. M. Meek, *Phys Rev.* **57** (1940), 722.
26. H. Kroemer, *Arch. Elektrotech.* **28** (1934), 703
27. L B. Snoddy and C. D. Bradley, *Phys. Rev.* **47** (1935), 541.
28. V. Nakaya and F. Yamasaki, *Proc. Roy. Soc. A,* **148** (1935), 446.
29. S Gorrill, see p. 437 of reference 68.
30. F. G. Dunnington, *Phys. Rev.* **38** (1931), 1506
31. F. G. Dunnington, ibid. 1535.
32. J. C Street and J. W. Beams, ibid 416
33. E. O. Lawrence and F. G. Dunnington, ibid. **35** (1930), 396.
34. L. von Hámos, *Ann. Phys.* **7** (1930), 857
35. H. J. White, *Phys Rev.* **46** (1934), 99.
36. R. R. Wilson, ibid. **50** (1936), 1082,
37. H. J. White, ibid. **49** (1936), 507.
38. A. E. J. Holtham and H. A. Prime, *Proc. Phys. Soc* B, **63** (1950), 561.
39. A. M. Zarem and F. R Marshall, *Rev. Sci Instrum.* **21** (1950), 514.
40. K. D. Froome, *J. Sci. Instrum.* **25** (1948), 371.
41. W. Kaye and R. Devaney, *J. Appl. Phys.* **18** (1947), 912.

42. A. C LAPSLEY, L. B SNODDY, and J. W. BEAMS, ibid. **19** (1948), 111.
43. J. W. BEAMS, *Rev. Sci. Instrum.* **1** (1930), 780.
44. H. J. WHITE, ibid. **6** (1935), 22.
45 C. V. BOYS, *Nature*, **118** (1926), 749.
46. B. F. J. SCHONLAND and H. COLLENS, *Proc. Roy Soc.* A, **143** (1934), 654.
47 B. F. J. SCHONLAND, D J. MALAN, and H. COLLENS, ibid. **152** (1935), 595.
48. T. E. ALLIBONE and B. F J. SCHONLAND, *Nature*, **134** (1934), 736.
49. T. E ALLIBONE and J. M. MEEK, *Proc. Roy. Soc.* A, **169** (1938), 246.
50. T. E. ALLIBONE and J. M. MEEK, ibid. **166** (1938), 97.
51. J. W. FLOWERS, *Gen. Elect. Rev.* **47** (1944), 9.
52. V. S. KOMELKOV, *Elektrichestvo*, No. 9 (1940).
53. V. S. KOMELKOV, *J. Tekhn. Fiz.* **10**, No 17 (1940).
54 V S. KOMELKOV, *Doklady Akad. Nauk U.S.S.R.*, **68** (1947), 57.
55. V. S. KOMELKOV, ibid. **47**, No. 4 (1945)
56. V. S. KOMELKOV, *Bull. Ac Sci. U.R.S.S. Dept. Tech. Sci.* No. 8 (1947), p. 955.
57. R F SAXE and J. M. MEEK, *Nature*, **162** (1948), 263.
58. R. F. SAXE and J M MEEK, *Brit Elect and Allied Ind. Res. Assoc.* Reports Ref L/T 183 (1947) and Ref L/T 188 (1947).
59. L. H FISHER and B. BEDERSON, *Phys. Rev* **75** (1949), 1615.
60 R STRIGEL, *Arch tech. Messen* **63** (1948), 52.
61. V. NEUBERT, *Z Phys.* **114** (1939), 705.
62. T. E ALLIBONE, *Journal I E.E.* **82** (1938), 513.
63. I STEKOLNIKOV and A. BELIAKOV, *Journal Exper. and Theor. Phys.* **8** No 4 (1938)
64. I. STEKOLNIKOV, *Elektrichestvo*, No. 2 (1940).
65. C. F. WAGNER, G. D MCCANN, and G. L. MACLANE, *Trans A I.E.E.* **60** (1941), 313
66. R STRIGEL, *Wiss Veroff. Siemens-Werk*, **11** (1932), 52
67. J M. MEEK and J. D CRAGGS, *Nature*, **152** (1943), 538
68. L. B. LOEB, *Fundamental Processes of Electrical Discharges in Gases*, J. Wiley, 1939.
69 H. M. GAUNT and J. D. CRAGGS, in preparation.
70. R. F. SAXE, Thesis, University of Liverpool, 1948.
71 H. A. PRIME and R. F. SAXE, *Journal I E.E.* **96** (II) (1949), 662.
72. H. A PRIME and R TURNOCK, ibid. **97** (II) (1950), 793.
73. R TURNOCK, in preparation.
74 W ROGOWSKI, E. FLEGLER, and R TAMM, *Arch. Elektrotech.* **18** (1927), 479; ibid 506.
75. W. ROGOWSKI and R. TAMM, ibid. **20** (1928), 107; ibid. 625.
76. R. TAMM, ibid. **19** (1928), 235
77. O. MAYR, ibid. **24** (1930), 15.
78. K. BUSS, ibid **26** (1932), 266; **27** (1933), 35.
79 A. TILLES, *Phys. Rev.* **46** (1934), 1015.
80. M. NEWMAN, ibid. **52** (1937), 652.
81. J. M. BRYANT and M. NEWMAN, *Trans. A I.E E.* **59** (1940), 813
82. R. C. FLETCHER, *Phys. Rev.* **76** (1949), 1501.
83. R. STRIGEL, *Wiss. Veroff. Siemens-Werk*, **11** (1932), 52.
84. J. W. FLOWERS, *Phys Rev.* **48** (1935), 955.
85 M. MESSNER, *Arch Elektrotech* **30** (1936), 133.

86 P L. BELLASCHI and W. L. TEAGUE, *Elec. Journal*, **32** (1935), 120.

87. P L BELLASCHI and W. L TEAGUE, ibid 56.

88 J. H HAGENGUTH, *Trans. A I E E*. **56** (1937), 67

89. T. E. ALLIBONE, W. G. HAWLEY, and F. R PERRY, *Journal I E E* **75** (1934), 670.

90. R STRIGEL, *Elektrische Stossfestigkeit*, J Springer, 1939.

91. B. GANGER, *Arch. Elektrotech*. **39** (1949), 508.

92. F. H. MERRILL and A. VON HIPPEL, *J. App Phys*. **10** (1939), 873.

93. J. G. PLEASANTS, *Elect Engng* **53** (1934), 300.

94 C. E MAGNUSSON, *Trans A.I E E* **49** (1930), 756 , **51** (1932), 117

95. G PRAETORIUS, *Arch. Elektrotech*. **34** (1940), 83.

96. W. ROGOWSKI, O. MARTIN, and H. THIFLEN, ibid **35** (1941), 424.

97 E. MARX, ibid **20** (1928), 589.

98 E MARX, *Elektrotech Z*. **51** (1930), 1161.

99 T. HONDA, *Arch. Elektrotech* **33** (1939), 458

100. P. ROSENLOCHER, ibid. **26** (1932), 19.

101. T. E. ALLIBONE, *Nature*, **161** (1948), 970.

102 L. B. SNODDY and J W BEAMS, *Phys Rev* **55** (1939), 663.

103 H. SCHNEIDER, *Arch. Elektrotech* **34** (1940), 457

104 V S. KOMELKOV, *Izvestiya Akad. Nauk SSSR*. **6** (1950), 851.

105. V. S KOMELKOV and A. M LIFSHITS, ibid. **10** (1950), 1463

THE LIGHTNING DISCHARGE

LIGHTNING, which is a particular instance on a large scale of the electric spark, has been investigated extensively during recent years, with the result that a fairly clear picture of the mechanism of development of lightning discharges has now been established. The principal methods used in these investigations may be summarized as follows.

(1) Measurement of the voltage gradient at the ground, between cloud and ground, and in the thunder-cloud.

(2) Oscillographic studies of the electromagnetic fields radiated by lightning discharges.

(3) Photographic studies of lightning discharges by rotating-camera techniques.

(4) Measurements of the peak magnitudes and time variations of the currents in lightning discharges

Recent studies of the electrification of thunder-clouds have been made by Simpson and his colleagues [1, 2, 3] but their results will not be described here as this book is concerned primarily with breakdown processes in gases, of which lightning is an example. Summaries of the numerous investigations relating to atmospheric electricity are given in books by Fleming [4], Humphreys [5], Schonland [6], and Chalmers [7]. The contents of this chapter are based largely on a review by Meek and Perry [81].

Lightning discharges between cloud and ground

Evidence of the composite nature of lightning discharges was obtained in 1890 by Hoffert [8] who photographed a discharge with a moving camera. He found that the discharge consisted of a number of separate strokes which occurred at different intervals of time, of the order of 10 m.sec., and that the strokes followed the same track in space. This observation of the multiple strokes forming a single discharge led within a few years to further studies by Walter [9] and Larsen [10]. In one photograph obtained by the latter a single lightning discharge was resolved into forty separate strokes, the total duration of the discharge being about six-tenths of a second.

In order to determine the direction and speed of propagation of the strokes of a lightning discharge a special camera was devised by Sir Charles Boys [11]. This camera has two lenses, arranged to rotate at

opposite ends of a diameter of a circle, and images of a lightning discharge
are recorded on a stationary film behind the lenses. There is then a
distortion of the photographed image resulting from the relative movement between lens and film, and a distinction can be made, therefore,
between any portion of the discharge which becomes luminous before
another portion In a later form of the camera [12], the lenses are fixed
and the film is mounted on the inside of a circular rotating drum.

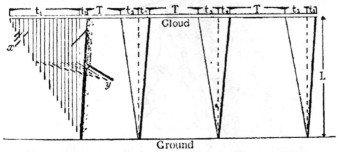

FIG 5.2 Representation of the development of a straight vertical discharge as
recorded on a Boys camera, in which there is relative movement between lens and
film in a horizontal direction, with time increasing from left to right of the record.
The average time intervals are as follows t_1, 0 01 sec , t_2, 0 00004 sec , t_3, 0 001 sec ;
T, 0 03 sec. The length L of the discharge is 2 km.

The first successful results with cameras of the above types were
obtained by Schonland and his associates in South Africa [13 to 19].
Subsequently similar records have been secured in the U.S.A. by
Workman, Beams, and Snoddy [20], Allbright [21], Holzer, Workman,
and Snoddy [22], McEachron [23], McCann [24], and Hagenguth [25]
and in the U.S.S R by Stekolnikov and his associates [26].

A typical photograph of a lightning discharge to the Empire State
Building in New York, as recorded by McEachron with a rotating-lens
type of camera, is given in Fig. 5.1, Pl. 7. This figure also includes, at
its centre, a photograph of the same discharge as recorded by a stationary
camera. The rotating-lens records show that the discharge consists of
a number of separate strokes, numbered 1 to 14 in Fig. 5.1, which take
place at different time intervals along the same track.

From the analysis of many rotating-camera photographs Schonland
has deduced the general mechanism of lightning discharges. This is
shown schematically in the drawing of Fig. 5.2, which represents an
idealized lightning discharge developing along a vertical track between
cloud and ground, as recorded on a film moving transversely to the

direction of the discharge. The process is initiated by a streamer which
develops downwards from the cloud in a series of steps which are
separated by time intervals varying between about 15 and 100 μsec.

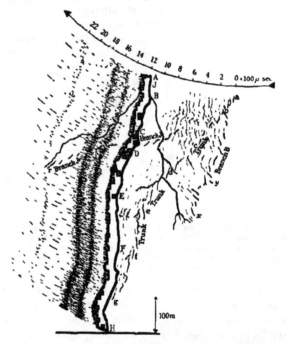

FIG. 5 4   Diagrammatic analysis of a stepped leader stroke,
illustrating the development of the individual step streamers.

Each step is revealed in the photograph by a sudden increased luminosity
of the channel between the tip of the streamer and the cloud. The
freshly ionized air at the tip of the streamer is more brightly luminous
than the remainder of the channel. The speed with which the process,
known as the stepped leader stroke, approaches the ground is generally
between 1×10^7 and 5×10^7 cm /sec., with a usual value of about $1 \cdot 5 \times 10^7$
cm./sec. The speed of the individual step streamers is of the order of
5×10^9 cm./sec.

A photograph of a stepped leader stroke recorded by Schonland [14]
is given in Fig. 5.3, Pl. 7. On account of the weak luminosity of the
leader stroke the details of its structure are not clearly distinguishable.

This is true of most photographs of stepped leader strokes, and, in some cases, no leader stroke has been recorded photographically, though confirmation of its presence is given by measurements of the electric field variations which it produces at the ground.

An analysis of the fine structure of one of the stepped leader strokes of Fig. 5.3 is shown in Fig 5.4. Each step follows the path traced out by

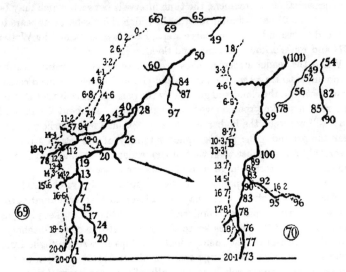

Fig. 5 5. Analyses of the times involved in the various stages of development of the stepped leader strokes and the return strokes in two separate discharges which took place from the same portion of a cloud, with discharge 69 preceding discharge 70 by about 0 0018 second. The leader-stroke times, given in light lettering, are in milliseconds, measured from the moment of appearance of the leader stroke of flash 69. The return stroke times, given in heavy lettering, are in microseconds

the preceding step, and branches which occur during the development of the leader stroke advance by the same stepped process. Details of the time taken for the leader strokes and the return strokes to develop are given in Fig. 5.5 for two separate discharges. The times for the leader stroke are in m.sec. and those for the return stroke are in μsec.

The above type of stepped leader development precedes the first stroke of a discharge in about 65 per cent. of Schonland's records. In most of the remaining records the stepped leader stroke develops more rapidly in its initial stages, with long bright steps. Its effective velocity is greater than 6×10^7 cm./sec., and in one case a velocity of $2 \cdot 6 \times 10^8$

cm./sec. has been recorded. After the leader stroke has proceeded part of the way between cloud and ground its speed suddenly falls, to a value which is usually about 1×10^7 cm./sec. The step length and the luminosity of the leader stroke also decrease, and the process can only be photographed with difficulty. In some instances, during this second slow stage of the development, the leader channel is brightly illuminated by occasional step streamers, the time intervals between which may be as large as 0·01 sec. Similar records, in which the discharge appears to proceed to ground in several large steps, have been obtained by Walter [27] and by Workman, Beams, and Snoddy [20].

When the leader stroke reaches ground, the main or return stroke commences to travel up the pre-ionized channel which has been established between the cloud and ground by the passage of the leader stroke. The return stroke proceeds with a speed of between about 2×10^9 and $1 \cdot 5 \times 10^{10}$ cm./sec., the higher speed being attained in its earliest stages near the ground. The photographed intensity of the return stroke is invariably much higher than that of the leader stroke, and it is during the return stroke that large current flows are observed.

Subsequent to the passage of the stepped leader stroke and the associated return stroke there may be an interval of several hundredths of a second, after which a second leader and return stroke takes place, as shown in Fig. 5.2. This leader to the second stroke of the discharge generally consists of a streamer which develops in a single flight from cloud to ground, at an average speed of about 2×10^8 cm./sec., and is termed a dart leader stroke. Occasionally, if the time interval since the previous stroke is exceptionally long, the dart leader may develop in one or two steps. Following the dart leader there is a return stroke which develops from ground to cloud. Later strokes of the same discharge consist of a similar sequence of dart leader and return stroke. In the majority of cases each of these strokes follows closely the original channel developed by the first stroke of the discharge. Branching is rarely observed on the second and subsequent strokes.

The sequence of events described above refers generally to discharges taking place in open country. A pronounced difference is observed for lightning striking high buildings, as shown by the experiments of McEachron [23] and later of McCann [24] in the U.S.A. The discharge is initiated, in most cases, by a stepped leader stroke which develops upwards from the building to the cloud, as indicated in Fig. 5.6 The speed of the upward-developing leader stroke, is about the same as that for the downward-developing leader stroke, and the time intervals between the steps are of the same order of magnitude. But, unlike the discharge to open country, the initial stepped leader is not followed by a return stroke. After the leader stroke has entered the cloud the luminosity of the channel persists, in what is known as a continuing stroke, during which current continues to flow. Downward-developing dart leaders are invariably observed preceding subsequent strokes of the discharge.

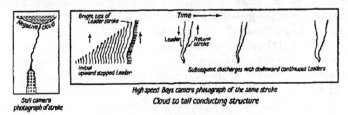

FIG 5.6 Schematic diagram showing discharges initiated by stepped leader strokes developing upwards from high structures to the cloud.

The majority of the photographs of lightning discharges have been obtained at night, as the usual photographic method requires that the shutter of the rotating camera should be open continuously before the flash occurs. Schonland [28] has devised a method whereby the camera shutter is opened by a mechanism triggered by the electromagnetic pulses which are radiated during the stepped leader process. This enables the shutter to open in sufficient time to record the later development of the stepped leader stroke as it approaches the ground, and also the return stroke and subsequent strokes. The shutter is arranged to close automatically after 0·5 sec. exposure. An alternative method, which gives less satisfactory results, makes use of a chlorine-bromine filter in conjunction with a quartz-fluorite lens. With this arrangement little fogging of the film takes place with the shutter open continuously in daylight, whereas light of short wave-lengths from the lightning discharge is transmitted by the filter-lens system, and photographs can be obtained in this manner of discharges at distances up to about three miles. Photo-electric recording of the light emitted by lightning discharges, at distances as great as twenty miles, has been successfully employed by Schonland [28].

The luminosity of the various processes comprising the lightning discharge is greatest near the tips of the streamers or leader strokes. This concentration of light may be considered to be caused by the greater fields existing at the tips, as compared with the stems, of the streamers, so that greater numbers of atoms and molecules are raised to excited states by electron impact. In general it is only the tips of the leader strokes which are sufficiently bright to be photographed In the dart leader the emitted light is largely confined to a distance of about 50 metres behind the tip and is only weakly discernible at greater distances behind the tip. Similar considerations apply in the step streamer. In the return stroke the emitting region is longer, but the luminosity near the ground has fallen appreciably before the bright tip of the stroke has reached half-way to the cloud.

Schonland [17] has postulated that the stepped leader process is preceded by a faintly luminous 'pilot' streamer, carrying a current of the order of several amperes and which travels continuously with a speed equal to the effective rate of advance of the leader stroke. The weakly ionized channel developed by the pilot streamer becomes unstable as the streamer advances, and after a period there is a rapid readjustment of the ionization in the channel, by which it is rendered more brightly luminous and more highly conducting. This process represents the passage of the step streamer, which advances rapidly down the channel to catch up with the advancing pilot streamer.

The existence of the pilot streamer still requires experimental confirmation. However, it is reasonable to suppose that even if it does exist, and there is circumstantial evidence in its favour, its luminosity would be so weak as to render it extremely difficult to photograph. By consideration of its existence Schonland [17] has been able to put forward an explanation of the remarkable regularity which is observed in the time intervals between the successive step-streamers, which, in 90 per cent. of the stepped leaders observed, lie between 50 and 90 μsec. The explanation is based on consideration of the decrease of electron density in the leader channel caused by electron capture and recombination processes. A decrease in electron density necessitates an increase in field strength if the current is to be maintained in the channel, and ultimately the field attains a value sufficient to cause breakdown and, therefore, fresh ionization in the channel, so that a step streamer is formed. This explanation has been extended quantitatively by Loeb and Meek [29, 30, 97], who have estimated the probable decrease of ionization in the channel between steps in terms of recombination processes.

An alternative proposal has been put forward by Bruce [31, 32], who draws attention to the low value of the mean electric field between cloud and ground [1, 2, 3]. This field is found to be of the order of several hundred V/cm. which is much less than that required to cause ionization by collision in a uniform field. In order to account for the propagation of lightning in such low fields Bruce suggests that as the leader stroke progresses from the cloud towards ground the potential difference between the discharge track and the surrounding space causes lateral corona currents to flow. The total corona current increases with the

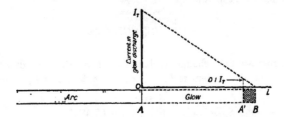

FIG. 5.7. Diagram illustrating formation of leader-stroke darts.
(I_T is the current required for glow to arc transition)

length of the channel, and it is postulated that when the current is about 1 A the character of the channel changes suddenly from a glow to an arc (see Chapter XII). The model of Fig 5.7 represents the conditions in the channel, showing how the total current flow increases to the critical value I_T. When the transition takes place the voltage gradient in the channel falls from a value of several hundreds of V/cm. to several tens of V/cm., and further, as an arc has a negative characteristic, the gradient in the arc phase will fall as the total current rises with increasing channel length.

Bruce presents experimental data in support of his theory. For example, if the streamer has started, its potential V_l relative to the surrounding space builds up at the rate of X_1 V/cm. (if the small voltage gradient X_a, necessary to maintain the arc, is neglected), where X_1 is the value of the pre-discharge electric field between cloud and ground. At distance l from the start,

$$V_l = V_0 + l(X_1 - X_a) \sim V_0 + lX_1, \qquad (5.1)$$

where V_0 is the initial voltage between the point at which the streamer originates and the surrounding space.

The corona current/cm. of channel at this point will be

$$I_c = k(V_0 + lX_1)^2, \qquad (5.2)$$

assuming that $I_c \propto V^2$ for the corona characteristic. The rate of change of electric moment μ at time t is therefore

$$\frac{d\mu}{dt} = \int_0^{vt} k(V_0+lX_1)^2 2l \, dl$$

if v is the velocity of propagation of the leader stroke. The total electric moment destroyed in time t is

$$\mu = \int_0^t \int_0^{vt} 2kl(V_0+lX_1)^2 \, dl dt. \qquad (5.3)$$

The electrostatic field change at distance r from the flash due to this change of electric moment is $E_s = \mu/r^3$, in which μ and E_s are in e.s.u. Hence

$$E_s = \frac{9\times10^{13}}{r^3} \int_0^t \int_0^{vt} 2kl(V_0+lX_1)^2 \, dl dt, \qquad (5.4)$$

l and r being in cm., v in cm /sec., V_0 and X_1 in volts and V/cm. respectively, and E_s in V/m.

Bruce, after discussing minor corrections to be made to the last formula because of branching from the channel, etc , then proceeds to solve for k, using E_s from the work of Appleton and Chapman [33] and various values of X_1 between 50 and 600 V/cm. k is found to be $\sim 10^{16}$ or 10^{17}, in agreement with the values quoted by Bruce from the early work of Watson [34] on D.C. corona, for which $k = 4\cdot2\times10^{17}$.

A criticism [35] has been expressed concerning Bruce's theory in that it is based on experiments on glow-to-arc transitions between metal electrodes for which there is evidence that electrode conditions are important (see Chapter XII) whereas the lightning mechanism must be governed by gas-dependent processes only. However, it may be, if Bruce's theory is correct, that a new concept of glow-to-arc transitions is required in which cathode mechanisms have little influence, and the experiments by Bruce [36] and by Suits [37] on various forms of discharges in hydrogen are of interest in this connexion. Nevertheless, apart from considerations of cathode effects, it is doubtful whether the experimental data on D.C. or 50 c/s A.C. arcs is directly applicable to the transient discharge plasma forming the leader stroke. There is evidence [38], for instance, that an arc subjected to an A.C. wave form of 1,000 cycles, or higher, has a positive volt-ampere characteristic, and the rapidity of the changes occurring in the leader stroke therefore make it unlikely that the channel has a negative characteristic as assumed by Bruce.

The usual speed of advance of the stepped leader stroke, which is about $1\cdot5\times10^7$ cm./sec , is closely that of the observed velocity of electrons in the minimum ionizing field in air. The much higher speed of about 5×10^9 cm./sec. which is attained by a step streamer may be explained by the presence ahead of its tip of a pre-ionized path in which appreciable ionization remains from the passage of the preceding streamer. Schonland [17] has given an expression for the velocity v of advance of a streamer along a pre-ionized path as

$$v = N^{1/3}\bar{v}d, \tag{5.5}$$

where N is the pre-existing electron density in electrons per cm.^{3} in the ionized path ahead of the tip, $\bar{v}$ cm./sec. is the electron drift-velocity in the field immediately ahead of the tip, and d cm. is the distance ahead of the tip in which the field exceeds the critical value for ionization by electron impact. Clearly values of v very much greater than the electron drift-velocity $\bar{v}$ may be attained in the presence of high values of N. Similar considerations apply in the case of the return stroke and the dart leader, but, because of the greater age of the ionized channel preceding the development of the dart leader, the speed of the latter is less than that of the return stroke or the individual steps of the initial stepped leader stroke.

Before the final explanation of the leader-stroke mechanism can be given it is probable that further experiments are required concerning the fundamental physical processes involved. Because of the random and uncontrolled nature of lightning discharges the progress of experimental work is necessarily slow, but much relevant information can be gained from studies of long sparks in the laboratory, as described in Chapter IV.

One feature of the leader-stroke development in lightning discharges to open country is that no positive leader strokes have been recorded to grow from the ground to meet the descending negative leader stroke, whereas, in the case of long sparks between a negative point and an earthed plane, positive leader strokes from the plane invariably occur. However, there is evidence that the ratio of the distance travelled by the positive leader stroke to the gap length decreases with increasing gap [99], and it is possible therefore that while positive leader strokes may develop from the ground in lightning discharges, they may be too short to be recorded under normal conditions where the camera is at a

considerable distance from the discharge. A still photograph, given by
McEachron [23], shows several streamers from the ground of lengths
about 200 cm. which may be upward-growing leader strokes.

According to Meek [30, 102] the initiation of the upward-growing
positive leader stroke is governed by the voltage gradient induced at
the earth's surface by the downward-growing negative leader stroke,
the magnitude of this gradient being inversely proportional to the
height of the charge on the leader stroke. This implies that, with increasing gap length, the negative leader has to approach relatively nearer to
the earth's surface before an upward-growing positive leader is initiated,
so that the latter travels a decreasingly smaller proportion of the gap.
Experimental evidence on sparks in the laboratory supports this view
[99]. The idea has been developed by Golde [100, 101] who has calculated
the height of the tip of the negative leader stroke when a positive leader
may be expected to develop from the earth.

The branching of a lightning discharge in open country appears to be
affected by the nature of the terrain. In flat, bare country McCann [24]
reports that discharges show little or no branching, whereas in hilly,
wooded country profuse branching is usually observed. It is probable
that in the latter case there is an increased point-discharge current,
which provides a positive-ion space-charge blanket near the earth's
surface, and encourages the development of branches from the descending negative leader stroke. Similar results have been observed in the
laboratory for a spark between a high-voltage negative point and an
earthed plate covered with a number of projecting points [38].

The development of the leader stroke is clearly an important factor
in governing the selection of the earthed object which is struck by the
lightning discharge. High buildings, which may initiate leader strokes,
are naturally more prone to be struck than the surrounding country
when thunderstorms are in the vicinity. A number of tests with models
in the laboratory have been carried out by Wagner, McCann, and
MacLane [39] who have studied the propagation of leader strokes for
various electrode arrangements which simulate on a small scale the
conditions obtaining for lightning discharges. Consideration of the
height of lightning rods and the degree of protection afforded to neighbouring objects have been studied, together with the influence of soil
resistivity and ground contours. Other recent investigations of the
protective value of lightning conductors, by means of scale models,
have been carried out by Zalesski [40], Akopian [41], Matthias [42],
Matthias and Burkardtsmaier [43], Wagner, McCann, and Lear [44].

Golde [45] has discussed the limited applicability of model tests in relation to the usual conditions obtaining during thunderstorms.

The localization of lightning discharges has been described by Goodlet [46]. It appears that the influence of soil conductivity is appreciable and that regions where there is a discontinuity in the geological formation are prone to be struck. A series of relevant experiments on sparks in the laboratory have been carried out by Stekolnikov and his associates [47 to 50]. In one of these investigations [50], a metal ball was buried beneath the flat surface of a layer of soil, and two small hills of soil were raised near the buried ball. The soil was made to form one electrode of a spark gap, and it was found that, on the application of impulse voltage to the gap, the sparks took place across the longer gap to the buried metal ball when the soil conductivity was low, in preference to the shorter gap to the hill-tops. When the soil conductivity was increased the sparks took place to the hills. A photograph obtained by Miranda [51] for a similar arrangement is shown in Fig. 5 8, Pl. 7, where sparks are seen to occur to a buried metal ball.

The return stroke

The high speed of the return stroke is accounted for by the existence of the pre-ionized channel ahead of its advancing tip, as explained in the previous section for step streamers and dart leaders. A gradual decrease in speed takes place as the return stroke proceeds up the channel, probably because of the greater age of the channel nearer the cloud.

Changes in luminosity of the main channel have been studied by Malan and Collens [15], and in Fig. 5.9 a photometric analysis of one of their photographs is given. They conclude that the main changes in luminosity, which may be associated with the current in the main channel, takes place when the tip of the return stroke reaches a branch. The luminosity of the channel between the ground and the branch then increases, while in its further progress up the channel the return stroke proceeds with diminished luminosity. The branch is illuminated to an extent comparable with the portion of the main channel between the branch and ground. In an explanation of the change of luminosity at branching points Schonland [17] has drawn attention to the relative ages of the ionized trails in the main channel and in the branches. This may be understood by reference to Fig. 5.10, which represents the time sequence of a vertical discharge with a single branch. The light lettering denotes the time, in μsec , at which the various parts of the main channel and the branch were formed The heavy lettering denotes the age of the ioniza-

tion at each point when the return streamer reaches it. For simplicity
the speed of the return stroke has been assumed to be uniform and to be
10^3 times that of the leader stroke. At the end of the branch the age of

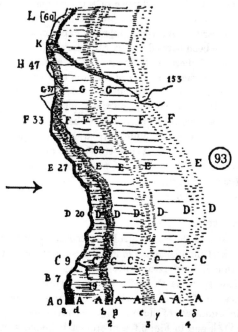

FIG 5.9 Diagrammatic analysis of the photographed
structure of a return stroke The figures at the leading
edge denote the times, in microseconds, at which the
return stroke first reaches adjacent points

the trail is little different from that at the base of the channel. The age
at E on the main channel is 8,080 μsec., whereas the age at F, which is
equidistant from the branching point A, is only 4,080 μsec. Therefore,
when the return stroke reaches A the portion travelling along the branch
develops faster than that along the main channel, and the branch is
illuminated more brightly than the portion of the main channel above
[17, 98].

By consideration of the rotating-camera photographs in conjunction
with the synchronized current oscillograms of lightning discharges to
the Empire State Building, Flowers [52, 53] has been able to obtain
approximate agreement between luminosity and current. A comparison

is made between the current oscillogram and the luminosity as determined with a micro-densitometer from the rotating photograph of the
discharge. While the general agreement between the two records is
evident, there are difficulties in establishing the calibration scale for

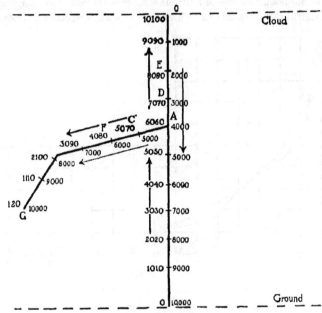

Fig 5 10   Analysis of the age of the ionized channel, for a hypothetical
discharge in a vertical direction, with one branch.

the estimation of currents from the luminosity records, as confirmed in
similar measurements by McCann [24, 54]. Some improvement may be
obtained by the use of a photo-electric electron-multiplier coupled to an
oscillograph, which records directly the variation with time of the
luminosity of lightning discharges, without the intermediate step of
a photographic record, this arrangement has been used by Meek and
Craggs [55] for the study of the light emission from spark channels
(see Chapter X).

Multiple strokes

The reason for the multiple-stroke sequence observed in lightning
discharges has been ascribed to the extension of the channel to new
charge centres within the cloud. Schonland [17] has suggested that this

extension may take place by the development of leader strokes from the new charge centres towards the existing charge centre. Alternatively, as pointed out by Bruce and Golde [56], the leader strokes may develop from the channel to the new charge centres.

Another possible cause of multiple strokes is the repeated charging of the original cloud centre from which the discharge originally developed.

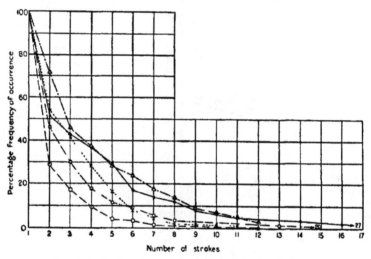

Number of strokes

FIG 5 12 Number of successive strokes in lightning discharges.

●———●	Schonland and others
⊙– – –○	McEachron.
△–––△	Stekolnikov and Valeev.
☐–··–☐	Noto.
×····×	Radio Research Board records

The points represent the proportion of flashes in which the number of component strokes is equal to, or greater than, the value shown on the abscissa-axis.

Some evidence in favour of this process is given by McCann [24], who has drawn attention to two rotating-camera photographs of multiple strokes, in one of which the time intervals between a series of eight strokes is remarkably regular, between 0 0093 and 0·0112 sec. McCann suggests, as a result of laboratory data on multiple-stroke discharges in the laboratory, that the continuous charging of the one cloud centre is the responsible factor in such instances. Further, the rate of charging, which is considered to proceed at a fairly uniform rate, is in accord with the measurements of voltage gradient at the ground.

While the component strokes of most lightning discharges take place along the same channel, exceptions are sometimes observed   A photograph obtained by Walter [57] of a discharge containing three strokes is shown in Fig. 5.11, Pl. 8. The first stroke on the left of the photograph is followed after 23 m.sec. by the second stroke, which retraces part of the path taken by the first stroke, but then develops along a

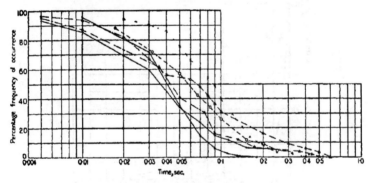

Fig 5 13   Time intervals between successive strokes.

●———●   Schonland and others
+———+   Schonland and others
○ — — ○   McEachron.
△—·—·△   Stekolnikov and Valeev.
▲—··—··▲   Stekolnikov
×··· ····×   Radio Research Board records.

The points represent the proportion of time intervals exceeding the value shown on the abscissa-axis.

separate track, the initial part of which has been formed as a branch of the first stroke   After a further 240 m sec. the third stroke develops, and follows the path of the second stroke. Similar observations are reported by Schonland, Malan, and Collens [16] who consider that changes in the lower portions of the channels used by multiple strokes are to be associated with long time intervals between the strokes, whereby the ionization existing in the channel has largely disappeared. A change in the upper portion of the channel may be attributed to the presence of a second charge-centre within the cloud.

A comparison of the number of multiple strokes in a discharge, as analysed by Bruce and Golde [56] from the results of a number of workers in different parts of the world, is given in Fig. 5.12. In exceptional records, not given in this diagram, as many as 40 strokes were observed by Larsen [10] and 42 strokes by Matthias [58]   Of 19 discharges photo-graphed by Allibone [68] 90 per cent. were composed of 2 or more strokes and the average number of strokes per discharge was about 4.

Further analyses given by Bruce and Golde [56] are given in Figs 5.13 and 5.14, and show respectively the distribution of the time intervals between successive strokes and the total duration of lightning discharges.

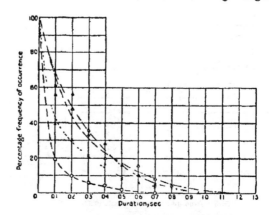

Fig. 5 14   Total duration of lightning discharges.

O— — —O   McEachron.
△—·—·—△   Stekolnikov and Valeev.
▲—·—·—▲   Stekolnikov
X ·········· X   Radio Research Board records.

The points represent the proportion of flashes in which the duration shown on the abscissa-axis is exceeded.

The average time interval between strokes is seen to be about 0·07 sec., whereas the average duration of a discharge is about 0 25 sec. Somewhat longer durations than those given in Fig. 5 14 are found by McEachron [23] for the discharges to high buildings.

Lightning discharges within the cloud

The preceding discussion has been mainly concerned with the lightning discharge to earth from the lower side of the cloud, as this is the type of flash which can most readily be observed and photographed. The visible discharge which emerges below the cloud may, however, be only part of a longer discharge track, the upper portion of which is concealed within the cloud itself and which originates at an internal charge-centre. It is probable that the lightning discharge to ground is initiated in the cloud as a 'mid-gap streamer', which develops from one end in a general downward direction towards ground, and from the other end in a general upward direction within the cloud   Only the downward-growing portion is recorded in the photographic studies, but it may be that an upward-growing portion would also be observed if the records were not obscured by the cloud itself. In the case of multiple-stroke flashes it has been suggested that nearby charge-centres may be involved in the strokes subsequent to the first so that the concealed portion of the discharge may vary both in direction and length during the progress of the discharge.

A large proportion of lightning discharges occur within the cloud itself, or may progress towards the earth without reaching it. Discharges of this latter type have been termed 'air discharges', and appear to be caused by discharges to heavy space-charge concentrations below the cloud [14, 17]. Alternatively, the development of the leader-stroke may be arrested by the collapse of the electric field, possibly because of the occurrence of other neighbouring discharges. The presence of discharges wholly within the cloud, and which, therefore, cannot directly be seen, is known by observation of the associated phenomena, such as field changes, general illumination of the cloud, and thunder. In certain types of discharge, however, part at least of the track is directly visible. Shipley [59] has attempted to classify nine different types of discharges frequently seen, of which four consist of discharges to ground and five consist of discharges within or about the cloud. Two of these latter types usually take place high up in the cumulus cloud and appear to involve localized charges of a portion of the cloud only. The other three types are shown as much longer discharges whose main direction is horizontal and which appear to traverse the major part of the cloud structure. Where these discharges occur near the lower edge of the cloud, they may be branched or non-branched

The relative proportions of cloud discharges and ground discharges are not accurately known but they appear to vary widely in different localities. Schonland [60], for storms observed at Somerset East, South Africa, found that the ratio of discharges in the cloud to discharges to ground varied from 50/1 to 5/1, with an average value of 10/1. Simpson [61], when measuring the charge on rain from thunderstorms at Simla, observed that 'most of the lightning discharges' took place within the cloud between the base and the summit. Halliday measured the field changes due to lightning discharges in Johannesburg and observed the type of discharge in 560 out of a total of 3,015 field-change records [62]. Of the observed discharges 277 were in the cloud and 283 were to ground, but an analysis of the total number of field changes would indicate that the ratio of total cloud discharges to total ground discharges was of the order of 1·7/1. The discrepancy is probably due to the greater difficulty of observing cloud discharges. According to Golde [63], Whipple and Scrase [64] observed that the ratio of cloud discharges to ground discharges at Kew was about 0·7/1.

The large differences in the observations by different workers of the ratio of cloud discharges to ground discharges are not fully understood. It is probable that they are influenced by the average height of the cloud base above ground, as with decreasing cloud height the number of discharges to ground may be expected to increase. There appears to be some correlation in the results obtained by various workers between the cloud height and the relative numbers of cloud and ground discharges [24]. However, other factors may be important, and it is anticipated that differences are to be expected in the results for thunderstorms in tropical and temperate zones, where the thermodynamical conditions producing the thundercloud vary, and where the distribution of charges in the cloud and the distance between charge-centres may differ.

McCann [65] describes a number of cloud-to-cloud discharges photo-graphed by a moving-film camera of sufficient resolution to separate the individual strokes. Some of his records show vertical or inclined dis-charges clearly involving the upper positive and the lower negative charge-centres of the cloud. The discharges are multiple in character and generally have a larger number of components, shorter time intervals between components, and a shorter total duration of the flash than is the case for discharges to ground. One photograph shows a multiple-stroke air-discharge with pronounced branching [65], and a second shows a multiple-stroke discharge originating at the top of the cloud and penetrating the cloud to reach the earth [65, 66]. Evans [67] shows a still photograph of a ground discharge which was seen by visual observa-tion to be linked to the discharge in the cloud-top above it. The photo-graph given in Fig. 5.15, Pl. 8 shows both a cloud discharge and a ground discharge, which may be associated, but there is no evidence concerning the sequence of the processes involved.

The photographic analysis of cloud discharges is not in all cases sufficiently exact to show the mechanism of their propagation, but this may be deduced from consideration of the electrical field changes produced. For 'air-discharges' Schonland, Malan, and Collens [14] showed that the discharge streamers progress in a series of steps in a similar manner to that of the first stroke of a multiple-stroke discharge but that no return strokes occur In measuring the field changes due to discharges within the cloud, Schonland, Hodges, and Collens [18] found that the usual field change from such a discharge was a simple slow rise to a final maximum, the rise carrying superimposed pulsations of the same kind as those found for the stepped leader process and with the same time interval between pulsations. There was an entire absence of the sudden change of field which occurs with a stroke to ground at the time of the development of the main return stroke. From this observation, Schonland with Hodges and Collens [18] and with Malan and Collens [16] deduced that for discharges within the cloud the mechanism was the same as for air discharges, i.e. a stepped leader process but no return stroke.

Bead and ball lightning

Bead lightning, which is observed fairly frequently, is caused by the persistence of luminosity of the discharge channel, which remains visible for longer times in certain sections than in others. Consequently the channel, before it disappears, may appear to break up into a number of detached light-sources. This subject has been discussed in a recent summary by Wolf [69]. In a description of the phenomenon, an observer has estimated the presence of twenty to thirty luminous beads remaining in the path of the discharge and spaced at about 2 ft. [70, 96]. The diameter of these beads of light was estimated at about 3 in., and their lifetime as about 0·5 sec., as compared to a duration of about 1 sec. for the preceding main discharge. The persistence of these local regions of light is witnessed by the streaks observed in rotating-camera photographs, as, for instance, in Fig. 5.11. It frequently occurs at bends in the channel, and has been explained by various authorities as an optical effect caused by the end-on view of the channel in the region of the bend, so that an effectively deeper length of ionization is in the line of sight. The persistence of the light appears to be too long to be accounted for in this manner, and it is probable that it is caused by more fundamental, as yet unexplained, processes [71, 72] (see p 414).

The evidence concerning ball lightning, sometimes termed a thunderbolt or fireball, is less definite, and many persons are sceptical as to its existence. But there is an appreciable amount of information available from the various accounts which have been given by the many observers, and the number of these independent accounts, together with their general agreement, would appear to substantiate ball lightning as a true phenomenon. Brand [73] has collected some 600 accounts of ball lightning, from about one-third of which he has been able to deduce the principal characteristics of the phenomenon. Other accounts have been investigated by Humphreys [74], who concludes that in many of the observations the phenomenon can be explained as a subjective effect, influenced by persistence of vision. The subject has been discussed more recently by Goodlet [46]. In general, ball lightning appears as a reddish luminous ball of 10 to 20 cm. diameter, which lasts for several seconds, and which sometimes, though not always, appears after an ordinary lightning discharge. It may disappear silently or with an explosion. Photographs of unusual discharges which may represent ball lightning have been given by Holzer and Workman [75]

Various theories have been put forward to account for ball lightning. Thornton [76] has suggested that it is formed largely of a mass of ozone and dissociated oxygen, produced by a normal lightning discharge, and that this reverts suddenly to molecular oxygen with explosive violence. The energy balance is calculated to be such as to account for the process. In a more recent theory, proposed by Neugebauer [78], it is shown that the stability of a gaseous sphere containing electrons and positive ions can be explained in terms of quantum mechanics, by consideration of exchange forces, if the electron density is sufficiently high, as may be the case in lightning discharge channels. The theory suggests that ball lightning is stable if

$$\tfrac{3}{2}kT < \frac{e^2nh^2}{8\pi mkT}, \tag{5.6}$$

where T is the electron temperature (probably not very different from the gas temperature) and e, n, and m are the charge, concentration, and mass of the electrons. h and k are Planck's and Boltzmann's constants. If the gas is 1 per cent. ionized at 1 atm. pressure then $T < 63°$ for stability, and it would be expected that most discharges would be unstable as this value is much lower than usual in such discharges (see Chapter X). The theory also predicts two modes of decay, the quiet and the explosive types, both of which appear to be observed.

The magnitude of lightning currents

Because of the complex nature of the lightning discharge and the limitations of the various measuring devices used to measure the stroke currents, it is certain that most of the results are to some extent incomplete and inaccurate. Sufficient data have been collected, however, to show that there is a wide range of currents involved, and the main problem is to obtain a correct statistical analysis of the frequency of distribution of the current magnitudes and wave-shapes.

The frequency of occurrence of lightning discharges as a function of their peak current is shown in Fig 5.16. A peak current of 160,000 A has been recorded by a fulchronograph [79] and a still larger current of 220,000 A has been deduced from records given by magnetic links

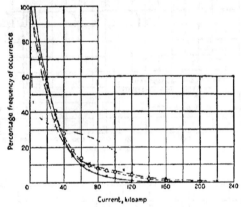

Fig. 5 16. Maximum currents observed in lightning discharges.

● ···· ··● Grunewald.
⊙ ····· ⊙ Lewis and Foust.
✕ —·—· ✕ Stekolnikov and Landon.
· · · · · · McEachron.

The points represent the proportion of current values exceeding that shown on the abscissa-axis.

installed on transmission-line towers. The chances of such large currents are remote, and a peak value of about 20,000 A is the average observed. These peak currents are of short duration, as shown in the oscillograms given in Fig. 5.17. Some investigators [14, 80] have found that the first stroke of the discharge is usually strongest and that subsequent strokes show a general tendency to become progressively weaker, though individual strokes in a series may contradict the general trend. On the other hand, McEachron [23] has recorded multiple-stroke discharges to the Empire State Building in which the largest peak-current value occurs after a number of preceding weaker strokes, and a similar result is shown in certain fulchronograph records [79].

The question of the polarity of the discharge current has long been a subject of controversy, and has been linked, naturally, with the problem of the distribution of charges in the thundercloud. The measurement of the discharge current by direct and indirect methods has to some extent clarified the issue, and it is now known that currents conveying

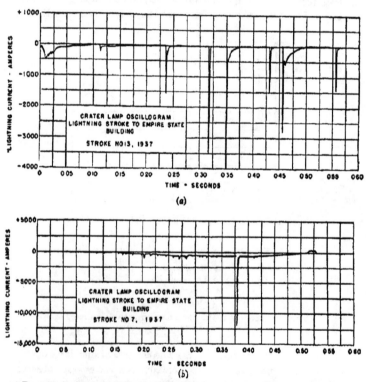

FIG 5 17 Oscillograms, of the variations of current with time in two separate lightning discharges, as recorded with a crater lamp oscillograph

negative charge to ground predominate. According to Bruce and Golde [56] a distinction is observed between discharges in the tropics and those in temperate zones. For the former it appears that the number of negative discharges exceeds the number of positive discharges in the ratio of 17 to 1, whereas in temperate zones the ratio is only 3 to 1. Out of forty-nine lightning discharges to the Empire State Building, McEachron [23] observed positive currents in only eight cases The positive current pulse is invariably preceded or followed by negative current pulses during the same discharge.

In addition to the short-time current pulses, low amplitude currents of long duration are usually observed in the lightning discharge. These currents range in value from tens of amperes to a few thousand amperes, and the term 'continuing stroke' has been used to denote their presence [23]. In some cases the continuing stroke persists for the duration of the discharge, so that current flow is maintained at a comparatively low level, with superimposed high current pulses. In other cases the current-flow ceases and is followed by isolated current pulses of higher magnitude. These observations were made for discharges to a high structure, but Wagner, McCann, and Beck [79] have shown that similar results may be obtained for discharges to transmission lines, and the persistence of luminosity in the discharge channel, as shown in rotating-camera photographs [15] would indicate the probable existence of continuous currents in the discharge to open country.

The data accumulated from various sources show that the usual wave-shape of the current pulses is such that the time taken to the peak value may be as short as 1 μsec. [23] or as long as 19 μsec. [79] with an average value of about 6 μsec. [56] The time taken for the current to fall to half the peak value ranges between about 7 to 115 μsec. with an average value in the region of 24 μsec. [23].

The foregoing discussion of the amplitude and the duration of the various components of the lightning-discharge current leads directly to a consideration of the amount of charge neutralized by each of these components, and thence directly to the total charge neutralized in a multiple-stroke discharge. The short-time pulses, of large peak current, generally carry only a small proportion of the total charge conveyed to earth. In measurements by McEachron [23] the charge for each current-pulse varies from less than 0·1 coulomb up to 5 coulombs. Only 50 per cent. of the individual strokes in a discharge are associated with a charge of 0·13 coulomb or more, and only 5 per cent. with more than 1·5 coulombs. On the other hand, the analysis given in Fig. 5 18 of the total charge involved shows that at least 50 per cent. of the discharges convey a charge of 25 coulombs or more. In several of the discharges a reversal of current flow took place, and an analysis of the amounts of charge of both polarities which are conveyed is included in Fig. 5.18, where it will be observed that the negative charge predominates. The maximum charge observed in McEachron's measurements was 164 coulombs, of which by far the greater proportion was conveyed to earth by a continuing low current, as shown in the oscillogram given in Fig. 5.17 (*b*). Similar observations, in which the greater proportion of charge neutralization takes place during the continuing low-current phase, are reported by Wagner, McCann, and Beck [79].

No explanation has been given for the mechanism governing the high

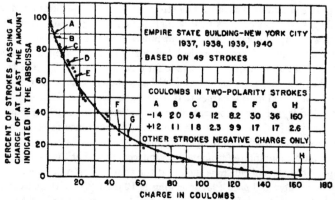

FIG. 5.18. Total charge neutralized by lightning discharges.

currents observed to flow in lightning discharges. In general, the ground forms the anode end of the discharge while the cathode end is in the atmosphere. The high current flowing in the main stroke cannot therefore be explained in the normal manner as for a discharge between metal electrodes, where, once the gap has been bridged by the leader stroke, the discharge is maintained by electrons from the cathode. A possible explanation is that during the growth of the leader stroke the voltage gradient along the leader channel is small compared with that ahead of the leader stroke, and consequently a high proportion of the cloud-ground voltage is available between the head of the leader stroke and ground to cause the propagation of the leader stroke. When the leader stroke reaches ground the whole of the cloud-ground voltage is suddenly impressed along the leader channel and produces a rapid multiplication of ionization in the channel with a consequent increase in current. A drop in the cloud-ground voltage then occurs and hence the current is in the form of a pulse of short duration, as observed experimentally. In the case of the laboratory spark a similar process may be expected, except that the current is then limited by the external circuit, whereas the lightning current may be considered to be produced by the discharge of a non-inductive capacitance through the channel without the introduction of any series resistance other than that effectively offered by the channel itself.

Additional information concerning the electrical characteristics of lightning discharges has been obtained from the study of the electric field changes produced by lightning [19, 33, 56, 60, 64, 77, 81 to 95].

REFERENCES QUOTED IN CHAPTER V

1. G. C. SIMPSON and F. J. SCRASE, *Proc. Roy. Soc.* A, **161** (1937), 309.
2. G. C SIMPSON and G D. ROBINSON, ibid. **177** (1941), 281.
3. G C. SIMPSON, *Quart. J. Roy. Met. Soc.* **68** (1942), 1.
4. J A. FLEMING (editor), *Terrestrial Magnetism and Electricity*, McGraw-Hill, 1939.
5 W J. HUMPHREYS, *Physics of the Air*, McGraw-Hill, 1940
6 B. F, J SCHONLAND, *Atmospheric Electricity*, Methuen, 1932.
7. J. A CHALMERS, *Atmospheric Electricity*, Oxford University Press, 1949.
8. H. H. HOFFERT, *Proc. Phys. Soc.* **10** (1890), 176.
9. B WALTER, *Ann. Phys.* **10** (1903), 393
10. A. LARSEN, *Annual Report Smithsonian Institute*, p 119, 1905.
11 C V. BOYS, *Nature*, **118** (1926), 749
12 C. V. BOYS, ibid. **124** (1929), 54
13. B F. J. SCHONLAND and H. COLLENS, *Proc Roy. Soc* A, **143** (1934), 654.
14. B F J. SCHONLAND, D J. MALAN, and H. COLLENS, ibid. **152** (1935), 595.
15 D J MALAN and H COLLENS, ibid **162** (1937), 175
16. B. F. J. SCHONLAND, D. J. MALAN, and H. COLLENS, ibid. **168** (1938), 455.
17. B F J. SCHONLAND, ibid. **164** (1938), 132.
18. B. F. J. SCHONLAND, D. B. HODGES, and H. COLLENS, ibid. **166** (1938), 56
19 D. J. MALAN and B F. J. SCHONLAND, ibid. **191** (1947), 485.
20. E. J. WORKMAN, J. W BEAMS, and L. B SNODDY, *Physics*, **7** (1936), 375.
21 J C ALLBRIGHT, *J Appl. Phys.* **8** (1937), 313.
22. R. E. HOLZER, E. J. WORKMAN, and L. B. SNODDY, ibid. **9** (1938), 134.
23 K B McEACHRON, *J. Franklin Inst* **227** (1939), 149
24 G. D McCANN, *Trans. Amer. Inst. Elect. Engrs.* **63** (1944), 1157.
25. J. H HAGENGUTH, ibid. **68** (1949), 1036.
26. I S. STEKOLNIKOV, *Lightning*, U S S R. Academy of Publications, Moscow, 1940.
27. B. WALTER, *Ann. Phys. Lpz* **22** (1935), 421.
28. B F. J. SCHONLAND, *Nature*, **141** (1938), 115
29. J. M MEEK, *Phys Rev* **55** (1939,) 972
30 L B LOEB and J. M MEEK, *The Mechanism of the Electric Spark*, Stanford University Press, 1941
31. C. E. R. BRUCE, *Nature*, **147** (1941), 805.
32. C. E R BRUCE, *Proc. Roy. Soc.* A, **183** (1944), 228.
33 E. V APPLETON and F W. CHAPMAN, ibid **158** (1937), 1.
34. E. A. WATSON, *Electrician*, **64** (1910), 707.
35. J M. MEEK, *Nature*, **148** (1941), 437.

36. C. E R. BRUCE, ibid. **161** (1948), 521.

37 C. G. SUITS, *J Appl. Phys* **10** (1939), 648.

38. B. F. J. SCHONLAND and T. E. ALLIBONE, *Nature*, **128**, (1931), 794

39. C. F. WAGNER, G. D. McCANN, and G. L. MACLANE, *Trans. Amer. Inst. Elect Engrs* **60** (1941), 313

40. A M ZALESSKI, *Conf Int. Gr. Rés Élect* , Paper No. 317, 1935.

41. A. AKOPIAN, ibid , Paper No. 328, 1937.

42. A. MATTHIAS, *Elektrotech. Z.* **58** (1937), 881, 928, 973.

43. A. MATTHIAS and W. BURKARDTSMAIER, ibid. **60** (1939), 681, 720.

44 C. F. WAGNER, G. D. McCANN, and C. M. LEAR, *Trans Amer. Inst. Elect. Engrs.* **61** (1942), 96.

45. R. H GOLDE, *J Instn. Elect Engrs.* **88** (II), (1941), 67.

46 B. L. GOODLET, ibid. **81** (1937), 1.

47. I. S. STEKOLNIKOV and V. V. YAVORSKY, *Elektrichestvo*, **57** (section 8) (1936), 13

48 I. S. STEKOLNIKOV and A P BELIAKOV, ibid (section 22) (1936), 16.

49. A. P. BELIAKOV and B HANOV, ibid. (section 22) (1936), 20.

50 I. S. STEKOLNIKOV, *Conf Int Gr Rés. Élect.*, Paper No 327, 1936

51 F. J. MIRANDA, *J. Instn. Elect. Engrs* **81** (1937), 46.

52. J. W FLOWERS, *Phys. Rev* **64** (1943), 225.

53 J. W FLOWERS, *Gen Elect. Rev* **47** (1944), 9

54. G D. McCANN, *J Instn Elect Engrs* **89** (1942), 646.

55. J. M MEEK and J. D. CRAGGS, *Nature*, **152** (1943), 538

56 C. E R BRUCE and R H GOLDE, *J Instn. Elect. Engrs* **88** (II) (1942), 487

57. B. WALTER, *Ann. Phys Lpz* **25** (1936), 124

58. A. MATTHIAS, *Elektrotech Z* **50** (1929), 1469

59. J. F. SHIPLEY, *Distrib Elect* **12** (1940), 413

60. B. F. J. SCHONLAND, *Proc Roy. Soc.* A, **118** (1928), 233.

61. G. C. SIMPSON, *Philos Trans.* A, **209** (1909), 412.

62. E C. HALLIDAY, *Proc. Roy Soc.* A, **138** (1932), 205.

63. R H. GOLDE, *Brit. Elect Allied Ind Res. Ass Report*, No. S/T 49 (1945), 10.

64. F. J. W. WHIPPLE and F J. SCRASE, *Meteor. Off Geophys. Mem* **7** No 68, 1936

65. G D. McCANN, Paper presented to Inst. of Aeronautical Sciences (America), December 1942.

66. G. D McCANN and D. E MORGAN, *Trans. Amer. Inst. Elect Engrs.* **62** (1943), 345.

67. F. A. EVANS, ibid **427**

68. T E. ALLIBONE, *Quart. J Roy. Met. Soc.* **70** (1944), 161.

69. F. WOLF, *Naturwiss.* **31** (1943), 215.

70 D. G. BEADLE, *Nature*, **137** (1936), 112

71. J. D CRAGGS and J. M. MEEK, *Proc. Roy Soc.* A, **186** (1946), 241

72. J. D. CRAGGS, W. HOPWOOD, and J. M. MEEK, *J. Appl. Phys.* **18** (1947), 919.

73. W. BRAND, *Der Kugelblitz*, H. Grand, Hamburg, 1923.

74. W. J. HUMPHREYS, *Proc. Amer. Phil Soc.* **76** (1936), 613.

75. R. E HOLZER and E. J WORKMAN, *J. Appl. Phys.* **10** (1939), 659.

76. W. M. THORNTON, *Phil. Mag* **21** (1911), 630.

77 E. J. WORKMAN and R. E. HOLZER, *Rev. Sci. Instrum.* **10** (1939), 160,

78. T NEUGEBAUER, *Z. Phys.* **106** (1937), 474.

79 C. F WAGNER, G. D. McCANN, and E. BECK, *Trans Amer. Inst Elect Engrs.* **60** (1941), 1222.

80. I. S STEKOLNIKOV and C. VALEEV, *Elektrichestvo*, **59** (section 1) (1938), 11.

81. J M MEEK and F. R PERRY, *Rep. Phys. Soc. Progr. Phys.* **10** (1946), 314.

82. T W. WORMELL, *Philos. Trans.* A, **238** (1939), 249

83 C. T. R. WILSON, ibid. **221** (1921), 73.

84 B. F. J. SCHONLAND and J CRAIB, *Proc Roy. Soc.* A, **114** (1927), 229.

85. F. R. PERRY, G H WEBSTER, and P. W. BAGULEY, *J. Inst. Elect. Engrs.* **89** (II) (1942), 185.

86. T. W WORMELL, *Proc. Roy. Soc.* A, **127** (1930), 567

87. R. A. WATSON WATT, F. J. HERD, and F. E. LUTKIN, ibid. **162** (1937), 267.

88 H. NORINDER, *J Franklin Inst* **220** (1935), 63.

89 H. NORINDER, *Proc. Phys. Soc.* **49** (1937), 364

90. F. E. LUTKIN, *J Inst. Elect. Engrs.* **82** (1938), 289.

91. F E LUTKIN, *Quart J. Roy. Met. Soc.* **67** (1941), 347

92. B F. J SCHONLAND, J. S ELDER, D B HODGES, W. E. PHILLIPS, and J. W VAN WYK, *Proc. Roy Soc.* A, **176** (1940), 180.

93 F. E LUTKIN, ibid **171** (1939), 285.

94. T. H. LABY, J. J. McNEILL, F G. NICHOLLS, and A F. B. NICKSON, ibid. **174** (1940), 145.

95 H NORINDER, *J Franklin Inst* **244** (1947), 109, 167

96. F. SCHEMINSKY and F. WOLF, *S. B. Akad. Wiss. Wien*, **156** (1948), 1.

97. L. B. LOEB, *J. Franklin Inst.* **246** (1948), 123.

98 B. WALTER, *Ann Phys Lpz* **6** (1947), 65.

99 T. E. ALLIBONE and J. M. MEEK, *Proc. Roy. Soc.* A, **166** (1938), 97, **169** (1938), 246.

100 R. H GOLDE, *Nature*, **160** (1947), 395

101. R. H. GOLDE, *Trans Amer. Inst Elect Engrs* **64** (1945), 902.

102 J M MEEK, *J Inst Elect. Engrs.* **89** (I) (1942), 335.

THEORY OF THE SPARK MECHANISM

THE Townsend criterion for the formation of a spark

$$\gamma e^{\alpha d} = 1 \qquad\qquad (6.1)$$

has been discussed in detail in Chapter II, where it is shown to be in general agreement with experiment for the breakdown of gases at reduced pressure. On the basis of a secondary mechanism dependent on positive ion bombardment of the cathode a formative time corresponding to the time of movement of positive ions across the gap is to be expected. This is observed for gaps at low gas pressures, and Schade [1] has deduced an expression, based on the Townsend theory of the spark, which relates the formative time with the voltage applied to the gap and the externally produced photo-current at the cathode. Much higher speeds of formation are recorded in the oscillographic studies of the development of long sparks at atmospheric pressure, and are of the order of those corresponding to electron movement across the gap (see Chapter IV). These short formative times occur when the gap is subjected to impulse voltages of only 2 or 3 per cent. above the D.C. breakdown voltage, but recent work [62, 87] shows that with smaller overvoltages longer formative times can occur, possibly corresponding to the time of transit of positive ions across the gap (see p. 279). Various modifications of the Townsend theory have been put forward by different investigators to account for the observed high speeds of formation of sparks in gaps subjected to small overvoltages. Loeb [2] considered the possibility of a series of electron avalanches forming in line across the gap. von Hippel and Franck [3] calculated the general redistribution of the field in a uniform gap by the positive space charge developed by the passage of successive electron avalanches, and thereby showed that the ionization in the later avalanches is enhanced, and that the breakdown of the gap may then occur in a time depending on the speed of electrons across the gap. Further calculations extending these proposals have been made by Schumann [4], Sammer [5], Kapzov [6], and Rogowski [7].

A further difficulty in the interpretation of the mechanism of breakdown of long gaps in gases at atmospheric pressure on the basis of the Townsend theory is that the impulse breakdown voltage of gaps between a positive point and a negative plane appears to be independent of the cathode material Similarly a cathode effect can hardly be invoked to explain the growth of a lightning discharge from a cloud. Such experimental results tend to throw doubt on a breakdown mechanism involving a cathode-dependent γ. The branched and irregular growth of sparks in long gaps is also difficult to reconcile with the original Townsend theory.

This chapter is concerned mainly with the streamer theory of the spark which has been developed from about 1939 onwards. Because of the lack of adequate data concerning photoionization and other fundamental processes the exact mechanism of streamer propagation is still by no means fully understood. Also many of the phenomena observed in spark breakdown, as described in Chapter IV, have not been explained as yet on the basis of the streamer or any other theory

The streamer theory of the spark

As a consequence of the many experimental observations on spark development, including principally the cloud-chamber results of Raether, as described in Chapter IV, a new theory of the spark was proposed in 1940 by Meek [8] and independently by Raether [9]. This theory has been elaborated by Loeb, Meek, Raether, and others [10 to 25] and is now generally referred to as the streamer theory of the spark. It is based on considerations of individual electron avalanches, the transition from an avalanche into a streamer, and the mechanism of advance of streamers. The theory involves ionization processes dependent on the gas only, including ionization by electron collisions according to the Townsend α-mechanism, photoionization, and space-charge field effects caused by avalanches and streamers.

The density of ionization in a single electron avalanche has been considered by Slepian [26], who suggested that a spark might develop when the ionization attains a density sufficient to cause thermal ionization of the gas, but it appears unlikely that this mechanism could be effective in the short time involved in spark formation (see the discussion of accommodation times on pp. 52 and 417). Slepian has also drawn attention to the space-charge fields produced in the track of an individual electron avalanche and its effect on the growth of the avalanche. In the streamer theory [8, 10] attention is again drawn to the space-charge field caused by an electron avalanche, the field being calculated in the following approximate manner.

Consider the application of a voltage gradient of E volts per cm. across a gap of length d cm. between parallel plane electrodes in a gas at a pressure of p mm. Hg. If the ratio E/p is sufficiently high, an electron leaving the cathode will ionize the gas molecules, and the additional electrons so formed will also be accelerated in the applied field and cause further ionization. When the original electron has moved a distance x in the direction of the applied field, the number of electrons created is $e^{\alpha x}$. The process is rapidly cumulative, and is appropriately termed an electron avalanche. In a field of the magnitude required to cause breakdown, the electrons travel at a speed of the order of 2×10^7 cm./sec., while the positive ions from which the electrons have been detached have a speed of about 2×10^5 cm /sec The positive ions may therefore be considered stationary in comparison with the more rapidly moving electrons, and the avalanche develops across the gap as a cloud of electrons behind which is left a positive ion space charge, in the manner indicated in Fig. 6.1.

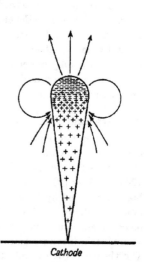

Fig. 6.1. Representation of the distribution of electrons and positive ions in an electron avalanche The arrows denote the direction of the resultant electric field surrounding the avalanche.

The number of ion pairs formed after an avalanche has developed for various distances across a 1-cm. gap in air at atmospheric pressure is shown in Table 6 1 The voltage gradient between the plates is that corresponding to spark breakdown, for which $\alpha = 18\cdot4$ from Sanders's values [27] About five times as many ion pairs are produced in the final 1 mm. of travel of the avalanche as in the preceding 9 mm.

The space charge produced by the electron avalanche produces a distortion of the field in the gap, as shown in Fig. 6.1. The distortion is greatest in the region of the head of the avalanche, where the ion density reaches its highest value. The space-charge field E_r augments the externally applied field E, and also creates a field in the direction radial to the axis. If it is assumed that $E_r = KE$ then the space-charge field increases the magnitude of the field along the axis to a maximum value of $(1+K)E$.

When the avalanche has crossed the gap, the electrons are swept into the anode, the positive ions remaining in a cone-shaped volume

TABLE 6.1

Distance in cm. travelled by avalanche	Number, e^{ax}, of ion pairs produced
0 4	$1\ 6 \times 10^3$
0 6	$6 \cdot 3 \times 10^4$
0 8	$2\ 5 \times 10^5$
0 9	$1\ 6 \times 10^7$
1 0	$1\ 0 \times 10^8$

extending across the gap, as shown in Fig. 6 2 (a). The ion density is relatively low except in the region near the anode, and therefore the presence of the positive ions does not in itself constitute breakdown of the gap. However, in the gas surrounding the avalanche photo-electrons are produced by photons emitted from the densely ionized gas constituting the avalanche stem. These electrons initiate auxiliary avalanches which, if the space-charge field developed by the main avalanche is of the order of the external field, will be directed towards the stem of the main avalanche. The greatest multiplication in these auxiliary avalanches will occur along the axis of the main avalanche where the space-charge field supplements the external field. Positive ions left behind by these avalanches effectively lengthen and intensify the space charge of the main avalanche in the direction of the cathode, and the process develops as a self-propagating streamer, shown in Fig. 6 2 (b). The streamer proceeds across the gap to form a conducting filament of highly ionized gas between the electrodes. This filament constitutes the initial stage of the spark channel through which the external circuit discharges (see Chapter X).

When a voltage gradient is applied to the gap in excess of the minimum breakdown value, the space-charge field developed by the avalanche attains a value of the order of the external field before the avalanche reaches the anode. In this case mid-gap streamers may be expected, and are, in fact, observed [11, 12].

The transition from an electron avalanche into a streamer is considered to occur when the radial field E_r produced by the positive ions at the head of the avalanche is of the order of the externally applied field E. Unless this is so, there will be no appreciable enhancement of ionization in the region of the avalanche or diversion of subsidiary electron avalanches to the main avalanche.

In order to calculate the approximate magnitude of the space-charge field it is assumed that the positive ions are contained in a spherical volume of radius r at the head of the avalanche. The field E_r produced by this space charge, at the radius r, is given by

$$E_r = \tfrac{4}{3}\pi r N \epsilon, \tag{6.2}$$

where N is the ion density and ϵ is the electric charge. In a distance dx

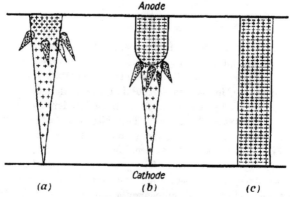

Anode

Cathode

(a) (b) (c)

FIG. 6.2. Transition from an electron avalanche to a streamer and the subsequent growth of a streamer across the gap.

at the end of a path x the number of ion pairs produced is $\alpha e^{\alpha x}\, dx$, and therefore

$$N = \frac{\alpha e^{\alpha x}}{\pi r^2}, \tag{6.3}$$

so that

$$E_r = \frac{4\epsilon \alpha e^{\alpha x}}{3r}. \tag{6.4}$$

The radius r is that of the avalanche after it has travelled a distance x and is given by the diffusion expression

$$r = \sqrt{(2DT)}. \tag{6.5}$$

Consequently

$$E_r = \frac{4\epsilon \alpha e^{\alpha x}}{3\sqrt{\{2D(x/v)\}}}$$

$$= \frac{4\epsilon \alpha e^{\alpha x}}{3\sqrt{\{(2D/k)(x/E)\}}}, \tag{6.6}$$

where v is the velocity of the avalanche and therefore of electrons in the applied field E, and k is the electron mobility. From consideration of the ratio D/k for electrons, Loeb and Meek [10] proceed to show that it is possible to obtain the following expression for E_r in the case of avalanches in air

$$E_r = 5 \cdot 27 \times 10^{-7} \frac{\alpha e^{\alpha x}}{(x/p)^{\frac{1}{2}}} \text{ V/cm}. \tag{6.7}$$

The criterion for onset of a streamer is that

$$E_r = KE, \tag{6.8}$$

where $K \sim 1$ The minimum breakdown voltage of a gap of length d cm in air at pressure p can then be calculated. A value of E is chosen and the corresponding value of α, as determined from the $\alpha/p = f(E/p)$ curves, is inserted in equation (6.7) If the resultant value of E_r is less than E, a higher value of E is chosen and the calculation is repeated until a value of E is obtained for which E_r and E are approximately equal. This value of E is then the breakdown voltage gradient for the gap.

If, for example, a gap of 1 cm. in air at 760 mm Hg is considered, Table 6.2 shows the values for α and E_r at various values of E.

TABLE 6.2

E V/cm	.	31,000	31,500	32,000	32,500
α		14 6	16 0	17·6	19·2
E_r V/cm.		480	2,100	11,500	62,000

Because of the exponential character of the ion multiplication in the gap the calculated breakdown voltage gradient, at which E_r is equal to E, is defined within fairly narrow limits for appreciable variations in the values of certain of the quantities assumed. From Table 6.2 the calculated breakdown voltage is about 32,200 V as compared with the measured value of 31,600 V.

In the above calculation it is assumed that the streamer formation occurs according to equation (6.8) when $K = 1$, as in Meek's original paper. In later work [13 to 17] the exact value to be assigned to K has been discussed in some detail and values down to 0·1 or less have been considered as adequate The difference caused in the calculated breakdown voltage by a change in K from 1 to 0·1 is not great, the value varying only from 32·2 to 31 6 kV for the breakdown of a 1-cm. gap. The prime consideration is that K should be of a value approaching the order of 1. The matter has been considered by Hopwood [17], who concludes that for avalanches approaching the anode K is never likely to be greater than about 0·5, as space-charge conditions in the avalanche would then be such as to inhibit further advance.

TABLE 6.3

Gap d cm.	Calculated breakdown voltages at 760 mm. Hg				Measured breakdown voltages
	E/p V/cm /mm. Hg	αd	N per cm^3	Voltage	
0 1	68 4	15 7	$1\,9 \times 10^{13}$	5,190	4,600
0 5	48 1	17 7	$6\,4 \times 10^{12}$	18,250	17,100
1·0	42 4	18 6	$3\,7 \times 10^{12}$	32,200	31,600
2 5	37 0	19 7	$2\,3 \times 10^{12}$	70,500	73,000
5 0	34 6	20 7	$1\,5 \times 10^{12}$	132,000	138,000
10 0	32 8	21 5	$8\cdot8 \times 10^{11}$	240,000	265,000
15 0	31 8	22 0	$7\,2 \times 10^{11}$	363,000	386,000
20 0	31 2	22 4	$5\,6 \times 10^{11}$	474,000	510,000

A comparison between the calculated and the measured breakdown voltages is given in Table 6.3 for several gaps in air at atmospheric pressure [10]. The table also includes the calculated values of E/p, αd, and N, the ion density at the head of the avalanche when it reaches the anode. For short gaps the calculated breakdown voltages are higher than the measured values and the deviation increases with the decreasing gap and therefore with increasing E/p The deviation becomes appreciable under conditions where E/p is greater than about 60 V/cm /mm. Hg, for which values of the secondary ionization coefficient γ become detectable (see p. 72). Breakdown at the shorter gaps may then be expected to occur by the Townsend mechanism rather than by the streamer mechanism [10].

The same trend is observed if the breakdown voltage of a given gap is calculated for various gas pressures, the calculated breakdown voltages exceeding the measured values to an increasing extent with decreasing gas pressure. However, it has been pointed out by Fisher [14, 15] and by Loeb [13] that the departure of the calculated breakdown curve from the measured breakdown curve occurs at a lower value of the product pd when the curve is calculated for a given gap with decreasing gas pressure than for a given pressure with decreasing gap. In these calculations by Fisher [14, 15] values of 0·1 or less for K have been used.

The calculated breakdown voltage falls below the measured value to an increasing extent as the gap length is increased above about 10 cm., so that in the fields required to cause breakdown of the longer gaps it appears that the avalanche-streamer transition occurs before the avalanche reaches the anode. A possible reason for this effect, suggested by Loeb and Meek [10], is that not only must the criterion that $E_r = E$ be satisfied, but also the ion density in the avalanche head must not be less than a certain value if photo-electrons are to be produced in sufficient numbers around the avalanche to initiate secondary avalanches. A tentative value for this critical minimum value of N is given as $N \sim 7 \times 10^{11}$ ions/cm.^{3} Reference to Table 6.3 shows that the value of the calculated breakdown field for a 20-cm. gap, on the basis of space-charge field considerations alone, only produces a value of $N = 5\cdot6 \times 10^{11}$ ions/cm.^{3} in the avalanche, and it is suggested that the field has therefore to be increased to produce a higher value of N if a streamer is to be formed to produce breakdown. No particular reason is given for the arbitrary choice of 7×10^{11} ions/cm.^{3} for N, and it is probable that as the deviation between calculation and measurement is appreciable for a gap of 10 cm., for which $N = 8\cdot8 \times 10^{11}$ ions/cm.^{3}, the latter value for N might equally well have been chosen The subject has been discussed further by Loeb and his associates [19, 21, 31] and also by Hopwood [17], and is considered again on p 268. If in fact there is such an arbitrary lower limit to N, then, as pointed out by Loeb and Meek [10], the slope of the curve relating breakdown voltage with gap length should decrease with increasing gap until a gap is reached for which the lower value of N obtains. For longer gaps the slope should increase to a constant value. An unsuccessful attempt to record this effect has been made by Fisher [15], who has measured breakdown voltages of gaps up to 2 cm. in air at pressures up to 2 atm., the maximum voltage used being 60 kV. The results were influenced at the longer gaps by field distortion caused largely by the walls of the discharge chamber. Other measurements [28, 29, 30] of breakdown in uniform fields, for gaps up to 16 cm., again fail to show any such effect (see p. 292).

Calculations of the breakdown voltages of different gaps in air at different pressures, such that the product of gap length and gas pressure is constant, show deviations from Paschen's law. Some results as calculated by Miller and Loeb [31] for the product $pd = 2,627$ mm. Hg × cm. are given in Table 6.4, where they are compared with experimental values [32]. The deviations from equality in the calculated breakdown voltages are appreciably smaller than those observed experimentally. This is accounted for by Loeb [18], who points out that the Meek equation neglects photon production and absorption in the gas If such factors are considered, the pressure dependence of the absorption coefficient for ionizing photons will have the effect of emphasizing pressure variations in the sparking equation to a still higher degree than does the Meek criterion. Thus departures from Paschen's law should be even greater than those predicted by the original Meek criterion. However, it may also be that the measured departure from Paschen's law at high gas pressure, as shown in Table 6.4, can be accounted for partly in terms of the pre-breakdown currents flowing in the gap and attendant space-charge field distortions (see p. 298).

TABLE 6.4

Gas pressure p mm Hg			775 7	10,084
Gap length d cm.		.	3 378	0 264
Measured breakdown voltage kV			95	81
Calculated breakdown voltage kV			94 3	93 4

The fact that the streamer theory enables the breakdown voltage of a uniform field in air to be calculated in rough agreement with the measured value is not considered to be a justification of the theory. Greater emphasis is placed on the general explanation given by the streamer theory of the physical processes leading to spark breakdown, and in particular the fact that it is based on gas mechanisms only, without the introduction of secondary cathode effects. This is in accordance with experimental observations, which show the decreasing influence of the cathode material on the sparking voltage as the gap length and gas pressure are increased. The gas ionization mechanisms involved in the streamer theory are ionization by electrons and photoionization, the latter replacing ionization of the gas by collision of positive ions, as envisaged by Townsend but which appears to be improbable (see p. 286).

It is important to bear in mind that many of the assumptions made in the derivation by Meek of the avalanche field are only approximate, and this in itself should preclude too great attention being paid to the degree of agreement between calculated and measured breakdown voltages. For instance the values taken for the avalanche diameters in these calculations are based on Raether's cloud-chamber measurements (see p. 181), and there is some variation in these values according to the expansion ratio used. The effect of the necessary condensable vapour may also be important. Further, Meek assumes the space charge to be concentrated in a sphere, which is clearly a crude approximation. Later investigators [23, 24, 25] have attempted more rigorous calculations of the distribution of space charge in the avalanche (see p. 275).

The approach made by Raether [11, 12] to the problem of avalanche-streamer transition is essentially similar to that by Loeb and Meek, but certain differences arise in the calculation of the space-charge field produced by the avalanche. The radius of the avalanche is calculated from considerations of diffusion processes to be

$$\bar{r} = (3Dt)^{\frac{1}{2}} = \left(\frac{9}{2}\frac{Ux}{E}\right)^{\frac{1}{2}},\qquad(6.9)$$

where x cm. is the distance travelled by the avalanche in the field E V/cm. during the time t sec., and U is the thermal energy of the electrons in eV. D is the diffusion coefficient for electrons, but it does not seem certain, in view of the presence of positive ions, that radial diffusion rates in avalanches are determined only by the value of D. The value to be taken for U is not known theoretically, but from a comparison between the experimental results for $\bar{r}$ and those expected from (6 9) Riemann [22] has deduced the values of U, denoted by U_b, required to give agreement. The values of U_b for various gases are listed in Table 6 5. The column headed U_u, in the same table, gives the values of the thermal energies as calculated from the observed speeds of advance of the avalanches. In the latter case the speed of the avalanche is assumed to be the same as the drift speed u of electrons in the applied field. Then, as

$$u = \frac{\epsilon E \lambda}{m v}, \qquad (6.10)$$

the thermal energy of the electrons in eV is given by

$$\left. \begin{aligned} \epsilon U_u &= \tfrac{1}{2} m v^2 \\ U_u &= \frac{\epsilon}{m} \frac{\lambda^2 E^2}{2 u^2} \end{aligned} \right\}. \qquad (6.11)$$

The values of U_u are appreciably higher than those of U_b, and therefore give unsatisfactory results for the avalanche radius if substituted in equation (6.9). The divergence between U_u and U_b may be attributed partly to the fact that equation (6.9) for the radius is inexact, and also to the oversimplified relation given by equation (6.11) and used to determine U_u. The latter relation assumes, for instance, that collisions are inelastic, whereas many more elastic than inelastic collisions will occur. This is pointed out by Raether [12] and by Riemann [22], who calculate the proportion f of the energy transferred by an electron per collision with results as shown in Table 6.5. More complete data on λ and f are given by Healey and Reed [33].

In order to obtain an approximate value for the field at the head of the avalanche Raether [11, 12] assumes the positive ions to be contained in a sphere of radius $r = (3Dt)^{\frac{1}{2}}$. The electric field at the sphere surface is

$$E_r = \frac{\epsilon e^{\alpha x}}{r^2}. \qquad (6.12)$$

The transition from an avalanche to a streamer occurs when $E_r = KE$, where $K \sim 1$, and hence

$$KE = \frac{\epsilon e^{\alpha x_c}}{r^2} = \frac{\epsilon e^{\alpha x_c}}{(9/2)(Ux_c/E)},$$ (6.13)

where x_c is the distance travelled by the avalanche.

TABLE 6.5

Gas	p mm. Hg at $0°$ C.	λ cm at $0°$ C, 1 mm. Hg ($\times 100$)	α/p ($\times 100$)	U_u eV	U_b eV	f ($\times 100$)
CO_2	285	1 71	2·4	1·1	0 5	1 5
O_2	290	2 78	1 28	3 0	1 0–1 5	5
Air	285	2 62	0 93	3 9	0 5–1·0	3
N_2	280	2 53	0 71	5 7		
	143		0 86	5 6	1 0–1 5	1 5
	94		1·4	6 0		
H_2	467	4 82	2 8	21 4		
	305		4·2	21·7	0 1–0 2	0 1
	121		8 3	23 2		
A	528	2 72	1 3	5 1	~ 0 5	0 2
	290		3 85	6 0		

Raether takes a value of 1·5 eV for U in air, as determined from his cloud-chamber experiments, and therefore equation (6.13) can be written

$$\alpha x_c = 17·7 + \log_e x_c + \log_e K.$$ (6.14)

Table 6.6 gives the values of the critical avalanche lengths x_c which have been determined from (6.14) for various gaps at the observed breakdown voltage for each gap. Values of x_c for $K = 1$ and for $K = 0·1$ are given but in most of the calculations it is assumed that $K = 1$. The exact value of K does not greatly influence the results provided K is in the region of 1. The critical avalanche lengths calculated by Meek's criterion for the various gaps in the measured breakdown fields are also given in Table 6.6. Raether [12] points out that the growth of the spark may be expected to occur according to the Townsend mechanism when $x_c > d$, but that the streamer theory obtains for $x_c < d$.

Calculations have been made by Raether [12] of the breakdown field for gaps in which $x_c < d$ on the assumption that the transition from an avalanche to a streamer occurs when the avalanche just crosses the gap, so that the avalanche length is d. The breakdown criterion is then

$$\alpha d = 17·7 + \log_e d.$$ (6.15)

Table 6.7 compares the values of the breakdown field calculated by Raether on this basis with the measured values. In the case of the 10-cm. gap a field strength of 24·9 kV/cm. is sufficient theoretically to

TABLE 6.6

Critical Avalanche Lengths x_c for Various Gaps, as calculated from Observed Breakdown Fields [12]

Gap d cm	x_0 (Raether)		x_c (Meek)
	$K = 1$	$K = 0.1$	
1 0	1 2	1 03	1 27
1 2	1 3	1 14	1·4
1 4	1 4	1 23	1 5
1 6	1 5	1 33	1 6
1 8	1 58	1 36	1 68
2 0	1·60	1 42	1·7

cause the critical multiplication of an avalanche when it reaches the anode, but the measured breakdown field is 26·4 kV/cm. for which the critical multiplication occurs when the avalanche is in the mid-gap region. Raether suggests that this increasing divergence between the calculated and measured breakdown voltages for the larger gaps may be accounted for by the consideration that the conditions governing the propagation of streamers may necessitate a higher field than that required to cause the avalanche-streamer transition   In Table 6.7 values are given, also by Raether, for αd in the calculated breakdown field, for the breakdown field estimated on the basis that breakdown occurs according to the empirical relation $\alpha d = 20$, and the breakdown field according to Meek's criterion.

TABLE 6 7

Measured and Calculated Values of Breakdown Fields for Several Gaps in Air at Atmospheric Pressure [12]

Gap d cm	Breakdown field in kV/cm				αd (Raether)
	Measured	Calculated (Raether)	Calculated (Meek)	Calculated ($\alpha d = 20$)	
2 0	29 8	28 9	29 0	29·2	18 4
6 0	27 4	25 7	25 8	25 8	19 5
10 0	26 4	24 9	25 0	24·9	20·0
16 0	25 8	24 1	24 2	23 8	20 5

The details of the criterion for the transition between an avalanche and a streamer have so far only been elaborated for air In other gases different values must be taken for the constants used in computing the space-charge fields surrounding an avalanche. Again, it is probable that the conditions of gap length and gas pressure at which the streamer mechanism has to be invoked will differ between gases. Reference to this matter is made by Loeb and Meek [10] in a discussion of the results obtained by White [97] in Kerr-cell measurements of spark growth (see p. 185). These measurements reveal the presence of streamers preceding sparks in hydrogen and in nitrogen, whereas diffuse glow discharges are recorded in the early stages of sparks across a 1·3-cm. gap in helium at 1 atm. Loeb and Meek suggest that this is possible evidence for the occurrence of the Townsend mechanism in helium at higher values of pd than in air. A similar result may be expected to occur in other gases, such as argon, where high values of γ are obtained at low values of E/p. This matter is discussed further by Loeb [98].

The fact that reasonable values for breakdown voltages in air can be calculated on the assumption that breakdown occurs when αd attains a particular value, for example $\alpha d = 20$ as used above by Raether [12], has been realized by a number of investigators [34 to 38]. If, however, the values of αd are calculated for the measured breakdown conditions the values are found to increase greatly with increasing gap length, as shown in Table 6.8 The reason why a formula of the type $\alpha d = 20$ gives fair agreement between the observed and calculated breakdown voltages is attributable largely to the rapid charge in α for a small change in the voltage gradient E across the gap, e g. a change in E from 31,000 to 32,000 V/cm. causes a change in α from 14 6 to 17 6. Consequently it is not surprising that the breakdown voltages calculated according to $\alpha d = 20$, as given in Table 6 7, agree to within about 10 per cent. with the measured values for the wide range of gaps selected of from 2 to 16 cm. Further reference to the semi-empirical formulae of Jørgenson [36] and Pedersen [37] is given on p. 273.

TABLE 6 8

Values of αd at Breakdown for Various Gaps in Air

Gap length, d cm		0 1	0 5	1 0	2 5	5	10	20
αd	.	9 2	13 5	16 5	25	32 5	42 5	60

Consideration has been given by Hochberg and Sandberg [38] to the development of a semi-empirical relation governing breakdown in various gases. They base their views on the streamer theory and point out that, because of the exponential multiplication in an avalanche, equality between the space-charge field of the avalanche and the external field is practically attained for a given gap at more or less the same value of α for different gases. They therefore assume that this nearly constant value for α, necessary for breakdown, is attained in the gases at different electric fields, which determine the breakdown strengths of the gases.

To test this supposition Hochberg and Sandberg have measured α/p as a function of E/p in a number of gases. They have then found the values of E/p in each of these gases for which $\alpha/p = 0.0215$, the latter corresponding to the breakdown of a 1-cm gap in air at atmospheric pressure. The ratio of the field at this constant value of α in a given gas to the field in air is then determined and is given in the second column of Table 6.9. The third column of this table shows the relative breakdown strength of the gases compared with air, as measured by Hochberg and Sandberg. The latter consider that the close agreement between the values in the two columns of Table 6.9 corroborates their assumption that an express condition for breakdown is the achievement of a definite value for the coefficient α. In gases of high dielectric strength this value is attained at correspondingly high fields. Hochberg and Sandberg do not give any reason for their arbitrary choice of the value 0.0215 for α/p in order to obtain their results, and do not discuss the effect of a variation in this quantity on the magnitude of the ratio of the electric fields. Also no reference is made to the manner in which the relative dielectric strength of the gases varies with gap length and gas pressure.

TABLE 6.9

Gas	E/E_{air}	Relative electrical strength
C_2H_5Cl	1 22	1 23
C_2H_5Br	1 44	1 52
C_5H_{12} .	1 63	1 65
SF_6 . .	2 22	2.3
$CHCl_3$	$\gtrsim$ 4 15	4 24
CCl_4	$\gtrsim$ 5 9	6 36
Ne	0 17	$\sim$0 14

The transformation from an avalanche to a streamer in a uniform field gap has been considered by Petropoulos [24], who has attempted to derive a more accurate expression for the radial tip field of the avalanche than that adopted by Meek [8, 10] and Raether [9, 11, 12]. It is assumed that all the electrons produced in the avalanche are contained in a sphere of radius r. This sphere also encloses a number of positive ions, the number being given by the product of the sphere volume and the positive ion density. Petropoulos assumes the latter density to be that given by Meek [8] The total number of electrons in the sphere is then

$$n_- = e^{\alpha x} \tag{6.16}$$

and the number of positive ions is

$$n_+ = \tfrac{4}{3} r \alpha e^{\alpha x}. \tag{6.17}$$

The radial field can then be calculated to give

$$X_r = \frac{14 \cdot 4 \times 10^{-7} (\tfrac{4}{3} r \alpha - 1) e^{\alpha x}}{r^2} \, \text{V/cm}. \tag{6.18}$$

Petropoulos calculates the conditions under which the radial field becomes an appreciable proportion of the applied field, when transformation from an avalanche to a streamer may be expected to occur. From consideration of the critical avalanche lengths he concludes that the limit of validity for the Townsend mechanism is $pd \sim 500$ mm. Hg×cm. Petropoulos also points out that because of electrostatic repulsion between the electrons in the head of the avalanche there will be a widening of the avalanche radius above that computed on the basis of thermal electron diffusion, but considers this effect to be negligible until the space-charge field is of the same order as the external applied field.

Other contributions to the streamer theory have been made by Teszner [25], Szpor [39, 40], and Fletcher [23]. Teszner has drawn attention to the influence of space-charge fields on the distribution of electrons at the head of the avalanche and proceeds to show that because of the difference in the fields acting at the front and rear of the electron cloud, the electrons at the front are accelerated and those at the rear are retarded, so causing the shape of the electron cloud to assume an ovoid form. The distribution of space charge in an avalanche has also been considered by Fletcher in an analysis of the formative time of a spark, and his conclusions are discussed again on p. 275.

There appears to be only one published attempt to apply the streamer theory to a gas other than air, namely sulphur hexafluoride. The value of α/p as a function of E/p in this gas has been measured by Hochberg and Sandberg [41, 92 to 94], who proceed to show that when breakdown occurs the field produced by the avalanche is of the order of the externally applied field. They also point out that this condition implies that α should attain a value which varies for different gases over a narrow range (see p. 264). This view has been criticized by Dobretzov [95], who considers that the errors in the measured relation between α and E, as recorded by Hochberg and Sandberg, are too large to enable conclusions to be drawn concerning the constancy of α at breakdown.

The propagation of streamers

Raether [9, 11, 12] has discussed the mechanisms of advance of both the anode-directed and the cathode-directed streamers which develop when an avalanche initiates streamer formation in the mid-gap region of a gap. Sketches illustrating the proposed mechanisms are given in Figs. 6.3 and 6 4. In the case of the positive streamer, or cathode-directed

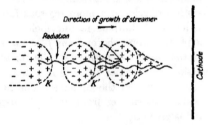

Fɪɢ 6 3   Growth of positive streamer

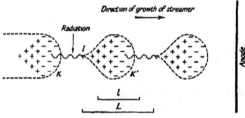

Fɪɢ. 6.4.   Growth of negative streamer.

streamer, shown in Fig. 6.3, photons are thought to produce photo-electrons ahead of the streamer and these initiate avalanches which develop towards the tip of the streamer because of the attraction of the positive space-charge field surrounding the tip. Each successive in-coming avalanche causes an extension of the tip of the streamer which advances in this manner to the cathode. The negative streamer, as illustrated in Fig. 6.4, grows as the result of successive avalanches developing away from the streamer tip in the space-charge field surround-ing the tip. There avalanches may be initiated by electrons from the tip or by photoelectrons produced in the gas ahead of the tip. Calcula-tions have been made by Raether [9] to show the feasibility of such mechanisms on the basis of approximate values of the space-charge fields and the photoionizing absorption coefficient.

Similar proposals relating to the propagation of streamers have been put forward by Loeb [18, 19, 21], who suggests that the conditions necessary for streamer propagation are (i) that sufficient high-energy photons must be created in the initial avalanche to ionize some of the gas atoms or molecules present, (ii) that these photons must be absorbed to produce electrons sufficiently close to the streamer tip, and (iii) that the space-charge field at the rear of the avalanche tip shall be great enough to give adequate secondary avalanches in the enhanced field. On this basis Loeb has developed a more rigorous form of the Meek criterion by the inclusion of additional factors, including the absorption coefficient μ of the photoionizing photons, and f the ratio of the number of such photons to the number of positive ions in the avalanche head. The critical expression for streamer advance in a uniform field is then deduced on the assumption that the advance is due to the successive formation of secondary avalanches from single photo-electrons ahead of the streamer and is given by

$$\tfrac{4}{3}K^2\rho\alpha f e^{\alpha\delta} = 1, \tag{6.19}$$

where $K = (a/4\pi)\exp(-\mu x_1)$, ρ is the radius of the streamer tip (assumed hemispherical), x_1 is the distance over which the secondary avalanches are produced, δ is the distance from the cathode, a is the solid angle within which the photons must be effective.

Expression (6.19) may be written as

$$Cf e^{\alpha\delta} = 1 \tag{6.20}$$

which, though resembling Townsend's expression for breakdown,

$$\gamma e^{\alpha\delta} = 1, \tag{6.21}$$

differs from it in that Cf is a quantity depending on gas pressure. Loeb has used equation (6.20) to explain the variation of the probability of spark formation with applied voltage and to account for various discrepancies between earlier theories and experimental results at high values of $p\delta$. The bearing of the more rigorous theory on the recent work on corona discharges has also been stressed by Loeb [99].

In a further elaboration of the theory, Loeb [19] has deduced an equation from which it is possible, in principle, to calculate the sparking voltages for different gaps. This equation is as follows:

$$(A/p)\int_{x_1}^{p} [E+(U/x^2)]\,dx = \alpha\delta+\log_e\left(\frac{\alpha}{\bar{\alpha}_1}\right)\left(\frac{x_1}{\rho}\right)^{\frac{1}{2}}, \tag{6.22}$$

where α is the Townsend ionization coefficient and is given by the approximate relation

$$\alpha = (A/p)E^2; \tag{6.23}$$

E is the externally applied electric field, p is the gas pressure, A is a constant; x is the distance from the centre of the space-charge zone, of radius ρ, to the point of formation of the photo-electron initiating a secondary avalanche; x_1 is the critical value of x beyond which secondary avalanches are not formed; $\bar{\alpha}_1$ is the value of α at the edge of the space-charge zone, δ is the gap length; $U = \frac{4}{3}\epsilon\rho\alpha e^{\alpha\delta}$, where ϵ is the electronic charge.

Loeb [19] proceeds to show how the value of E required to produce a spark may be determined by trial and error from a knowledge of δ, p, ρ, f, and μ. The value of ρ is determined from consideration of D, the coefficient of electron diffusion, and $\bar{v}$, the average electron drift velocity.

Because of the lack of knowledge of the values of some of the quantities involved, such as f and μ, equation (6.22) has little immediate practical value in the calculation of sparking potentials, but is of importance in that it includes a more detailed consideration of the possible physical processes occurring in streamer growth. It has been modified again recently by Loeb and Wijsman [21] to include the possibility that several electron avalanches develop simultaneously into the streamer tip. But even though this modified sparking criterion has now become so complex as to make calculations of sparking voltage on this basis virtually impracticable, the criterion cannot yet be considered as entirely rigorous as there are still other factors to be taken into account, such as the distribution of ionization in avalanches and streamers, and the variations of α in highly divergent fields [58, 59].

The introduction of the photon absorption coefficient into the sparking criterion has been discussed by Miller and Loeb [31] and again by Loeb [13], who suggest that it can account for the departures from Paschen's law observed by Trump, Safford, and Cloud [32] and other investigators of the breakdown of gases at high pressures. These departures are appreciably greater than those predicted by the original Meek criterion, but on the other hand it is possible that they may be accounted for by space-charge fields produced in the gap by pre-breakdown currents, as shown by Howell [42].

In the various proposals by Loeb and his colleagues [10, 18 to 21] concerning the streamer mechanism it has been assumed that the photoionizing radiations from the avalanche head, or from the streamer tip, are produced by electron excitation processes. However, Hopwood [17] has recently pointed out that the requisite photons could be produced in the avalanche as it approaches the anode by radiative electron-ion recombination. For such recombination to occur at a sufficient rate, the electrons at the head of the advancing avalanche must be retarded by the space-charge field of the positive ions left behind in the avalanche stem, that is when the space-charge field is of the order of the externally applied field in the gap. The Meek criterion, put forward originally to account for the production of secondary avalanches feeding into the primary avalanche as the mechanism governing streamer propagation, then automatically satisfies also the further proposal put forward by Hopwood concerning the production of the electrons initiating the secondary avalanches, namely that such electrons can be produced by photons emitted by electron-ion recombination.

Hopwood [17] has also pointed out that, if the space-charge field mechanism is valid, the effective value of α near the anode and the ion densities there will be less than those calculated for an undistorted field, as was assumed for simplicity in Meek's earlier work. Consequently, the actual ion densities will be less than the ion densities originally calculated by Meek. If consideration is given to this effect in the calculations the results give breakdown voltages in closer agreement with the experimentally determined values for long gaps at higher gas pressures.

Although in the various theoretical treatments of the streamer theory it is generally assumed that the secondary electrons in the gas are produced photo-electrically by photons from the parent avalanche, there is no actual proof that this occurs. Another possible mechanism of production of these secondary electrons would be the ionization of gas atoms or molecules by positive ions. However, it seems to be improbable that such a process could be effective at the voltage gradients involved (see p. 285).

Townsend's sparking criterion, given in equation (6.1), for the breakdown of gaps at low pd, when cathode mechanisms become important, has been discussed in two recent papers by Loeb [13, 18] with particular reference to the statistical factors involved. Loeb points out that the shape of the curve relating the probability of sparking to the applied voltage, as shown in Fig. 6.5, can be accounted for if fluctuations of α and γ are considered, and quotes in support the work of Schade [1], Braunbek [44], and Hertz [45]. Curves similar to that of Fig. 6.5 have also been obtained by Wilson [46] at higher values of pd, where it is now thought that breakdown occurs through streamer formation. Loeb discusses the latter results in terms of the sparking criterion given by equation (6.20). The fluctuations causing the variations in breakdown voltage are now those in C and f, but, as data concerning these two quantities are lacking, no detailed quantitative account of the sparking threshold region can be given for the condition of high pd.

Although a number of the characteristics of avalanches and streamers as described in Chapters III and IV appear to be in general accordance with the streamer theory, there are a number of experimentally observed phenomena which have not been interpreted on this basis. No complete account has yet been given, for instance, of the mechanism of growth of streamers in point-plane gaps, one unexplained feature being the

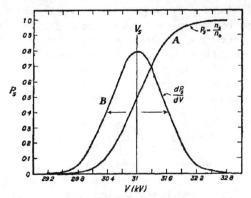

Fig 6 5 (*A*) Probability of sparking P_s plotted against applied potential in kV. (*B*) Derivative of sparking probability against voltage indicating the distribution of sparking events about the threshold plotted in terms of voltage

nature of the head of the leader stroke as described in Chapter IV. The head of the leader stroke is formed of a narrow leader channel which branches out at its tip into a voluminous shower of discharge, as noted by Allibone and Meek [43] and studied in detail by Komelkov [47, 48] and by Saxe and Meek [49, 50]. Measurements by the latter of the light distribution in the discharge show that the visible diameter can be as large as 20 cm. (see Chapter IV). According to Komelkov [47, 48] the discharge consists of numerous filamentary streamers, and on this assumption Kukharin [90] has made an attempt to calculate the current flowing in the leader stroke. Komelkov [91] gives a theoretical analysis of the leader-stroke mechanism based on considerations of the growth of successive avalanches in a radial and axial direction at the head of the leader stroke. He suggests that, as a streamer develops, the temperature in the streamer channel rises until thermal ionization eventually takes place and is followed by a transition from the streamer into the leader channel. The existence of the shower of discharge is considered by

Bruce [51] to substantiate his glow-to-arc transition theory for the spark and lightning mechanisms, the glow condition corresponding to the shower and the arc condition to the core of the leader channel. Investigations on D.C. corona (see Chapter III) also show the presence of the luminous shower discharge.

Breakdown in non-uniform fields

The breakdown of gaps in which the field varies across the gap, as for instance a gap between spheres, has been discussed by Meek [10, 52] and Raether [12] in terms of the streamer theory. In this case the number of ion pairs formed by an electron avalanche crossing the gap is given by the exponential of the integral of α across the gap. The space-charge field E_r at the head of the avalanche when it has travelled a distance x in a non-uniform field is given by Meek [8, 10] as

$$E_r = 5 \cdot 3 \times 10^7 \frac{\alpha_x \exp\left[\int_0^x \alpha \, dx\right]}{(x/p)^{\frac{1}{2}}} \text{ V/cm} , \qquad (6.24)$$

where α_x is the value of α corresponding to the external field at the head of the avalanche. The minimum breakdown voltage of the gap is then computed on the assumption that the transition from an avalanche to a streamer should occur when the avalanche reaches the anode. The magnitude of the radial field E_r should then equal the external field at the anode surface. For long gaps between spheres the mid-gap field may fall to such low values that the value of α is negligible except in the region of the sphere surfaces. It is then suggested [10, 52, 53] that electron avalanches do not cross the entire gap but that streamers are initiated by avalanches growing across the high field region surrounding one of the spheres, so that the number of ion pairs in an avalanche is given by the exponential of the integral of α across half the gap only. As in long gaps between spheres the field at the high-voltage sphere is generally higher than that at the earthed sphere, because of the presence of stray fields, the electron avalanche giving rise to a streamer may be expected to occur at the high-voltage sphere.

From such considerations the breakdown voltages at different gaps between spheres of several sizes have been calculated by Meek [10, 52, 53]. The resultant curves for spheres of diameter 25 cm. are shown in Fig. 6.6. The full curve is that measured. The dashed curves are calculated on the assumptions (I) that the electron avalanche crosses the whole gap and (II) that the electron avalanche crosses only the high field part of the gap near the high-voltage electrode. Then in the former case E_r as given by (6.24) includes the term $\exp\left[\int_0^d \alpha \, dx\right]$, where d is the

gap length, and in the latter case E_r includes the term $\exp\left[\frac{1}{2}\int_0^d \alpha \, dx\right]$.

A region of discontinuity may be expected to occur for gap lengths in the border-line region between mechanisms I and II, and this has been observed experimentally [52 to 56] In measurements with spheres of

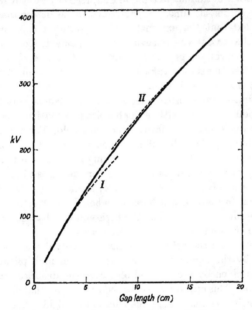

Fig. 6.6 Calculated and observed curves for the breakdown voltage between 25 cm. diameter spheres.
— — — — calculated.
——————— observed.

25 cm diameter Dattan [54] recorded a scattering in breakdown voltages of greater than 2 per cent., and rising to a maximum of 6 per cent., for spacings between 5·25 and 7·75 cm Reference to Fig. 6.6 shows that this is in reasonable agreement with that expected theoretically.

With still longer gaps between spheres, in excess of about four diameters, corona is observed at a voltage lower than the breakdown value. In this case the transition from an avalanche to a streamer defines the onset of corona and not the spark-over of the gap, Similar considerations apply to point-plane and other asymmetrical gaps, where streamers are formed in the intense fields near the electrodes but are not propagated across the gap until the voltage is raised to a value sufficient to make such propagation possible. For such gaps the breakdown voltage cannot be computed theoretically, as it depends on the conditions governing the propagation of streamers for which no quantitative mechanism has yet been proposed. (Some of the factors influencing streamer growth are discussed earlier in this chapter and in Chapter III.) However, the onset voltage for corona-streamer propagation can be determined in these gaps provided that the external field distribution is known, as for example in the coaxial wire and cylinder arrangement. Corona streamers form when the voltage gradient is sufficient to cause E_r, as given by equation (6.24), to attain a value of the order of the applied field at the head of the electron avalanche.

With a high-voltage positive electrode the avalanche travels towards this electrode, whereas with a high-voltage negative electrode the avalanche travels outwards, into a decreasing field. For a given voltage applied to the gap the value of E_r as given by equation (6.24) is then higher at the head of the avalanche moving into the electrode than for that moving away from it, as in the former case α_x is higher while the other factors involved in equation (6.24) are unchanged. This means that streamer formation, which occurs when $E_r \sim E$, can be produced at a slightly lower voltage when the high-voltage electrode is negative than when it is positive, as is generally observed.

In gaps in which the field distribution is highly divergent errors are introduced by the assumption that the electron multiplication in an avalanche is given by the integral of α over the distance travelled by the avalanche. This appears to be caused by the fact that electrons may not reach their terminal energy distribution in highly divergent fields, and the effect becomes appreciable when the divergence is greater than about 2·5 per cent. over an electron mean free path [16, 58, 59].

Semi-empirical relations for the calculation of breakdown voltages in non-uniform fields have been proposed by Peek [57], ver Planck [35], Jørgenson [36], and Pedersen [37]. According to ver Planck breakdown occurs when the number of electrons $\int_0^d \alpha \, dx$ in an avalanche exceeds a value $N(E, p)$ which is characteristic of the gas and depends on the cathode field and gas pressure. The value of $N(E, p)$ is determined from measurements of breakdown in uniform fields. The criterion has been applied by ver Planck to the calculation of breakdown between coaxial cylinders, with results in reasonable agreement with experiment. A similar proposal is made by Jørgenson, who states that breakdown occurs when

$$\int_0^x \alpha \, dx = \phi(x, p),$$ (6.25)

where x is the length of the electron avalanche, and $\phi(x, p)$ is a function of x and the gas pressure p, independent of the field distribution. The function $\phi(x, p)$ is deduced from measurements of breakdown voltage in uniform fields. The application of equation (6.25) to the calculation of the breakdown voltage of non-uniform fields gives values fairly close to the measured values. However, the effect of polarity on the breakdown voltage is not given by equation (6.25) and the expression has subsequently been modified by Pedersen [37] accordingly. Pedersen assumes that a streamer leading to breakdown develops when the density of positive ions at the head of the avalanche attains a certain value, and on this basis has formulated the breakdown condition

$$\log_e \alpha_x + \int_0^x \alpha \, dx = g(x, p),$$ (6.26)

where α_x is the value of α at the head of the avalanche. With this semi-empirical expression Pedersen has calculated the breakdown voltages for a gap between coaxial cylinders of radii 3·81 cm. and 1·11 cm. respectively. With the inner cylinder positive the calculated breakdown voltage is 54·6 kV, and with the inner cylinder negative it is 56·1 kV. The measured values are given as 54·5 and 55·5 kV respectively. The corresponding values calculated by Meek's criterion are 56·4 and 58·5 kV.

A further theory for the breakdown of non-uniform gaps, based on considerations of the corona current flowing in the gap and glow-to-arc transition phenomena, has been proposed by Bruce [51]. This theory, which was evolved initially to explain the growth of lightning discharges, is discussed in Chapter V.

The time of formation of spark discharges

The formative time of a spark in a uniform field has been calculated by Raether [11, 12] on the assumption that this time consists mainly of the time taken by an avalanche to grow to the size where avalanche-streamer transition occurs. The subsequent streamer develops so much more rapidly than the avalanche that its time of growth across the gap can be neglected in comparison with that of the avalanche. In Fig. 6.7 curves are given by Raether for the variation with applied overvoltage of the calculated time of formation of sparks across gaps of 2 cm. and of

3 cm. in air at atmospheric pressure. The calculated curves give time lags of the same order of magnitude as those measured.

In the above calculations Raether has assumed that the transition from an avalanche to a streamer occurs when $\alpha d = 20$. A more rigorous approach to the problem has been made by Fletcher [23], who has first

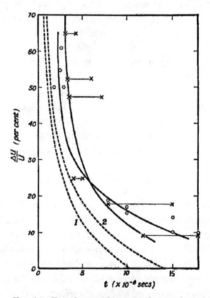

FIG. 6.7. Time lag as a function of overvoltage for gaps in air at atmospheric pressure. ○ for 2 cm gap, × for 3 cm gap The dashed curves 1 and 2 give the calculated values for 2 and 3 cm gaps.

considered in some detail the distribution of space charge in an avalanche. In his analysis Fletcher assumes that when the avalanche first develops the distributions are dominated mainly by the diffusion of electrons, the densities of electrons and positive ions being small enough to make the influence of the space-charge field negligible. The rates of growth of the electron and positive ion densities, n_- and n_+, are then given by

$$\frac{\partial n_-}{\partial t} = \alpha v n_- - D\nabla^2 n_- + v\frac{\partial n_-}{\partial z}, \qquad (6.27)$$

$$\frac{\partial n_+}{\partial t} = \alpha v n_-,$$

where v is the drift velocity in the direction of the field E taken in the z direction and D is the diffusion coefficient. Equations (6.27) and (6.28) have the following solutions:

$$n_- = (4\pi Dt)^{-\frac{3}{2}} \exp\left[-\frac{x^2+y^2+(z-vt)^2}{4Dt} + \alpha vt\right],$$ (6 29)

$$n_+ = \alpha v \int_0^t (4\pi Dt')^{-\frac{3}{2}} \exp\left[-\frac{x^2+y^2+(z-vt')^2}{4Dt'} + \alpha vt'\right] dt'.$$ (6.30)

The electrons are distributed in a normal Gaussian manner, viewed in a coordinate system moving in the z direction with the velocity v. For this distribution the average distance of an electron from the centre is $(6Dt)^{\frac{1}{2}}$ which can be considered the breadth of the avalanche. The integral determining the positive ion distribution cannot be evaluated in terms of well-known functions.

By making several simplifying assumptions Fletcher has deduced the values of n_- and n_+ along the axis of the avalanche ($x = y = 0$). For convenience the values are expressed in terms of the maximum electron density
$$n_0 = (\pi a z_0)^{-\frac{1}{2}} e^{\alpha z_0},$$ (6.31)

where $z_0 = vt$ and $a = 4D/v$. Fletcher also writes

$$w = \frac{z-z_0}{(az_0)^{\frac{1}{2}}} \quad \text{and} \quad b = \alpha(az_0)^{\frac{1}{2}}.$$

Then in the vicinity of the avalanche head, i.e. for $z \sim z_0$, the values of n_- and n_+ are determined as

$$n_- = n_0 e^{-w^2},$$ (6.32)

$$n_+ = \begin{cases} \frac{1}{2}bn_0[1+I(-w)]e^{bw}, & w < 0 \\ \frac{1}{2}bn_0[1-I(w)], & w > 0 \end{cases}.$$ (6.33)

These values are plotted in Fig. 6.8 for different values of the parameter $b = \alpha(az_0)^{\frac{1}{2}}$, which is approximately the ratio of the avalanche breadth to the ionizing distance $1/\alpha$. The centre of mass of the electrons is at $z = z_0$ while that of the positive ions is at $z = 1/\alpha$ approximately. As $\alpha(az_0)^{\frac{1}{2}}$ increases the cloud of electrons tends to overlap more and more the cloud of positive ions.

As the densities of electrons and positive ions increase due to ionization the space-charge field alters the calculated distribution An approximate method for determining the conditions under which this occurs has been developed by Fletcher, who assumes that the distribution is unaffected up to a time t after the initiation of the avalanche. For times greater than t the space-charge field is considered to be proportional to the number of electrons N and inversely proportional to the avalanche radius r_s, as would be true for a spherical distribution of charge. The radius r_s is obtained by integrating the drift velocity due to the space-charge field E_s. Then

$$E_s = E_t \frac{N}{N_t} \left(\frac{r_t}{r_s} \right)^2, \qquad (6.34)$$

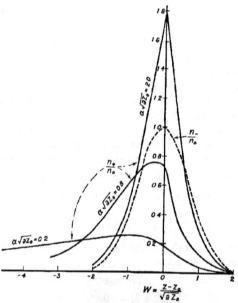

FIG 6.8. Distribution of electron and positive ion densities along the axis of the electron avalanche for different values of the ratio of avalanche radius $(3az_0/2)^{\frac{1}{2}}$ to ionization distance $1/\alpha$. Densities are plotted in terms of the maximum electron density n_0 and distance from the centre of the electron cloud is plotted in terms of $(az_0)^{\frac{1}{2}}$, approximately the avalanche radius.

where $E_s = E_t$ and $N = N_t$ at time t. Fletcher derives an expression for the space-charge field E_s as follows:

$$E_s = \tfrac{3}{16} a\alpha E \left(\frac{N}{N_t} \right)^{\frac{1}{2}}, \qquad (6\ 35)$$

where

$$N_t = \frac{9\pi\epsilon_0}{4e} E a^2 \begin{cases} \{1 + \tfrac{3}{2}a\alpha \log N_t - [\tfrac{3}{2}a\alpha \log N_t]^{\frac{1}{2}}\}^{-1}, & \tfrac{3}{2}a\alpha \log N_t < 1, \\ [\tfrac{3}{2}a\alpha \log N_t]^{\frac{1}{2}}, & \tfrac{3}{2}a\alpha \log N_t > 1. \end{cases} \qquad (6\ 36)$$

The critical avalanche size N_c which drops the internal field in the avalanche to zero is obtained when the space-charge field E_s is equal to the external field E. From equation (6.35) this occurs when

$$N_c = \frac{256}{9} \frac{N_l}{(a\alpha)^2}. \tag{6.37}$$

An attempt to calculate the formative time lag t_f of sparks is then made by Fletcher on the assumption that this time lag consists mainly of the time taken for the avalanche to grow to the critical size, so that

$$t_f = \frac{1}{v\alpha} \log N_c. \tag{6.38}$$

To obtain numerical values for the formative time lag Fletcher takes the value $a = 3 \times 10^{-4}$ cm. as calculated from Raether's observations of the avalanche breadth. He also assumes $v = 0.224(E)^{\frac{1}{2}}$ cm./sec , which gives the avalanche velocity observed by Raether [65] at fields near the minimum breakdown voltage and is extrapolated to the higher fields by the expected dependence on the square root of the field E. By the use of Sanders's values [27] for α Fletcher has calculated the curve relating time lag with applied field as shown in Fig. 4.34 where it is compared with his measured values.

Fletcher's calculation of the formative lag shows that it should be a function of field alone and independent of gap width or applied voltage. This is found to be true for the high but not for the low fields. As shown in Fig 4 34, for a given gap the time lags are longer for the smaller gaps when the field is below a critical value. This critical field is believed to be that for which the initial avalanche just crosses the gap to achieve the critical avalanche size. For a lower field a single avalanche is unable to create a high enough space-charge field to cause breakdown, and more than one avalanche is required to bring about this condition with a consequent increase in the formative time. The value of the critical field can be computed from the value of α_c at which N_c is obtained:

$$V = E_c d = E_c \frac{\log N_c}{\alpha_c}. \tag{6.39}$$

A curve relating the calculated critical field as a function of applied voltage is given in Fig. 4.34. The arrows show the expected critical fields for the three applied voltages used in the measurement of the longer times. The agreement with observation is sufficiently close to enable Fletcher to conclude that this interpretation of the behaviour of the formative lag in the lower fields is a valid one However, Fletcher points out that the results are affected only slightly if a constant value of 10^8 is taken for N_c rather than the computed value as given in equation (6 37). This is because of the insensitivity of the log N_c term in equation (6.39), a change in the value of N_c by a factor of 10 changes the calculated time lag by only 12 per cent. The principal value of the computation is to establish that the correct order of magnitude for N_c is given by considering the space-charge field of a single avalanche to be the dominating factor governing breakdown.

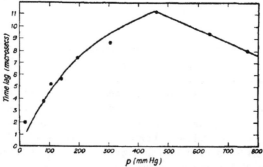

FIG 6 9   Formative time lag as a function of gas pressure for a 1-cm. gap.

An attempt to determine the transition region from the streamer mechanism to the Townsend mechanism by a study of the formative time lags of spark breakdown has been made by Fisher and Bederson [61]. Experiments were made with voltages close to the D.C. breakdown value, in contrast with the measurements by earlier workers where generally overvoltages of at least 1 per cent. appear to have been used. The cathode of a 1-cm. gap was illuminated by ultraviolet light, and an approach voltage below but close to the D.C. breakdown value was applied. Then an additional small square voltage pulse, with a rise time of 0·1 μsec., was applied across the gap. The primary photo-current was about 50 electrons/μsec. A curve showing the variation of formative time lag as a function of gas pressure is shown in Fig. 6.9. The points plotted are average values, and are about twice the minimum values recorded. The distribution of the results at any given pressure and overvoltage shows that the measured lags are not statistical. The formative time lags decrease rapidly with increasing overvoltage, and with an overvoltage of about 1 per cent. are a small fraction of a microsecond at all pressures studied.

In later measurements by Fisher and Bederson [87] the formative time lags of uniform-field gaps of 0 3 to 1·4 cm in length in air, at pressures from 1 atm. down to a few centimetres Hg, have been studied for small overvoltages. For pressures above 200 mm. Hg the formative time lags close to the sparking threshold voltage are ~ 100 μsec or more. At all

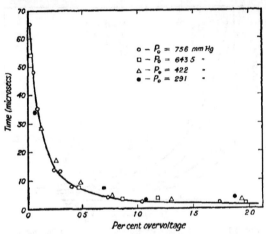

Fig 6 10. Time lag as a function of overvoltage for a 1-cm. gap.

the pressures studied the time lags decrease as the overvoltage increases until at about 2 per cent. overvoltage above the sparking threshold the formative time lags are ~ 1 μsec. A curve showing the time lags as a function of overvoltage for a 1-cm. gap at different gas pressures is given in Fig 6 10 Other curves showing the time lags as a function of overvoltage for four different gaps are given in Fig 6.11. From the results it appears that the curves are independent of pressure from atmospheric pressure down to 200 mm. Hg, and that for a given percentage overvoltage the time lags increase linearly with gap length. The number of initiating electrons released per second from the cathode has been varied by a factor of about seven without any material effect on the results.

The existence of long time lags at low overvoltages could imply that the motion of positive ions plays a part in the breakdown process. The fact that the time lags decrease continuously and smoothly as the overvoltage is increased indicates that the role of positive ions gradually declines in importance as the field strength is raised until finally the time lags become so short that positive ions may be considered to remain stationary throughout the formative time of spark breakdown.

Fisher and Bederson [62, 87] conclude that the Townsend mechanism, based on the emission of secondary electrons from the cathode by positive ion bombardment, cannot give an adequate explanation of the above results as the time lags decrease much more rapidly with increasing overvoltage than can be accounted for by the variation of positive ion

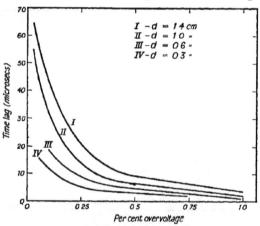

FIG. 6 11   Time lag as a function of overvoltage for four gap lengths. The curves represent the average of data for all pressures between atmospheric and 200 mm. Hg.

velocities. They suggest, however, that the positive ions may assist the formation of the spark by producing a general distortion of the field in the gap [88, 89] Electrons produced in the cathode region subsequent to the first avalanche multiply more intensely than in the undistorted field, as α/p increases more rapidly than linearly with E/p. These electrons may be produced by secondary mechanisms, the most probable mechanisms being photoionization in the gas and photoelectric emission from the cathode As the positive ions, originally created in the anode region, move toward the cathode, the field strength in the cathode region increases more and more rapidly with time. Therefore, because of the non-linearity of the curve relating α/p with E/p, the quantity $\int_0^d \alpha \, dx$, which represents the number of ion pairs produced by an electron starting from the cathode, increases with time. If conditions in the gap are not satisfactory for the formation of a spark when the first avalanche has crossed the gap, it is possible that, as the positive ions created in the first avalanche approach the cathode, subsequent avalanches may produce the charge densities necessary for the onset of a streamer. With such a mechanism times corresponding to many transit times of a positive ion may occur before the spark develops. As the overvoltage is increased the enhanced field necessary for the production of a spark requires a shorter distance of travel of the positive ions and so the time lag decreases. Eventually, with a sufficiently high overvoltage, no motion of the positive ions is required and a streamer develops from the anode region to cause a spark. With still higher overvoltages an adequate space-charge field can be built up by the avalanche before it reaches the anode and a mid-gap streamer develops.

From such considerations Fisher and Bederson [62, 87] raise the question as to whether it is necessary to invoke the Townsend mechanism even at low gas pressures, and indicate that the streamer mechanism may obtain, with such modifications as described above, over the whole range of gap lengths and gas pressures. The opposite view is held by Ganger [63] as the result of his oscillographic measurements of the times of formation of sparks in air, for various pressures and gap lengths, as described on p. 214. Gánger was unable to detect any discontinuity in the curves relating time lag with overvoltage, and hence considers that there is no transition from a Townsend mechanism to a streamer mechanism. He concludes that a Townsend mechanism involving ionization by positive ions is possible for the range of gaps studied, for voltages up to 50 kV. In all Gánger's measurements overvoltages of several per cent. or more were used.

Criticisms of the streamer theory

The streamer theory has been criticized by Zeleny [64], who considers that radial space-charge fields are unlikely to divert large numbers of electrons into the main avalanche. He also states that unreasonably large values of the factor K are required in order to bring the calculated value of breakdown voltage into conformity with the measured values.

These criticisms have been discussed by Hopwood [17], who points out that in the derivation of Meek's equation the ion densities at the head of the avalanche are calculated on the basis that α corresponds to the external field E which is assumed to be operative during the whole time of avalanche growth. This assumption is clearly only approximate, as the voltage distribution in the gap will be distorted when the voltage applied to the gap is about the breakdown value, because of the high ion densities created at the avalanche tip. The effect of this space charge is such that electrons in the rear of the advancing electron cloud, ahead of the positive space charge, will experience a steadily decreasing field and consequently the mean effective value of α will be reduced. Ion densities close to the anode will then be somewhat smaller than those calculated on the assumption that the value of α corresponding to the externally applied field E obtains across the whole of the gap in the track of the avalanche. Consequently voltages higher than those calculated will be necessary to ensure that the ionization produced in an avalanche is sufficient to cause streamer formation according to Meek's criterion, that $E_r = E$. Hopwood considers that the effect should increase at longer gaps, since as the gap is increased the value of E/p for breakdown decreases and at the lower E/p, where α is small, a small change in E reduces α to an almost negligible value, whereas at higher E/p, as in the breakdown of shorter gaps, a similar proportional change in E will reduce α by only a few per cent. It is possible therefore that discrepancies detected in the straightforward application of Meek's equation to large gaps can be largely attributed to oversimplified assumptions in the derivation of the equation rather than to a failure of the general concepts on which the mechanism is based.

Attention has been drawn by Zeleny [64] to the fact that the breakdown voltages of gaps in air can be calculated, in good agreement with the measured values, by the use of the simple assumption that the total number of ion pairs produced in an avalanche during its passage across the gap should reach a particular value. The value mentioned by Zeleny is

$$\exp \alpha d = 10^9. \tag{6.40}$$

The fact that a simple expression of this type gives reasonable values of breakdown voltages has been realized by a number of investigators (see p. 273). For instance, Schumann [34] proposed that breakdown of medium gaps, of the order of 1 to 10 cm., could be calculated on the assumption that

$$\alpha d = 20. \tag{6.41}$$

Zeleny concludes that no particular significance should be assigned to the degree of agreement with which the streamer theory predicts breakdown voltages. However, the fact that equations (6.40) and (6.41) give breakdown voltages in close accord with the measured values is now understandable from consideration of the streamer theory. Further, whereas the streamer theory gives an interpretation of the spark mechanism in terms of fundamental physical processes, no explanation is given by Zeleny for the physical basis underlying equations (6.40) and (6.41) which are formulated on an empirical basis only.

The view has been expressed by Llewellyn Jones and Parker [66] that the streamer theory is unnecessary and that the growth of sparks at atmospheric pressure can be explained quantitatively by a mechanism of the Townsend type. The Townsend equation for the current flowing in a gap is expressed in the form

$$i = i_0 \frac{\exp \alpha d}{1 - (\omega/\alpha)(\exp \alpha d - 1)}, \qquad (6.42)$$

where ω is a generalized coefficient of secondary ionization given by

$$\omega = \delta + \gamma + \beta; \qquad (6.43)$$

δ represents electron emission from the cathode caused by impact of photons and excited atoms from electron avalanches, γ represents cathode emission caused by positive ion bombardment, and β represents ionization of gas atoms resulting from direct or indirect action of positive ions. Spark breakdown occurs when the denominator of equation (6.42) is zero.

Llewellyn Jones and Parker point out that although the relative importance of δ, γ, and β is not known with certainty the values of α and ω can be measured. For constant E, p, and i_0, and for small values of ω/α,

$$\log_e i/i_0 = \alpha d. \qquad (6.44)$$

The slope of the curve relating $\log_e i/i_0$ with d then gives α As α approaches the sparking distance the curve departs from linearity in accordance with equation (6.42), and the value of ω may then be calculated. This process is known to occur for discharges in short gaps at low gas pressure, as described in Chapters I and II.

It has been widely assumed that the probability of ionization of gas atoms by collision with their own positive ions is negligible at values of E/p obtaining at spark breakdown. This view which is based on classical momentum considerations and on a few low-pressure experiments with alkali ions is discussed again on p. 286. The factor β is therefore generally ignored and the explanation of secondary ionization is usually sought in terms of δ and γ only.

A series of experiments have been made by Llewellyn Jones and Parker [66] to record the variation of $\log_e i$ with d for longer gaps and higher gas pressures than those studied by previous investigators. In some results obtained for air at 200 mm. Hg the relation between $\log_e i$ and d is linear for d up to about $1 \cdot 25$ cm. and for values of E/p from 40 to 45. For greater values of d the curves depart from linearity, thus indicating measurable values of ω/α. The values calculated for ω/α are of the order of 10^{-5} for nickel electrodes and increase with E/p. From these measurements Llewellyn Jones and Parker conclude that the

Townsend mechanism still obtains for values of pd up to 769 mm. Hg × cm.

By observation of the influence of cathode surface on the values of i_0 and i an indication is given of the nature of the secondary process. It appears that the γ-process is important, even at gaps up to 3·7 cm. at 200 mm. Hg, while the δ-process is relatively unimportant. However, Llewellyn Jones and Parker also draw attention to the fact that for positive point corona discharges it is difficult to understand how cathode ionization due to positive ion bombardment can be significant and hence, they state, a β-process must then become important. From studies of corona Medicus [67] has concluded that a β-process is effective in discharges in non-uniform fields in diatomic gases, in accordance with other experimental data [68] on the action of positive ions in hydrogen in a uniform field. Werner [69] also considers that ionization by positive ions in the gas may be effective in Geiger counter discharges.

In their analysis Llewellyn Jones and Parker neglect consideration of the possibility of photoionization in the gas as a factor contributing to the secondary ionization process. In the absence of data showing that photoionization in the gas is unimportant there would appear to be no reason why a fourth factor taking into account such photoionization should not be added to the factors δ, γ, and β forming the secondary ionization term defined by Llewellyn Jones and Parker. Further experimental evidence to distinguish between the relative contributions made by these factors is required before any final conclusions can be drawn.

The role of ionization by positive ions in spark breakdown is discussed by Townsend [70], who has calculated the probability of a positive ion acquiring an energy sufficient to cause ionization in a gap. In these calculations Townsend assumes that (a) air molecules of ionization potential $V_i = 15$ V can be ionized by positive ions of energy $V_0 = 2V_i$, (b) the mean free path of an ion can be calculated by extrapolation of the viscosity theory to high temperatures, and (c) the free paths lie entirely in the field direction. The number of times a positive ion acquires an energy of at least V_0 in travelling across a gap of length L is then

$$n = \frac{d}{\lambda}\exp\left[-\frac{V_0\,d}{V\lambda}\right], \qquad (6.45)$$

where V is the voltage across the gap and λ is the mean free path of a positive ion. Townsend proceeds to consider the breakdown of a 1-cm. gap in air between parallel plates, for which he assumes the breakdown field to be 33,000 V/cm. The mean free path λ of the positive ions is taken to be 7×10^{-6} cm., 7·3 times greater than that of the gas molecules. The energy acquired by an ion in moving a mean free path in the direction of the field is then 2·3 eV. As the number of mean free paths in a distance of 1 cm. is 14,300, the number of free paths in which an ion acquires energies greater than 30 eV is $14,300e^{-13} = 0·03$, as calculated from equation (6.45). For a breakdown field of 33,000 V/cm the value of α is 19·8 so that $\beta = 19·8e^{-19·8} = 5 \times 10^{-8}$. As the magnitude of β is appreciably lower than the number of free paths in which ions acquire energies greater than 30 eV, Townsend concludes that the breakdown of the gas can be explained in terms of ionization of gas molecules by collisions of positive ions on the hypothesis that a molecule is ionized in a small proportion of the total number of collisions of ions with energies greater than 30 eV.

The views expressed by Townsend have been criticized by Varney, Loeb, and Haseltine [71], who state that the value of the mean free path assumed by Townsend for the positive ions is seriously in error. From considerations of the experimental work by Hershey [72] and by Ramsauer and Beeck [73], Varney, Loeb, and Haseltine conclude that the mean free path of fast ions is probably about 9×10^{-6} cm. in air, the same as that of the gas molecules, and is certainly not 7·3 times larger as supposed by Townsend. Insertion of this value for λ in equation (6 45) gives $n = 2·8 \times 10^{-38}$ when $V_0 = 30$ eV. This is lower than Townsend's calculated value of n by a factor of about 10^{-36}. Varney, Loeb, and Haseltine also draw attention to the fact that the breakdown field of a 1-cm. gap in air is more correctly 31,600 V rather than 33,000 V as adopted by Townsend. In the former case α is approximately 17 and the value of β is then $7·1 \times 10^{-7}$. It is apparent therefore that even if ionization could occur at $V_0 = 30$ eV the chance of positive ions acquiring such energy is trivial and could not explain the observed ionization.

The possibility of ionization occurring for positive ions with energies as low as 30 eV is disputed by Varney, Loeb, and Haseltine [71], who state that there is no experimental evidence despite the numerous measurements which have been made to detect such ionization [68, 74 to 78, 83 to 86, 96, 104]. Resonance ionization phenomena appear to occur in atomic gases such as Ne, A, Kr, Xe, and Hg, when atoms are struck by alkali ions or by their own fast neutral atoms. The ionization efficiency exceeds 1 per cent. in several cases and is characterized by a sharp onset. Varney and his co-workers [76, 77, 78] have observed ionization to commence when alkali ions traverse these gases at the energies listed in Table 6.10. All the onsets occur at energies of five or six times the ionization potential. No ionization was detected below 500 eV by any alkali ion in H_2, He, N_2, CO_2, or CH_4.

TABLE 6.10
Onset Potentials in Volts

	Ne	A	Kr	Xe	Hg
Na	130	.	.		80
K		82	69	114	.
Rb . . .		135	97 5	150	.
Cs . . .	.	338	200	77	.
Electrons . . .	21 6	15 8	14·0	12 1	10 4

Related to the ionization by alkali ion collision is the ionization produced in these gases by their own fast neutral atoms [79 to 82, 100, 102, 103, 105]. When positive ions pass through the gas from which they have been formed a number of the ions capture electrons from neutral atoms and proceed with unchanged velocity as neutral atoms [101]. If these fast neutral atoms collide with other atoms they can cause ionization. Recent measurements by Horton and Millest [103] of the ionization of helium by neutral helium atoms show that ionization begins when the kinetic energy of the bombarding atoms is twice as great as the minimum kinetic energy which electrons must possess in order to ionize helium.

Although there is still considerable uncertainty about the spark mechanism, and differing views are expressed on the subject, it is evident that the mechanism at low gas pressures is influenced by the cathode while that at atmospheric and higher gas pressures is not noticeably dependent on the cathode material (provided large pre-breakdown currents, and attendant space-charge distortions, do not exist; see p. 298). The cloud-chamber experiments of Raether and his colleagues have shown that the breakdown of gaps in air, for which the product of pressure and gap length is greater than 1,000 mm. Hg × cm., is attributable to the transition from an individual avalanche to a streamer, but at low gas pressures it appears that the breakdown on the basis of the Townsend mechanism is caused by the build-up of the current flowing in the gap and therefore by a succession of avalanches. The lack of evidence of a sharp transition between the classical Townsend régime and the streamer régime tends to indicate that in due course a more generalized theory of the spark may be forthcoming, in which some factors predominate at low gas pressures and others at high gas pressures.

In fact the streamer theory might well be considered in certain respects as a special case of the Townsend theory in which the secondary ionization coefficient γ denotes the production of secondary electrons in the gas and not at the cathode; in this event it seems reasonable to suggest that, whereas the Townsend theory explains the occurrence of a spark in terms of a discontinuity arising in a mathematical equation involving the ionization coefficients, the streamer theory endeavours to present in greater detail the physical picture of the mechanism of spark growth.

REFERENCES QUOTED IN CHAPTER VI

1 R. Schade, Z. Phys. **104** (1937), 487

2 L. B Loeb, Science, **69** (1929), 509.

3. A. von Hippel and J. Franck, Z Phys. **57** (1929), 695

4. W. O. Schumann, Z. tech. Phys. **11** (1930), 194.

5. J. J Sammer, Z Phys **81** (1933), 440.

6 N. Kapzov, ibid **75** (1932), 380.

7. W. Rogowski, Arch. Elektrotech. **20** (1928), 99

8. J. M. Meek, Phys. Rev. **57** (1940), 722.

9. H. Raether, Arch. Elektrotech. **34** (1940), 49.

10. L. B. Loeb and J M. Meek, The Mechanism of the Electric Spark, Stanford University Press, 1941.

11 H. Raether, Z. Phys. **117** (1941), 394, 524.

12. H. Raether, Ergebnisse d. exakten Naturwiss **22** (1949), 73

13. L. B. Loeb, Proc Phys. Soc. **60** (1948), 561.

14. L. H. Fisher, Phys Rev **69** (1946), 530

15. L. H. Fisher, ibid **72** (1947), 423.

16 L H Fisher and F. L. Weissler, ibid. **66** (1944), 95.

17. W. Hopwood, Proc Phys. Soc. **62**B (1949), 657.

18 L. B. Loeb, Rev. Mod. Phys. **20** (1948), 151.

19. L B. Loeb, Phys. Rev. **74** (1948), 210.

20. L. B. Loeb and J. M. Meek, J. Appl. Phys. **11** (1940), 438, 459.

21. L. B Loeb and R. A. Wijsman, ibid **19** (1948), 797.

22. W. Riemann, Z. Phys. **122** (1944), 216.

23. R. C. Fletcher, Phys Rev. **76** (1949), 1501.

24. G. M Petropoulos, ibid. **78** (1950), 250.

25. S. Tesznek, Bull. Soc. Franc. Elect. **6** (1946), 61.

26 J Slepian, Elect. World, **91** (1928), 768.

27. F. H Sanders, Phys Rev **44** (1933), 1020

28. H. Ritz, Arch. Elektrotech. **26** (1932), 219.

29 W Holzer, ibid. **865**.

30. F. M Bruce, J. Instn. Elect. Engrs **94** (II) (1947), 138.

31. C. G. Miller and L B Loeb, Phys Rev **73** (1948), 84.

32 J G Trump, F. J Safford, and R. W Cloud, Trans. Amer. Inst Elect. Engrs **60** (1941), 112.

33. R. H. Healey and J W. Reed, The Behaviour of Slow Electrons in Gases, Amalgamated Wireless, Sydney, 1941.

34. W. O. Schumann, Elektrische Durchbruchfeldstärke von Gasen, J. Springer, 1923.

35. D. W. VER PLANCK, *Trans. Amer. Inst. Elect. Engrs.* **60** (1941), 99.
36. M. O. JØRGENSON, *Elektrische Funkenspannungen*, E. Munksgaard, Copenhagen, 1943.
37. A. PEDERSEN, *Appl Sci. Res.* **B1** (1949), 299.
38. B. M. HOCHBERG and E. Y. SANDBERG, *Comptes Rendus Doklady U.R.S.S.* **53** (1946), 511.
39. S. SZPOR, *Bull. Ass. Suisse Elect.*, No. 1, 1942.
40. S. SZPOR, *Recueil d. travaux d. Polonais int. en Suisse*, **2**, 1944.
41. B. M. HOCHBERG and E. Y. SANDBERG, *J. Tech. Phys. U.S S.R.* **12** (1942), 65.
42. H. H. HOWELL, *Trans. Amer Inst. Elect. Engrs* **58** (1939), 193.
43. T E. ALLIBONE and J. M. MEEK, *Proc. Roy. Soc. A,* **166** (1938), 97 ; **169** (1938), 246.
44. W. BRAUNBEK, *Z. Phys.* **107** (1937), 180.
45. G. HERTZ, ibid. **106** (1937), 102.
46. R. R. WILSON, *Phys Rev.* **50** (1936), 1082.
47 V. S. KOMELKOV, *Doklady Akad. Nauk, U.S S.R.* **68** (1947), 57.
48. V. S KOMELKOV, *Bull Ac Sci. U S S R*, Dept Tech. Sci **8** (1947), 955.
49. R. F. SAXE and J. M MEEK, *Nature,* **162** (1948), 263.
50. R F SAXE and J. M. MEEK, *Brit. Elect. Allied Ind Res Assoc Reports* L/T 183 (1947), L/T 188 (1947).
51. C. E. R. BRUCE, *Proc Roy Soc A,* **183** (1944), 228.
52. J. M. MEEK, *J. Franklin Inst.* **230** (1940), 229.
53. J. M MEEK, *J. Instn. Elect. Engrs* **89** (1942), 335.
54. W. DATTAN, *Elektrotech Z* **57** (1936), 377.
55. J. CLAUSSNITZER, ibid. 177.
56. E. HUETER, ibid. 621.
57. F. W. PEEK *Dielectric Phenomena in High Voltage Engineering*, McGraw-Hill, 1929.
58 P. L. MORTON, *Phys. Rev.* **70** (1946), 358.
59. G. W. JOHNSTON, ibid. **73** (1948), 284.
60 L H. FISHER and B. BEDERSON, ibid. **75** (1949), 1324.
61. L. H. FISHER and B BEDERSON, ibid. 1615.
62. L H. FISHER, *Elect. Engng.* **69** (1950), 613.
63. B GANGER, *Arch. Elektrotech.* **39** (1949), 508
64. J. ZELENY, *J. Appl Phys.* **13** (1942), 103 and 444.
65. H. RAETHER, *Z. Phys.* **107** (1937), 91.
66. F. LLEWELLYN JONES and A. B. PARKER, *Nature,* **165** (1950), 960.
67. H MEDICUS, *Z. angew. Phys.* **1** (1948), 106 ; **1** (1949), 316.
68. J. S. TOWNSEND and F. LLEWELLYN JONES, *Phil. Mag.* **15** (1933), 282.
69. S. WERNER, *Nature,* **165** (1950), 1018.
70 J. S. TOWNSEND, *Phil. Mag.* **28** (1939), 111
71. R N. VARNEY, L. B. LOEB, and W. R. HASELTINE, ibid **29** (1940), 379.
72 A. V. HERSHEY, *Phys Rev.* **56** (1939), 908.
73. C. RAMSAUER and O. BEECK, *Ann. d. Phys.* **47** (1928), 1.
74. R. M. SUTTON and J. C MOUZON, *Phys Rev* **37** (1931), 379.
75. K. T. COMPTON and I LANGMUIR, *Rev. Mod. Phys.* **2** (1930), 134.
76. R N. VARNEY, *Phys. Rev* **47** (1935), 483.
77 R N. VARNEY and W. C. COLE, ibid. **50** (1936), 261.
78. R. N. VARNEY, A. C. COLE, and M. E. GARDNER, ibid. **52** (1937), 526.

79. R. N. Varney, ibid. **50** (1936), 159.

80. H. Wayland, ibid. **52** (1937), 31.

81. C. J. Brasefield, ibid. **42** (1932), 11; **43** (1933), 785.

82. R. Dopel, *Ann. d. Phys.* **16** (1933), 1.

83. O. Beeck, ibid. **6** (1930), 1001; **18** (1933), 414.

84. O. Beeck, *Phys. Zeits.* **35** (1934), 36.

85. A. Rostagni, *Phys. Rev.* **53** (1938), 729.

86. R. N. Varney, ibid. **732.**

87. L. H Fisher and B. Bederson, ibid. **81** (1951), 109.

88. R. N. Varney, H. J. White, L. B. Loeb, and D. Q. Posin, ibid. **48** (1935), 818.

89. J. M. Meek, *Proc. Phys. Soc.* **52** (1940), 547.

90. E. S Kukharin, *Electrichestvo,* No 10 (1950), 35.

91. V. S. Komelkov, *Izv. Akad. Nauk* (Otdel Tekhn. Nauk), No. 6 (1950), 851.

92. B. M. Hochberg and E. Y. Sandberg, *Doklady Akad. Nauk, U S S R.* **53** (1946), 515.

93. B. M. Hochberg and E. Y. Sandberg, *Izv. Akad Nauk, U S.S.R , Ser. Fiz.,* **10** (1946), 425.

94. B. M. Hochberg and E. Y. Sandberg, *J. Tech. Phys.* **17** (1947), 299.

95. L. N. Dobretsov, *Doklady Akad Nauk, U.S S.R.* **59** (1948), 1547.

96. F. Wolf, *Z. Phys.* **74** (1932), 575.

97. H. J White, *Phys. Rev.* **46** (1934), 99

98. L. B. Loeb, ibid. **81** (1951), 287.

99. L. B. Loeb, ibid. **73** (1948), 798.

100. A. Rostagni, *Nuovo Cim.* **11** (1934), 621.

101. H. Kallmann and B Rosen, *Z. Phys.* **58** (1929), 52, **61** (1930), 61.

102 H. W. Berry, *Phys Rev* **61** (1942), 378.

103. F. Horton and D. M Millest, *Proc. Roy. Soc.* A, **185** (1946), 381.

104. J. C Mouzon, *Phys. Rev.* **41** (1932), 605.

105. A. Rostagni, *Nature,* **134** (1934), 626.

BREAKDOWN VOLTAGE CHARACTERISTICS

Uniform fields

THE breakdown in uniform fields by 50 c./s. A.C. voltages has been studied by Ritz [1] for gaps of up to 12 cm. in air The electrodes used to produce the uniform field are of the form suggested by Rogowski [2, 3] and have plane sparking surfaces with the curvature of the edges so designed that sparks occur only in the uniform-field section of the gap. Ritz shows that the variation of breakdown voltage with gap length, for air at 20° C. and 760 mm. Hg can be expressed closely by the relation

$$V = 24 \cdot 55d + 6 \cdot 66\sqrt{d} \text{ kV} \qquad (7.1)$$

The breakdown voltages measured by Ritz for gaps up to 1 cm. are listed in Table 7.1, which also includes the breakdown voltages calculated according to equation (7.1). The breakdown voltages for gaps above 1 cm. are given in Table 7.2, where they are compared with the values given by Schumann [4], Holzer [5], Bruce [6], and those recorded for sphere gaps [7] when the sphere diameter is large compared with the sphere spacing so that the gap is nearly uniform. The curve obtained by Ritz [1] for the breakdown voltage gradient in gaps up to 3 cm is plotted in Fig. 7.1, which also includes values given by other investigators [8 to 12]

The breakdown voltages given in Tables 7.1 and 7.2 have been measured with A.C. voltages, but the values may be considered to apply equally well to D.C. breakdown. No divergence is recorded between D.C. and A.C. breakdown voltages for small gaps between spheres [13], and the D.C. breakdown voltages determined by Trump, Safford, and Cloud [14] for gaps up to 3·5 cm. appear to agree closely with the values given in Table 7.2. The results given by Klemm [10], and plotted in Fig. 7.1, are for D.C. breakdown and again show no noticeable deviation from the other values plotted which have been obtained with A.C. voltages. It may therefore be assumed that within the margins of experimental error no difference has yet been detected between the D.C. and A.C. breakdown voltages. The erratic results obtained in some measurements of A.C. breakdown in short gaps, of the order of 1 mm., has been shown to be due to lack of adequate irradiation [15].

The general form of equation (7.1) is discussed by Ritz [1] and by other workers [5, 8, 16, 17, 18], who find that their breakdown measure-

TABLE 7.1

A.C. Breakdown Voltages in Uniform Fields in Air at 20° C. and 760 mm. Hg (Absolute Humidity of 10 mm Hg)

Gap in cm.	Breakdown voltages in kV		Measured breakdown gradient in kV/cm
	Measured	Calculated	
0 06	3 13	3 10	52·16
0 07	3·49	3 48	49·86
0 08	3 84	3 85	48 06
0 09	4 18	4 21	46 44
0·10	4·54	4 56	45 40
0 12	5 23	5 26	43 58
0 15	6 25	6 26	41 67
0 2	7 90	7 89	39 50
0 3	11 02	11 01	36 73
0 4	14 01	14 03	35 03
0 5	17 0	16 98	34 0
0 8	25 7	25 60	32 13
1 0	31 35	31 21	31 35

TABLE 7.2

A.C. Breakdown Voltages in Uniform Fields in Air at 20° C. and 760 mm. Hg

Gap in cm	Breakdown voltages in kV				
	Schumann (16)	Ritz (1)	Holzer (5)	Bruce (6)	Sphere gap (7)
1	31·7	31 35	31 66	30 30	31 0
2	59 6	58 7	61 2	57 04	58
3	87 0	85 8	86 94	83 19	85
4	114 0	112 0	113 04	109 0	112
5	140 0	138 5	137 8	134 7	137
6	166 2	163 8	163 44	160 2	164
7	191 8	189 9	187 74	185 6	190
8	216 8	215 0	212 88	211 0	215
9	241 2	240 0	237 78	236 3	240
10	266 0	265 0	263 0	261 4	265
11	290 4	290 0	288 2	286 6	288
12	.	315 5	313 2	311 6	312
13		.	338 1	.	336
14	..	..	363 2		362
15	..		387 7	..	388
16	..		412 6		412

ments give results which conform to expressions of this type, as shown by equations (7.3), (7.4), (7.5) below. The explanation is based on the suggestion by Schumann [16] that breakdown occurs in a uniform field when the following condition obtains (see p. 263):

$$\alpha d = K, \tag{7.2}$$

where K is a constant. If it is assumed that

$$\alpha = A(E-E_0)^2 \tag{7.3}$$

then it follows that

$$E = E_0 + B/\sqrt{d} \tag{7.4}$$

or

$$V = Cd + B\sqrt{d}, \tag{7.5}$$

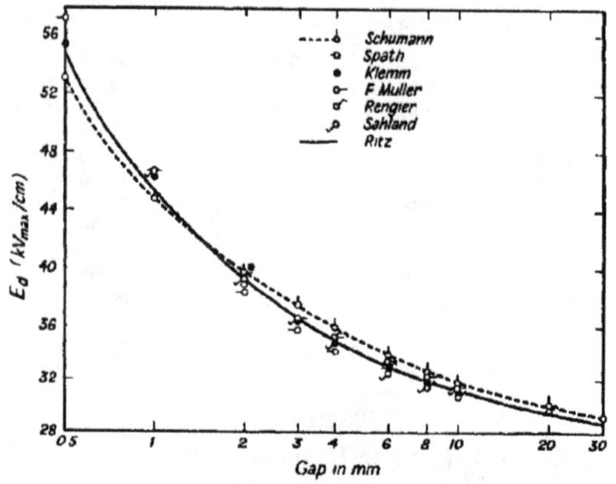

Fig 7.1 Breakdown voltage gradient in air at 1 atm as a function of
gap length between parallel plates

where A, B, and C are constants Experiments by Posin [19] show that
for a wide range of voltage gradients the value of α in nitrogen can be
expressed in the form given in equation (7 3).

Ritz [1] has studied the influence of gas density on his results for gas
densities varying from $\rho = 0.8$ to 1.1, where $\rho = 1$ is the density
corresponding to air at 20° C. and 760 mm Hg, and finds that the
breakdown voltage is constant for a given value of the product ρd, in
conformity with Paschen's law. The accuracy of the measurements is
stated to be within 0.5 per cent. The influence of a variation of gas density
on the breakdown voltage can be expressed by a modification of equation
(7.1) as follows: $V = 24.55\rho d + 6.66\sqrt{(\rho d)}$ kV. (7.6)

Electrodes of the Rogowski type [2, 3] have also been used by Holzer
[5] in measurements of the A.C. breakdown voltage of uniform fields in
air for gaps up to 16 cm. Some of his results are given in Table 7.2.
According to Holzer the breakdown voltage can be represented by

$$V = 23.85\rho d + 7.85\sqrt{(\rho d)} \text{ kV.} \tag{7.7}$$

In later experiments by Bruce [6] the A.C. breakdown of gaps from
0.2 to 12 cm. in air has been studied. The uniform field is obtained by
the use of electrodes with plane sparking surface and curved edges

constructed to a design suggested by Stephenson [18]. The breakdown voltages observed by Bruce, and given in Table 7.2, appear to diverge appreciably from those of other workers, particularly at the shorter gaps. The relation between the breakdown voltage and the gap length is stated by Bruce as

$$V = 24\ 22d + 6 \cdot 08\sqrt{d}\ \text{kV.} \tag{7.8}$$

The accuracy of eq (7 8) is quoted by Bruce as within $\pm 0 \cdot 2$ per cent.

The differences between the breakdown voltages given in Table 7.2 have been discussed by Meek [162] and by Bruce [163] Meek points out that in Bruce's investigations the influence of humidity has been neglected whereas the earlier workers have shown that humidity affects the results. Although the effect is small it appears to be noticeable in measurements involving high accuracy (see p. 295). Variations in the electrode shape may also cause differences In Ritz's measurements three sets of electrodes were used, with diameters of 8·5, 18, and 45 cm. for gaps up to 2, 5, and 12 cm. respectively. In Holzer's studies the electrodes were of 60 cm diameter. The diameter of the plane surfaces of Bruce's electrodes were 5·7, 11·2, and 19·8 cm., with overall diameters of 16 5, 33 0, and 54·0 cm , for maximum measured voltages of 140, 280, and 420 kV (peak) respectively.

From considerations of the measuring techniques greater reliability must be placed on Bruce's values for the breakdown voltages as compared with the others given in Table 7 2. For example, Bruce [163] points out that the maximum deviations in Ritz's measured breakdown voltages are about 0·8 per cent., compared with about 0·15 per cent. recorded by him.

Other measurements of breakdown voltage in uniform gaps in air at atmospheric and lower pressures have been made by Fitzsimmons [20], Haseltine [21], and Fisher [22]. Some of Fisher's results for the D C. breakdown of air at atmospheric pressure are given in Table 7.3, where they are compared with the earlier results obtained by Schumann [16] and Spath [9] with A.C. voltages. Fisher's results are in good agreement with those of Spath for gaps up to 0·6 cm. For gaps > 1 cm Fisher's values are lower than those obtained by other investigators and this is attributed by Fisher to the influence of the walls of the discharge chamber on the field in the gap. Measurements by Fisher of breakdown voltages in air at reduced pressures, down to 261 mm. Hg, show that Paschen's law is obeyed though divergences occur for higher values of the product pd, for instance for $pd > 800$ mm. Hg $\times$ cm. at $p = 261$ mm. Hg, again because of the influence of the chamber walls on the gap field.

TABLE 7.3

d cm.	pd mm. Hg × cm.	Breakdown voltage in kV		
		Schumann	Spath	Fisher
0 4	306	14 4	13 8	14 1
0 5	382	17 4	16 8	16 9
0 6	459	20 3	19 8	19 6
0 7	536	23 2	22 75	22 3
0 8	612	26 1	25 6	24 9
0·9	688	28 9	28 6	27 5
1 0	765	31 7	31 55	30·1
2 0	1,530	59 6		55 7

Both Fitzsimmons [20] and Haseltine [21] observe a lowering of the breakdown voltage caused by the formation of oxides of nitrogen if repeated sparking is allowed to occur in the same chamber without a circulation of fresh air In earlier experiments by Goodlet, Edwards, and Perry [23] the breakdown voltage of a non-uniform field gap between coaxial cylinders in air is found to decrease from 88 kV to 64 kV as the result of successive sparking when the gap is enclosed.

Fisher [22] notes a small but definite spread of 300 to 500 V in the breakdown voltage with water vapour present in the chamber. All his measured voltages, as given in Table 7.3, refer to dry air. The influence of humidity on the breakdown voltage in uniform fields has been studied by several other investigators [1, 24, 25] According to Ritz [1] the breakdown voltage of a 1·0-cm. gap in air at 760 mm. Hg is increased by 2 per cent. for a change in the absolute humidity from 10 to 25 mm. Hg. The following expression is given by Ritz for the variation of breakdown voltage with humidity

$$V = 6 \ 66\sqrt{(d\rho)} + \left[24{\cdot}55 + 0{\cdot}41\left(\frac{e}{10} - 1\right)\right]d\rho, \qquad (7.9)$$

where e is the absolute humidity in mm. Hg. The breakdown voltages given by Ritz in Table 7.2 refer to an absolute humidity of 10 mm. Hg.

There appear to have been few investigations of the impulse breakdown of gaps in uniform fields [5, 25, 26], the results differing but little from those obtained for sphere gaps, which have been more widely studied (see pp. 305–312). In measurements of the impulse breakdown of air in a uniform field Holzer [5] has varied the rate of rise of the impulse voltage up to $9{\cdot}3 \times 10^8$ kV/sec. and finds that for this maximum rate of rise used the impulse breakdown voltage exceeds the A.C breakdown value by only about 2 to 4 per cent. for a 12-cm. gap, the difference becoming less as the gap length is reduced. In Cooper's measurements [26] impulse voltages of 1 μsec. duration are applied recurrently to the gap at the rate of 400 pulses per sec., and the peak value of the impulse voltage required to cause breakdown is measured for various time intervals between the application of the voltage and the occurrence of breakdown. A time interval of 15 sec. then implies that approximately 6,000 applications of the impulse voltage, with the same peak value, will have been made to the gap before breakdown takes place. A small transition region, not exceeding 2 per cent., is recorded by Cooper between the maximum voltage which fails to cause breakdown within 5 min. and the minimum voltage causing breakdown within 1 sec. Results are given in Table 7.4, for an unirradiated gap and for the gap when irradiated by 0·2 mg. of radium. The impulse breakdown voltage is noticeably higher than the A.C. breakdown voltage as given by Bruce [6] for the form of electrodes used by Cooper.

TABLE 7.4

Impulse Breakdown of Uniform Fields in Air

Gap length in cm	Impulse breakdown voltage in kV			A.C. breakdown voltage in kV
	Irradiated gap breakdown within 10 to 30 sec	Unirradiated gap breakdown within 15 to 30 sec	Unirradiated gap breakdown within 1 sec.	
0 05	2 6	3 1	3 4	2 6
0 10	4 6	5 3	5 8	4 4
0 15	6 4	7 0	7 5	6 0
0 20	8 1	8 3	8 8	7 6
0 30	11 4	11 5	12 2	10 6
0 40	14 6	14 7	14 9	13 5
0 50	17 6	17 8	18 1	16 4
0 60	20 6	20 6	20 6	19 2
0 70	23 4	23 4	23 5	22 0

Studies of the breakdown of uniform fields in air at pressures above atmospheric have been made by a number of investigators [14, 27 to 31]. Some curves given by Howell [27] for the variation of the D.C. breakdown voltage with the gas pressure for a number of gaps are given in Fig. 7.2. Similar results have been obtained by Trump, Safford, and Cloud [14] as shown in Fig. 7.5. In all these investigations deviations from Paschen's law are observed, as shown in Table 7.5, which gives the breakdown voltages measured for three values of the product pd [14]. The breakdown voltage decreases appreciably with increasing pressure for a given value of pd. While the reason for this effect may be partly a consequence of the spark mechanism (see p. 258) it is also possible that the presence of pre-breakdown currents in the gap, as discussed below, may be a contributory factor.

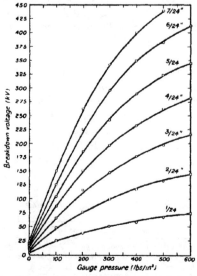

FIG 7.2   Breakdown voltages for plane electrodes
in air at various constant spacings.

TABLE 7.5

pd lb /inch² × inch	Breakdown voltages, in kV, for different gas pressures p in lb /inch²		
	p = 45	p = 105	p = 195
40	180	175	160
80	370	345	320
160		705	635

In the measurements of the breakdown of air at atmospheric pressure there is no noticeable difference between the breakdown voltages recorded for different electrode materials of the type normally used, such as brass or copper [7], provided that the surfaces are clean and smooth. However, the condition of the electrode surface affects the breakdown characteristic to an increasing extent as the gas pressure is increased. This is shown by the curves given in Fig. 7.3, as recorded by Howell [27] for a 0 25-in. gap in air between plane electrodes. The curves denoted 'rough electrodes' represent an extreme case, where the

electrodes have been sand-papered but not polished. When the electrodes
have been polished and have been subjected to a number of sparks the
curves labelled 'partially conditioned electrodes' are obtained. After
prolonged sparking the breakdown voltage rises to the 'conditioned
electrodes' curve. The differences between the various curves are
explained by Howell [27] in terms of small point discharges set up on the

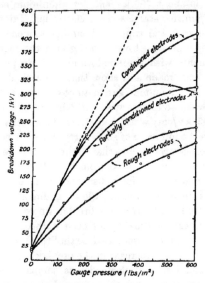

FIG. 7.3. Effect of electrode conditioning on the
breakdown voltage of a 0.25-inch gap between
plane electrodes

minute irregularities on the electrode surfaces. These discharges cause
an appreciable current to flow in the gap before the voltage reaches its
breakdown value, and it is probable that the resultant space-charge
distortion of the field is sufficient to cause a lowering of the breakdown
voltage (see p. 369). The presence of this pre-breakdown current is
confirmed in Howell's experiments. For the gap of Fig. 7.3, at 500 lb./in.²,
the pre-breakdown current at 200 kV is about 40 μA for the rough
electrodes but is less than 1 μA for the smooth electrodes. In the latter
case the current increases to 10 μA at 325 kV and to 30 μA at 375 kV,
breakdown occurring at 385 kV.

Studies of the influence of the surface roughness of the electrodes, and
of the electrode material, on the breakdown characteristics of gases at

high pressures have also been made by Felici and Marchal [30]. Both hydrogen and air have been investigated, at pressures up to 70 atm , for gaps of the order of 1 to 2 mm. and for voltages up to 250 kV. At the lower gas pressures the breakdown voltages for different electrode materials are closely the same, but differences occur at higher pressures and become clearly noticeable, e.g. the breakdown field for a gap of about 1·4 mm. is about 700 kV/cm. for aluminium electrodes and is about 1050 kV/cm. for stainless-steel electrodes. Intermediate values are recorded for iron and copper. A similar tendency is observed in hydrogen, the stainless-steel electrodes giving the highest breakdown fields. Experiments with the anode and cathode plates of different materials have been made and show that the cathode material is the dominating factor: for a 2-mm. gap in air at 60 atm. the breakdown field is 960 kV/cm. with a stainless-steel cathode and an aluminium anode, but is only 705 kV/cm when the electrode materials are reversed. Experiments on the influence of surface roughness on the breakdown of gaps between iron electrodes show that the breakdown field of a particular gap in air at 20 atm. is increased from 310 kV/cm. to 505 kV/cm. by careful polishing (see p. 118 for a discussion of field emission).

In later experiments by Trump, Cloud, Mann, and Hanson [74] the influence of electrode material on the breakdown voltages of compressed air, nitrogen, and carbon dioxide has been studied. Electrodes shaped to the Rogowski contour [3] were used, so that the field can be considered to be uniform. A marked increase in the breakdown voltage gradient is obtained with stainless-steel electrodes as compared with aluminium electrodes. Some curves giving the mean values are reproduced in Fig. 7.4. The curves correspond to the mean values obtained with gaps of $\frac{1}{4}$ in., $\frac{1}{2}$ in., and $\frac{3}{4}$ in., no dependence of gradient upon gap length being observed over this range of gaps within the limits of experimental error. In the measurements the electrodes were conditioned by repeated sparking. With stainless-steel electrodes the final breakdown voltages, as used in the curves of Fig. 7.4, were about 50 per cent. higher than the initial breakdown voltages and were recorded after several hundred sparks. With aluminium electrodes the final breakdown voltage was reached after fewer sparks had occurred. From experiments in which the anode and cathode materials were interchanged it is concluded that the emission characteristics of the cathode are more important than those of the anode in influencing the breakdown voltage, but that nevertheless the breakdown voltage is also affected by the anode material. While field emission from the cathode is considered to be a probable mechanism,

it is suggested that the high fields may also serve to enhance secondary
and photo-electric emission to many times their normal value [84].

The breakdown of uniform fields in gases other than air, at atmospheric and higher pressures, has been studied by Finkelmann [28] for
nitrogen, carbon dioxide, and hydrogen, Palm [32] for nitrogen and
carbon dioxide, Gänger [31] for nitrogen and carbon dioxide, Trump,
Safford, and Cloud [14] for freon, Felici and Marchal [30] for hydrogen,
Weber [33] for freon and freon-air mixtures.

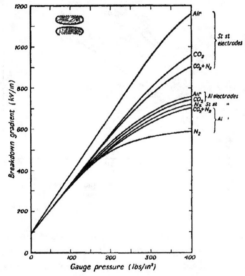

Fig. 7.4. Breakdown voltage gradients for several gases at
high pressures for uniform fields between stainless steel electrodes and between aluminium electrodes

Curves given by Finkelmann [28] for the breakdown voltage of
hydrogen as a function of pd are compared with curves for air in Fig. 7.5
(a). The number against each curve denotes the gap length in centimetres
used in obtaining the curve. For a given gap the deviation from Paschen's
law becomes more pronounced in hydrogen than in air as the pressure is
increased. Fig. 7.5 (b) shows the breakdown curves obtained by Finkelmann for the breakdown of carbon dioxide and nitrogen.

Curves recorded by Trump, Safford, and Cloud [14] for the breakdown
of freon in uniform fields are given in Fig. 7.6. Comparison of these
curves with those for air shows that the breakdown strength of freon at a

given pressure is approximately three times that of air at the same pressure. Measurements by Weber [33] of the impulse and A.C. breakdown characteristics of freon in mixtures show that a small amount of freon in air has a more than proportional effect in increasing the breakdown strength of the mixture (see also p. 340).

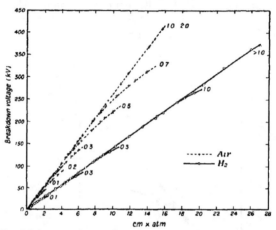

FIG. 7 5 (a) Breakdown voltage between parallel plates in air and hydrogen as a function of the product pressure × gap length The figures against each curve denote the corresponding gas pressure in atmospheres.

Measurements of the breakdown strength of a large number of gases and vapours and of nitrogen saturated with various vapours have been made by Charlton and Cooper [34]. The electrodes used had a spherical shape of 4 in diameter, and were spaced at gaps between 0·125 in. and 0·25 in. The field is therefore nearly uniform. The quantity chosen for comparison is called the 'relative dielectric strength' and is the ratio of the breakdown voltage of the gas or nitrogen-vapour mixture to the breakdown voltage of nitrogen at room temperature and atmospheric pressure, when both are measured in the same gap. The results for some of the gases or mixtures studied are given in Table 7.6.

Similar measurements have been made by Hochberg [35], again apparently for a nearly uniform field, though the exact form of the gap used is not specified. The results are given in Table 7.7.

The relative breakdown strengths of air, CCl_2F_2, and CSF_8 have been compared by Geballe and Linn [36]. Their results do not apply strictly to uniform fields, as although they used plane electrodes of 7·6 cm.

diameter spaced 1·5 cm. apart, the anode plane contained at its centre
a hemispherical boss of 1·6 mm. radius. Curves for the breakdown
voltage of CSF₈ are compared with those of other gases in Fig. 7.7.
It is found that the breakdown voltage increases steadily with each spark
until it reaches a steady value about 10 per cent. above the initial break-

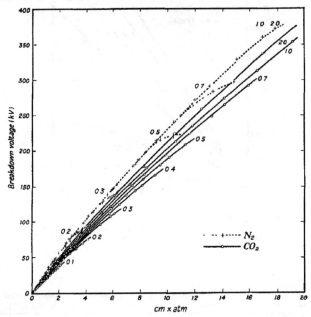

Fig 7 5 (b) Breakdown voltage between parallel plates in nitrogen and carbon
dioxide as a function of the product pressure × gap length. The figures against
each curve denote the corresponding gas pressure in atmospheres.

down voltage. The final or 'saturated' breakdown voltage is shown by
the dashed curve of Fig. 7.7. The reason for this progressive increase in
breakdown voltage is attributed by Geballe and Linn to the decomposition of the gas into CF_4 and SF_4 and the consequent rise in gas pressure.

The high breakdown strengths of gases such as CCl_4, CCl_2F_2, SF_6,
etc., would appear to imply that the ionization coefficient α must be
appreciably lower than in gases such as nitrogen at the same gas pressure
and for the same applied field. The only measurements of α in gases
containing chlorine or fluorine appear to be those of Hochberg and
Sandberg [35, 37], who have reported data obtained for a range of gases

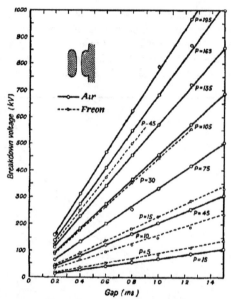

FIG. 7 6 Variation of D.C breakdown voltage of air and
of freon-12 with electrode separation at several absolute
pressures in pounds per square inch. Uniform field.

TABLE 7.6

Gas or N₂-vapour mixture saturated at 23° C and 760 mm. Hg total pressure	Vapour pressure in mm Hg	Boiling-point in °C	Relative dielectric strength†
CCl₃F	725	24 1	3 0
C₂Cl₂F₄	760	3 8	2 8
C₂Cl₃F₃	306	47 2	2·6
CCl₂F₂	760	−28	2 4
BCl₃	760	12 7	2·3
CH₃I	370	42·35	2 2
SO₂	760	−10 0	1 9
CCl₄	105	76 74	1 65
CHCl₃	180	61 28	1 58
Cl₂	760	−34 6	1 55
TiCl₄	12	135 8	1 17
N₂O	760	−89 5	1·14
CH₃Cl	760	−24 4	1 06
CO	760	−191·59	1 02
CH₄	760	−161 4	1 0
C₂H₅Cl	760	12·7	1 0
CO₂	760	−28 5	0 88

† With respect to nitrogen at 1 atm. pressure and room temperature.

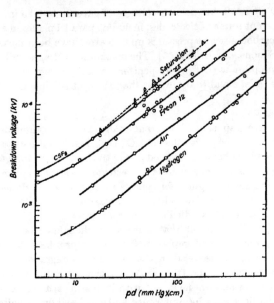

Fig. 7.7. Breakdown voltage curves in CSF$_8$, freon, air, and hydrogen. All data reduced to 0° C

TABLE 7.7

Gas	Relative dielectric strength†
CCl$_4$	6 3
SeF$_6$	4 5
CCl$_3$F	3–4 4
CCl$_3$H	4 3
C$_2$H$_5$I	3 0
C$_2$H$_4$I	2 9
C$_2$Cl$_2$F$_2$	2 8
SF$_6$	2·3–2·5
SOF$_2$	2·5
CCl$_2$F$_2$	2·4–2 5
SO$_2$	1 9–2 3
C$_2$H$_5$Br . . .	1·5
C$_2$H$_5$Cl	1 25
CF$_4$ . . .	1·1
N$_2$	1 0
CO$_2$ . . .	0 9

† With respect to nitrogen at 1 atm. pressure and room temperature.

of widely varying properties. Discharge mechanisms which would account for low values of α have been suggested by Warren, Hopwood, and Craggs [39].

The 60 c./s. A.C breakdown voltages of gaps up to 0·25 in. between three different forms of electrodes, including parallel planes in a number of fluorocarbon gases at pressures up to 3 atm. have been measured by Wilson, Simons, and Brice [160] The breakdown voltages for butforane (C_4F_{10}), propforane (C_3F_8), and pentforane (C_5F_{12}) were found in most instances to be equal to or greater than those of sulphur hexafluoride.

Sphere gaps

The sphere gap is used internationally as an instrument for the measurement of the peak value of A.C , D C., and impulse voltages, and calibration tables have been issued giving the breakdown voltages corresponding to different gap lengths between various sizes of sphere [7, 41] These tables include figures for spheres up to 200 cm. in diameter and voltages up to 2·5 million volts. The breakdown characteristics have therefore been widely studied by numerous investigators [10, 13, 15, 42 to 56, 60] with the result that many more data have been obtained concerning this type of gap than for the uniform-field gap between parallel plates, as discussed in the preceding paragraphs. For small gaps between large spheres the breakdown characteristics are closely the same as for the uniform field, but with increasing spacing between the spheres the field loses its uniformity and the breakdown voltage falls below that for the uniform field. The greater the diameter of the spheres the greater is the gap length to which the spheres can be separated before the breakdown voltage falls below that for the uniform field.

Values for the A.C. breakdown voltages of a number of gaps in air at 760 mm. Hg and 20° C. for several sizes of sphere, are given in Table 7.8 (see p. 313 for the calibration of spheres of 2 cm. diameter). This table is taken from the 'British Standard Rules for the Measurement of Voltage with Sphere Gaps' [7], hereafter referred to as B S.358, and applies to gaps in which one sphere is earthed. The values given are stated by B S 358 to be accurate to within 3 per cent. provided that the gap d is less than the sphere radius R. The error is likely to be greater for $R < d < 1·5R$ and, in the range $1·5R < d < 2R$ the breakdown voltages listed are regarded as of doubtful accuracy and are given in brackets. B.S.358 includes specifications for the mounting of the spheres, and for the position of the sphere gap in relation to neighbouring bodies, if the calibration tables are to apply. Details are also given concerning the surface and conditioning of the spheres. It is recommended that spheres of brass, bronze, steel, copper, aluminium, or light alloys be used.

The breakdown voltage of a sphere gap decreases with decreasing gas density, the variation being directly proportional to the density for $0.9 < \rho < 1.05$, where $\rho = 1$ is the gas density at 20° C. and 760 mm. Hg. For larger changes in gas density the direct relationship between gas density and breakdown voltage ceases to obtain, and the breakdown voltage is given by

$$V = KV_n, \tag{7.10}$$

TABLE 7.8

A.C Breakdown Voltages (in kV) of Gaps in Air at 20° C. and 760 mm. Hg between Spheres of Different Diameters [7]

Gap in cm.	Sphere diameter in cm.					
	6 25	12 5	25	50	100	200
1	31 9	31 5	31		. .	
1 5	45 9	45 6	45			.
2	58 2	59 2	59	58		.
2 5	69 6	72 0	72	72	71	
3	79 1	85 2	86	85	84	.
4	94·8	109	112	112	112	
5	.	129	137	137	137	137
6	.	146	161	164	163	163
8		174	205	214	215	215
10	.		243	243	266	265
15	. .		314	372	387	389
20			. .	461	503	510
25	.			532	611	630
30	.			591	709	745
35			.	640	797	858
40			.		876	965
45		.			949	1,070
50	. .	. .			1,010	1,180
75		.		.	1,240	1,600
100		.	.	.	. .	1,930
150		.			. .	2,350

where V_n is the breakdown voltage at 20° C. and 760 mm. Hg, and K is a factor the value of which is given in Table 7.9 [7]. The significance of the factor K in terms of gap length and sphere diameter has been discussed by Whitehead and others [51, 57, 58].

TABLE 7.9

ρ .	. 0 70	0 75	0 80	0 85	0 90	0 95	1 0	1 05	1·10
K .	0·72	0 76	0·81	0 86	0 90	0 95	1 0	1 05	1·09

In the specifications for the use of sphere gaps [7, 41] it is stated that the breakdown voltage is independent of the humidity of the atmosphere over the range of gaps covered by the calibration tables, i.e. for $d < R$, provided that dew is not deposited on the sphere surfaces. However, small variations may be expected with changes in humidity if the results obtained with uniform gaps are correct (see p. 295). In recent measurements of the breakdown of a 0·4-cm. gap between spheres of 2 cm. diameter Lewis [59] has found that the breakdown voltage increases by 0·1 per cent. per mm Hg increase of water vapour in the atmosphere.

The calibration given in Table 7.8 is stated [7] to apply to the breakdown of the gaps by negative impulse voltages, with an accuracy of 3 per cent., provided that $d < R$ and that the time to peak value of the impulse is at least 1 μsec and the time for the wave to decay from its peak to half-value is at least 5 μsec. Separate calibration tables are issued [7] for the breakdown of sphere gaps by positive impulse voltages, the positive impulse values being higher than the negative values for gaps greater than about 0·5R. For instance, with spheres of 25 cm. diameter, a difference is noted [7] in the breakdown voltages for a gap of 5 cm., when the positive and negative impulse breakdown voltages are given as 138 and 137 kV respectively. At a radius spacing, for the same spheres, the corresponding values are 298 and 282 kV.

This difference in the breakdown voltage characteristics of sphere gaps for positive and negative impulse voltages has been studied by a number of investigators [42, 46, 47, 56, 60, 61, 62]. McMillan and Starr [60] measured breakdown voltages as a function of gap length between spheres of 6·25 cm. diameter and of 25 cm. diameter when subjected to A.C. and to impulse voltages. For short gaps the breakdown curves coincided but with increasing spacing the positive impulse breakdown curve rose above the negative impulse breakdown curve, which was found to coincide with the A.C. breakdown curve. The difference between the positive and negative impulse breakdown voltages, for a gap of 15 cm. between 25 cm. diameter spheres, amounted to 20 per cent. The measurements were continued by McMillan [61], who explained the results in terms of space charges in the gap. No studies were made of the influence of irradiation on the behaviour of the gaps.

Experiments by Fielder [62], Meador [56], Bellaschi and McAuley [46], Dattan [47], and Davis and Bowdler [42] show that similar results are obtained with larger spheres, up to 200 cm. diameter. It was noticed that spheres of 6·25 cm. and 12·5 cm. diameter behave more erratically than spheres of larger diameter, and the suggestion is made by Meador [56] that this may be an indication of time lag, but the effect of irradiation was not studied. In the measurements by Davis and Bowdler [42], with breakdown voltages up to 1,000 kV, it was found that the negative impulse breakdown voltage may exceed the A.C. breakdown value by up to about 6 per cent for a gap $d = R$ and the corresponding difference for the positive impulse breakdown voltage may be as great as 10 per cent. The percentage difference between the positive and negative impulse breakdown voltages, at $d = R$, was not more than 4 per cent. as compared with a difference of up to 8 per cent in Meador's results. The difference between the positive and negative values given by Dattan [47] varies up to 15 per cent. for spheres spaced at $d = R$.

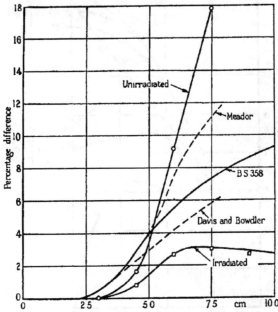

Fig 7 8 Variation with gap length of the difference between the positive and negative impulse breakdown voltages (50 per cent values), expressed as a percentage of the negative value, for a 12 5-cm. diameter sphere gap.

The curves labelled 'irradiated' and 'unirradiated' were obtained with a 1/50 wave.

Recently it has been pointed out by Meek [63, 64] that the difference between the positive and negative impulse breakdown characteristics is greatly reduced, for gaps between spheres of 25 cm. diameter or less, by the use of adequate irradiation (see also Chapter VIII). Meek's results for the percentage difference between the positive and the negative impulse breakdown voltages of gaps between spheres of 12·5 cm. diameter, when unirradiated and when irradiated by 0 5 mg. of radium, are plotted as a function of the gap length in Fig. 7.8, which

includes also the curves obtained by Meador [56] and by Davis and
Bowdler [42] for this size of sphere. Meador's results would appear to
have been obtained under conditions of weak irradiation, whereas it is

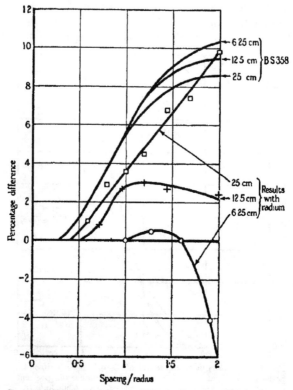

Fig. 7.9 Percentage difference between the positive and negative
impulse breakdown voltages as a function of the ratio of spacing to
sphere radius, for spheres of 6 25 cm., 12 5 cm., and 25 cm diameter.

possible that stronger irradiation, possibly by illumination from the
impulse generator spark gaps, may have been present in Davis and
Bowdler's experiments. The B S.358 curve [7] representing the mean
of a number of investigations is roughly intermediate between the unirradiated and the irradiated curves.

The results obtained by Meek [64] for the percentage difference
between the positive and negative impulse breakdown voltages for
irradiated gaps between these sizes of sphere are plotted in Fig. 7 9 as

a function of d/R, i.e the ratio of the gap to the sphere radius. The corresponding curves deduced from the British Standard Rules are also plotted. No polarity difference is noted by Meek for the 6.25-cm. spheres up to $d/R = 1$. For $1 < d/R < 1{\cdot}6$ the positive impulse voltage exceeds the negative value by a small amount but for $d/R > 1\ 6$ the positive impulse voltage falls below the negative value, by an amount of 6 per cent. at $d/R = 2$. At the latter value the values given in the British Standard Rules [7] show the positive impulse voltage to exceed the negative impulse voltage by about 10·4 per cent.

It appears therefore that even when precautions are taken to irradiate the gap and so to reduce the scatter in the positive impulse breakdown voltages there is still a difference between the positive and negative impulse breakdown voltages for spheres of 12·5 cm. diameter or more at a radius spacing. A possible explanation of this effect has been given by Meek [64]. In a sphere gap, with one sphere earthed, the field in the gap is asymmetrical when the spacing is about a sphere radius or more. Under such conditions the field distribution in the gap remains the same whether the high-voltage sphere is of positive or negative polarity The avalanche initiating the breakdown for a positive high-voltage sphere moves towards the sphere into a region of increasing field strength, and the criterion for breakdown is that the space-charge field at the avalanche head E_a should equal the field E_R at the sphere surface. With a negative high-voltage sphere the avalanche moves away from the sphere, and the criterion for breakdown is that $E_a = E_{R+l}$, the field at a distance l from the sphere. The breakdown for the negative sphere may then be expected to occur at a voltage slightly lower than for the positive sphere. With longer gaps where corona is observed before breakdown, the positive breakdown voltage is less than the negative breakdown voltage, as a higher voltage is required to propagate a negative streamer than a positive streamer across the gap. For further details of the above and related effects in sphere-gap breakdown the reader is referred to Chapters VI and VIII.

According to B.S.358 the breakdown calibration curve of a sphere gap for negative D.C. voltages is the same as for negative impulse and A.C. voltages. Similarly the positive D.C. breakdown voltage curve is the same as the positive impulse curve. There appears to be little experimental evidence [13, 64, 161] on which to base the D.C. calibration, and for this reason presumably a possible error of 5 per cent. is quoted in B.S.358 for the tables if applied to D.C. breakdown, provided the spacing does not exceed a sphere radius. According to measurements by Meek [64] the difference between the positive and negative D.C. breakdown voltages is negligible for gaps up to a radius spacing between 6·25 and 12·5 cm. diameter spheres Bouwers and Kuntke [161] have measured D.C. breakdown voltages for gaps between spheres of 100 cm. diameter, with one sphere earthed, and find little difference between the curves for positive and negative D C. breakdown for gaps up to 40 cm.; the D.C. breakdown voltages are slightly lower than the corresponding A.C. values given in Table 7.8.

For a given gap Table 7.8 shows that the breakdown voltage is generally higher for the larger spheres, the field then being more nearly uniform.   However, at certain low gap settings a slightly higher voltage appears to be required to cause breakdown between spheres of small diameter than for larger spheres at the same setting. It is possible that this may be the result of inadequate irradiation of the gap in the experiments on which the results of Table 7.8 are based. Irradiation becomes increasingly necessary as the gap length and sphere size are reduced because of the decrease in sparking area (see Chapter VIII).

Toepler [67] has shown that the measured breakdown curve for a sphere gap can be matched by two empirical curves, and that there is a discontinuity in the slopes of the two curves at their meeting-point For gap lengths shorter than that at which the discontinuity occurs the breakdown voltage is independent of the sphere diameter.   The critical gap length d_K is given by Claussnitzer [68] for several sphere diameters, the value being 7·15 cm. for spheres of 25 cm. diameter. For gaps up to about 0·5d_K Claussnitzer finds that the breakdown voltage corresponds closely to that for the uniform field. Between 0·5d_K and d_K the breakdown voltage is higher than that for the same gap between spheres of larger diameter, the deviation reaching a maximum at about 0·8d_K. For gaps greater than d_K the breakdown occurs at a voltage less than that for the more uniform field between larger spheres. A detailed analysis of the effect is given by Jørgensen [157]. Meek [63, 64, 65] has suggested a possible explanation of the effect, namely that for gaps up to 0·5d_K the electron avalanche leading to breakdown travels completely across the gap, whereas for gaps greater than d_K, because of the low field in the mid-gap region, the electron avalanche only travels across part of the gap in the high-field region near one of the electrodes (see Chapter VI). On this basis a transition region may be expected for gaps between 0 5d_K and d_K. Support for the suggested mechanism is given by the results of Hueter [43] and Dattan [47], who find an increase in the scattering of breakdown voltage measurements in the region between $0.5d_K$ and d_K. Dattan, in measurements with spheres of 25 cm. diameter, records a scattering of more than 2 per cent for gaps between 5·25 and 7·75 cm , the maximum scattering reaching 6 per cent.

A number of investigations have been made of the breakdown voltage required to cause breakdown of sphere gaps as a function of the time interval between the application of the voltage wave and the ensuing breakdown [46, 69, 70, 71]. In the measurements by Bellaschi and Teague [69] the breakdown of several gaps between three sizes of sphere has been studied, the impulse voltage having a wavefront of 1·5 μsec. and a wavetail of 40 μsec. Some of the results are given in Table 7.10. The impulse ratio as given in Table 7 10 is the ratio of the voltage causing breakdown at the stated time to the voltage causing breakdown after 2 μsec. This ratio increases gradually with decreasing time-to-breakdown (see Chapters IV and VIII). In similar measurements by Hagenguth [70, 71], for gaps between spheres of 25 cm and 50 cm. diameter, it is found that for a 4-cm. gap between 25 cm. spheres, with an impulse voltage rising at the rate of 1,000 kV/μsec , the respective overvoltages required to cause breakdown in times of 1, 0·5, and 0·2 μsec are 13 per cent., 54 per cent., and 91 per cent. Hagenguth has also shown that ultraviolet light from a mercury arc-lamp has no effect on the results, so that it may be assumed that the overvoltage required to cause breakdown in the shorter times is necessitated by the mechanism of formation of the spark rather than by a statistical time lag. These effects are discussed in further detail in Chapters IV and VIII, which also include additional data for shorter gaps

The impulse breakdown of short gaps between spheres of 2 cm. diameter has been studied by Cooper [72], who used impulses of duration 0·1 to 4·0 μsec. (see also p 296) The impulses were applied recurrently to the gap at repetition rate of between 100 and 3,000 pulses/sec Some of his results for gaps of various lengths when irradiated by 0·2 mg. of radium are given in Table 7.11, which includes also the D.C. breakdown voltage of the gap [72]. For gaps between 0·2 and 0 7 cm. the impulse breakdown voltage exceeds the D.C. breakdown value by almost 6 per cent., the difference increasing for shorter gaps. The D.C. breakdown voltage of gaps up to 1 cm. between spheres of this size is the same as the A.C. breakdown voltage [13].

The breakdown of gaps between spheres in air at pressures above atmospheric has been studied by Zeier [73], Reher [29], Skilling [75], Skilling and Brenner [76], Trump, Safford, and Cloud [14], and Gossens [77]. In these measurements the gaps used are generally small compared

TABLE 7.10

Ratio of the Breakdown Voltage at Different Times to Breakdown to the Breakdown Voltage at 2 0 μsec , for Different Sphere Gaps in Air at N.T.P.

Sphere diameter in cm.	Polarity	Ratio of gap to sphere diameter (expressed as percentage)	Ratio		
			1 0 μsec.	0 6 μsec	0 2 μsec.
6 25	+	16	1 17	1 35	1 71
		32	1 07	1 17	1 43
		64	1 06	1 22	1 47
25	+	16	1 13	1 22	1·35
		32	1 07	1 14	1 27
		76	1 06	1 13	1 27
200	+	4	1 03	1 10	1 64
		8	1 01	1 09	1 44
		16	1 01	1 09	1 35
		32	1 00	1 06	1 23
6 25	—	16	1 19	1 41	1 75
		32	1 10	1 27	1 54
		64	1 08	1 24	1 53
25	—	16	1 18	1 35	1 61
		32	1 12	1 19	1 39
		64	1 08	1 23	1 48
200	—	4	1 03	1 16	1 75
		8	1 03	1 17	1 57
		16	1 04	1 15	1 42
		32	1 01	1 09	1 27

TABLE 7.11

Comparison between the Impulse Breakdown Voltage, for Recurrent Impulses, and the D.C. Breakdown Voltage of a 2-cm. Diameter Sphere Gap

Gap length in cm .	0 05	0 10	0 20	0 30	0 40	0 50	0 60	0 70
Recurrent impulse breakdown voltage in kV .	3 5	5 1	8 5	11 7	14 9	18 1	21 2	24 1
D C breakdown voltage in kV	2 8	4 6	8 0	11 1	14 1	17 1	20 0	22 7
Per cent. difference in values	25 0	10 8	6 3	5 4	5 7	5 7	6 0	6 2

with the sphere diameter and the results are therefore closely the same as for uniform fields (see p. 292). Deviations from Paschen's law are recorded.

Zeier [73] has investigated the breakdown of nitrogen and of carbon dioxide, and notes deviations from Paschen's law at the higher gas pressures. In nitrogen, for a 1·0-mm. gap between spheres of 30 cm. diameter the deviation begins at a pressure of 20 kg./cm.^{2} approximately. Breakdown measurements for sphere gaps in nitrogen and freon, and in mixtures of these two gases have been made by Nonken [78], Skilling and Brenner [79], and Camilli and Chapman [80], with results in general

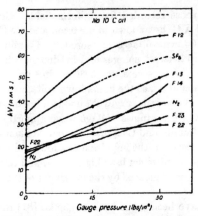

Fig. 7.10. Breakdown voltage curves at 60 c/s. for various gases between 1 in diameter spheres spaced at 0 25 in. The breakdown voltage for a transformer oil is included for comparison F 12, CF_2Cl_2, F 13, CF_3Cl, F 14, CF_4, F 22, $CHClF_2$, F 23, CHF_3.

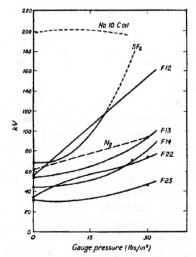

Fig. 7 11 Impulse breakdown voltage curves for various gases between 1 in. diameter spheres spaced at 0 25 in. The breakdown voltage for a transformer oil is included for comparison. F 12, CF_2Cl_2, F 13, CF_3Cl, F 14, CF_4, F 22, $CHClF_2$, F 23, CHF_3.

agreement with those recorded for uniform fields Gossens [77] has studied D.C. and A.C. breakdown in air-freon and air-CCl_4 mixtures

The A.C. breakdown voltages measured by Camilli and Chapman [80] for a $\frac{1}{4}$-in. gap between spheres of 1 in diameter in SF_6 and other halogenated gases, at pressures up to 30 lb./in.^{2} gauge, are shown in Fig 7.10, which includes also the breakdown voltage of a typical transformer oil. Fig. 7 11 gives the curves obtained for breakdown of the same gap with a 1·5/40 μsec. impulse voltage wave applied to the gap In the discussion of Camilli and Chapman's paper Nonken points out that the impulse ratio between the impulse breakdown voltage and the A.C. breakdown voltage in nitrogen is as high as 2·7, and considers that the results may have been influenced by the proximity of the insulating walls of the test-chamber.

Experiments have been made by Kowalenko [81] to determine the breakdown voltages of a 0·4-cm gap between spheres of 0·43 cm. radius in a number of gases at 30 mm. Hg and 300° K. The results are given in Table 7.12. An inconclusive attempt is made by Kowalenko to correlate the breakdown strengths with other characteristics of the gases such as molecular weight, mean free path, ionization potential, and dissociation potential.

TABLE 7.12

Gas	H_2	O_2	N_2	NO	HCl	HBr	HI
Breakdown kV	1·06	1 46	1 66	1 70	2 48	3 66	5 22

Sphere-plane gaps

The breakdown voltage characteristics of gaps in air between a sphere and a plane have been studied by Klemm [10], Goodlet, Edwards, and Perry [23], Bellaschi and Teague [82], Allibone, Hawley, and Perry [83] for spheres of several diameters. Typical curves [23] obtained with A.C. voltages are given in Fig. 7.12. The curves 'A' correspond to the voltage at which corona discharges can be detected. For the shorter gaps the corona-onset voltage cannot be distinguished from the breakdown voltage but for longer gaps the voltage required to cause breakdown, as given by the curves 'B' is appreciably higher than that for corona onset. With a sphere of given size the field distribution in the gap tends to that of the point-plane gap as the gap is increased, and consequently the breakdown voltage characteristics of the sphere-plane gaps are roughly the same as those for a point-plane gap at the larger gaps.

Curves for the impulse voltage breakdown of sphere-plane gaps are shown in Fig. 4.38 [83] With the 12·5-cm. diameter sphere there is a

discontinuity in the curve at a gap of about 20 in. but no theoretical explanation of this effect appears to have been given.

The breakdown of sphere-plane gaps in air and in nitrogen at pressures
above atmospheric has been studied by Ganger [31]. Curves for the
breakdown voltage of air as a function of gap length for two sphere sizes
and for various gauge pressures up to 40 atm. are given in Figs. 7.13 and
7.14. The breakdown voltages in compressed nitrogen are generally

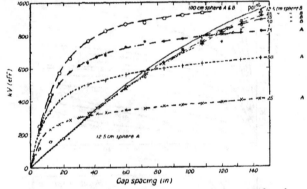

FIG. 7.12. Breakdown voltage and corona onset voltage curves for sphere-
plane gaps The point-plane breakdown voltage curve is added for comparison. (Curves *A* are for corona onset Curves *B* are for breakdown)

lower than in air under the same conditions. Measurements of sphere-
plane breakdown in compressed gases have also been carried out by Zeier
[73] for air, N_2, N_2-air, and CO_2.

Coaxial cylinders

There appears to have been little published work since about 1930
dealing with the breakdown of gaps in air between coaxial cylinders.
The experimental data before 1929 are well summarized by Whitehead
[112] and by Peek [113].

Fig. 7.15 shows some curves given by Uhlmann [86] for the D.C. and
A.C. breakdown voltages of gaps between coaxial cylinders in which
the radius of the outer cylinder is 5 cm. The results show that for an
inner cylinder of radius less than about 0·5 cm. corona is observed at
voltages lower than the breakdown value, the corona onset voltage
being indicated by the dashed curves. For inner cylinders of radii
greater than 0·5 cm. no pre-breakdown corona is detected and it may
be assumed that as soon as a corona streamer is formed the voltage
between the electrodes is sufficient to enable the streamer to cross the

gap and cause breakdown [66]. Other curves are given by Uhlmann [86] for outer cylinders of radii 3 cm. and 10 cm.

Some curves obtained by Howell [27] for the D.C. breakdown voltages of four different gaps in air at pressures above atmospheric are given in

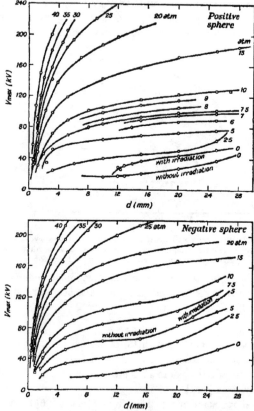

Fig. 7 13   Breakdown voltage curves in air, at various gauge pressures in atm , between a sphere of 5 mm. diameter and a plane.

Fig. 7.16. For the case where the inner cylinder is $\frac{1}{2}$ in. in diameter and the outer cylinder is $1\frac{1}{16}$ in. in diameter, there is a distinct polarity effect at the higher gas pressures For the larger cylinders a polarity effect is again noted, but there appears to be no definite positive D.C. breakdown voltage which can be reproduced, the value ranging from about 50 to

80 per cent. of the negative D.C. breakdown value. The positive breakdown voltage could be increased by conditioning of the electrodes (see
p. 297) but the improvement was often suddenly terminated by a burst

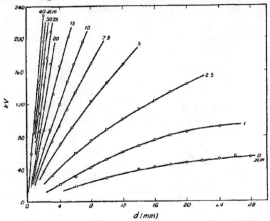

FIG. 7.14. Breakdown voltage curves in compressed air, at different
gauge pressures, between a 5-cm diameter sphere and a plane.

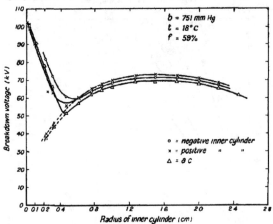

FIG. 7 15 Breakdown voltage curves in air at 751 mm Hg between
coaxial cylinders as a function of the radius of the inner cylinder.
The radius of the outer cylinder is constant at 5 cm.

of sparks and a great reduction in the breakdown voltage. With further
operation the breakdown voltage again increased, but the lowering
was ultimately repeated. For a negative inner cylinder the breakdown

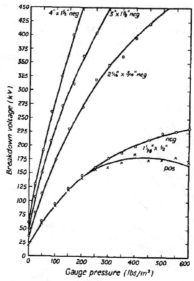

Fıg 7 16. Breakdown voltage curves for coaxial cylinders
in compressed air. The cylinder diameters, and the inner
cylinder diameter, are shown.

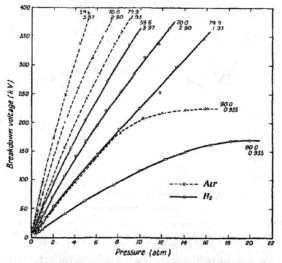

Fıg. 7.17. Breakdown voltage curves in compressed air and hydrogen between coaxial cylinders Tho figures against the curves give
the radius of the inner cylinder in mm. and the gap in cm.

voltage increased to a steady value with conditioning of the electrodes, as in the case of plane electrodes (see p. 297). Throughout the experiments tiny flashes could be detected on the inner cylinder at the instant of sparking. As with the plane electrodes, pre-breakdown currents were associated with roughened electrodes, and could be reduced by conditioning.

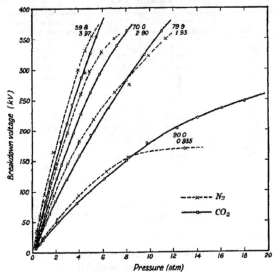

Fig 7.18. Breakdown voltage curves in compressed nitrogen and carbon dioxide between coaxial cylinders. The figures against the curves give the radius of the inner cylinder in mm, and the gap length in cm.

Other curves recorded by Finkelmann [28] for breakdown in air, and also in hydrogen, are given in Fig. 7.17. The curves refer to different sizes of inner cylinder, the outer cylinder having a constant radius of 9·97 cm. Finkelmann [28] has also measured the breakdown voltages between coaxial cylinders in compressed nitrogen and carbon dioxide. The curves for these gases are shown in Fig 7.18. Other measurements on compressed nitrogen and carbon dioxide have been made by Palm [32] and by Bölsterli [114]. The latter has also investigated the breakdown of compressed helium. Results for the D.C. and A.C. breakdown voltages between coaxial cylinders in compressed air, air-CCl$_4$, and air-CCl$_2$F$_2$ mixtures are given by Gossens [77].

Thornton [115] has measured the corona starting gradients in wire-cylinder gaps and has extrapolated the values to obtain the corresponding gradients for large wire diameters, i.e. effectively for parallel plane gaps. A 50 c./s. alternating voltage was used in the experiments. The gradients are given in Table 7.13 for a number of gases, at 760 mm. Hg and 0° C., and are selected from a longer list in Thornton's paper.

TABLE 7.13

Gas	Corona starting gradient kV/cm.	Gas	Corona starting gradient kV/cm
Air	35 5	CH_4	22 3
H_2	15 5	CH_2Cl_2	126
He	4	$CHCl_3$	162
Ne	4 5	CCl_4	204
A	7·2	CH_3Cl	45 6
Kr	9 5	C_2H_5Cl	109
O_2	29 1	C_3H_7Cl	161
N_2	38 0	C_4H_9Cl	200
Cl_2	85 0†	$C_5H_{11}Cl$	264
CO	45 5	CH_3Br	97
CO_2	26·2	C_2H_5Br	98
NH_3	56·7	C_3H_7Br	155
N_2O	55 3	CH_3I	75
H_2S	52 1	C_2H_5I	102
SO_2	67 2	$C_2H_4Cl_2$	240
CS_2	64 2		

† This value appears to be too high [34].

Point-plane gaps

Measurements of the breakdown voltages of short gaps in air between a point and a plane have been made by several investigators [85 to 89]. Curves of the D C breakdown voltages for gaps up to 1 cm. as recorded by Strigel [87, 88] are shown in Fig. 7.19. While these curves refer to gaps between a point and a sphere of 5 cm. diameter, the results may be considered to apply closely to point-plane gaps, particularly for the shorter gaps. Fig. 7.19 also includes curves for a point-point gap and for a gap between spheres of 5 cm. diameter. Curves given by Uhlmann [86] for the positive and negative D.C. breakdown voltages of gaps up to about 10 cm. between a point and a plane are shown in Fig. 7.20. If the point is replaced by a cylindrical rod of 0 4 cm. diameter with a hemispherical tip the curves of Fig. 7.21 are obtained.

With increasing gap length the exact shape of the point becomes of decreasing importance in its influence on the breakdown voltage, and in many investigations a rod with a $\frac{1}{2}$-in. or $\frac{3}{8}$-in. square section and a square-cut end is used [82, 83]. Curves giving the positive and negative impulse characteristics of gaps between a 0·5-in. square rod and a plane

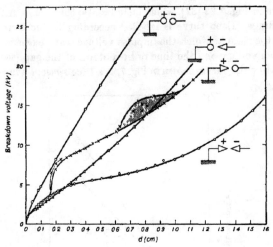

FIG. 7.19. D.C breakdown voltage curves in air for various combinations of point and sphere electrodes. The sphere diameter is 5 cm

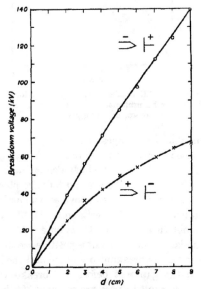

FIG. 7.20. D.C. breakdown voltage curves for air between a 30° conical point and a plane

are shown in Fig. 7.22 [82], which includes also the 60 c./s. A.C. break-down voltage. Each curve is labelled according to the time interval between the time at which the impulse voltage wave exceeds the A.C. breakdown voltage and the time of breakdown of the gap (see p. 214). In all the impulse curves shown in Fig. 7.22 a 1·5/40 μsec. wave was used.

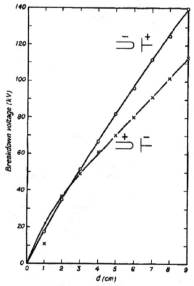

FIG. 7.21. D.C. breakdown voltage curves for air between a hemispherically-ended rod, of 0 4 cm. diameter, and a plane.

The impulse breakdown voltages of point-plane gaps in air for gaps up to 450 cm. have been measured by Gorev, Zalesky, and Riabov [90]. The results are given in Table 7 14 which also includes impulse break-down voltages for horizontal rod-gaps and for wire-plane gaps.

The breakdown of air at pressures above atmospheric in point-plane gaps has been studied by several investigators [27, 31, 75, 76, 77, 92, 93, 94]. Some results obtained by Howell [27] for a 0·25 in. gap between a needle point and a plane when the point is positive are shown in Fig 7 23. The upper curve gives the breakdown voltage and the lower curve the current flowing in the gap just before the breakdown occurs. The breakdown voltage rises until a pressure of about 100 lb./in.² is reached, when its value becomes uncertain and may have any value between the

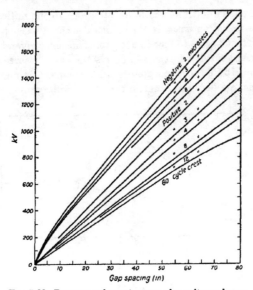

FIG. 7.22. Positive and negative impulse voltage characteristics of rod-to-plane gap.

TABLE 7.14

Positive and Negative Impulse Breakdown Voltages of Point-plane Gaps, Rod Gaps, and Wire-plane Gaps in Air at 20° C., 760 mm. Hg, and Absolute Humidity 11 gm./m.³, 1·5/40 μsec. impulse wave. Voltages given in kV

Gap in cm	Point-plane +	Point-plane −	Rod gap +	Rod gap −	Wire-plane +	Wire-plane −
100	539	945	662	720	570	912
125	673	1,140	803	885	705	1,140
150	807	1,305	945	1,045	841	1,328
175	941	1,502	1,090	1,205	976	1,515
200	1,075	1,670	1,230	1,370	1,112	1,682
225	1,209	1,830	1,370	1,530	1,247	1,845
250	1,343	1,980	1,515	1,695	1,383	2,000
275	1,477	2,125	1,655	1,855	1,518	..
300	1,611	2,260	1,800	2,020	1,654	..
325	1,745	.	1,940	2,155	1,789	..
350	1,879	..	2,080	2,285	1,925	.
375	2,013	..	2,210	2,400	..	.
400	2,147	.	2,320		.	..
425	2,281		2,425	..	.	..
450	2,415		2,512			

two curves shown until at 150 lb./in.² it drops to the initial potential curve. The initial potential is the corona discharge onset potential at which a current begins to flow in the gap. The curve given in Fig. 7.23 for the current flowing in the gap shows that at pressures above the initial value, 150 lb./in.², no pre-breakdown current is measured, i.e. corona discharge does not occur in the gap at voltages lower than the

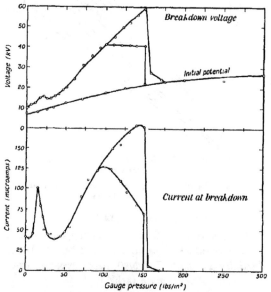

Fig. 7.23   Voltage and current characteristics for a 0 25 in. gap in air
between a positive point and a plane.

breakdown value. The sudden rise in the current curve at lower pressures, of about 10 to 15 lb./in ², is associated with a change in the appearance of the discharge   At atmospheric pressure the dull point maintains a glow just at the tip of the needle, with sometimes a faint beam extending across to the cathode. With a small increase in the gas pressure the discharge suddenly lengthens and the current rises. At first it extends to the cathode, but with further increase in gas pressure it shortens and the current drops rapidly. At still higher pressures the current again increases but the length of the visible discharge decreases   The shortened corona takes on a variety of forms and each gives a different result so that measurements are difficult to reproduce.

Visual observation of the sparks shows that at pressures below about 100 lb./in.^{2} they develop more or less straight across the gap from the positive point to the plane   At higher pressures the sparks follow increasingly longer paths and may terminate on the plane at a distance more than the gap length from the perpendicular to the plane through

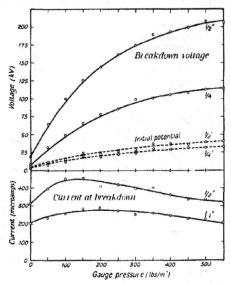

Fig. 7 24. Voltage and current characteristics for a 0 25 in. gap in air between a negative point and a plane.

the point. At pressures above the critical pressure the spreading of the sparks falls and they follow the direct path from the point to the plane.

The reasons for the observed behaviour in the breakdown of gaps in air between a positive point and a plane have been discussed by Howell [27] and also by other investigators [95]. It appears that the peak in the voltage characteristic may be explained by the redistribution of the field in the gap caused by space-charge accumulation. At the lower gas pressures, where corona discharge occurs at voltages lower than the breakdown value, the pre-breakdown current flowing in the gap produces space charges which tend to make the field more uniform, and the voltage required to cause breakdown of the gap is therefore increased. At higher gas pressures, above that corresponding to the voltage peak, it would appear that as soon as a streamer is formed at the point the voltage is

sufficient to enable the propagation of the streamer across the gap.
Breakdown consequently occurs before space charges can accumulate
in the gap and affect the field distribution; under such conditions no
steady corona discharge is observed to precede breakdown. The path
taken by the spark is also modified by the presence of space charge and,
at gas pressures below that corresponding to the voltage peak, tends to

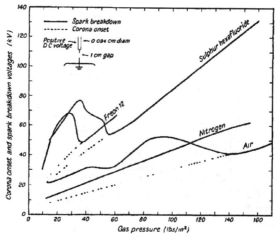

FIG. 7 25. Voltage characteristics for a positive point-plane gap
in compressed air, nitrogen, freon, and sulphur hexafluoride.

occur along the outside of the space charge rather than towards the
plane through the centre of the space-charge region where more uniform
field conditions appear to prevail.

Curves given by Howell [27] for the breakdown voltages between a
negative point and plane, for gaps of $\frac{1}{4}$ in. and $\frac{1}{2}$ in , are reproduced in
Fig. 7.24. The corresponding currents flowing before breakdown are
also shown. Sparks from a negative point were found to occur along
roughly the shortest path between the electrodes. Corona discharge
was detected at voltages below the breakdown value over the whole gas
pressure range, up to 550 lb./in.², and therefore appreciable space-charge
modification of the field must be present The peak observed in the
positive point-plane breakdown curve is absent in negative point-plane
breakdown. This may possibly be explained by the consideration that
in negative point corona over the gas pressure range studied, the corona
discharge is localized at the point and streamers do not immediately

develop across the gap at corona onset to cause spark breakdown. However, the exact influence of the mid-gap space charges in positive and negative point-plane breakdown appears as yet to be unknown.

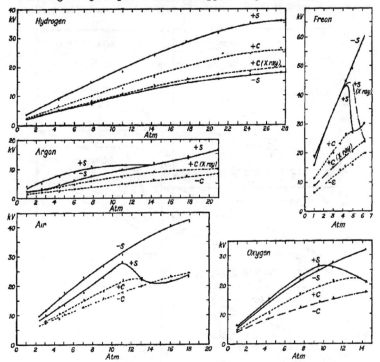

Fig 7 26 Voltage characteristics for various gases between a point and a plane spaced at 0 3 cm Curves S give the spark breakdown voltage. Curves C give the corona onset voltage.

The critical pressure at which the peak in the breakdown characteristic occurs, for gaps between a positive point and a plane, has been studied by Howell [27] for nitrogen and for nitrogen-helium mixtures for various gap lengths. The result for a ¼-in. gap gave a critical pressure of 190 lb./in.² for nitrogen, 500 lb./in.² for helium mixed with 3 per cent. of nitrogen, and 350 lb./in.² when equal partial pressures of the gases were used.

Other studies of the breakdown of point-plane gaps in nitrogen have been made by Goldman and Wul [96, 97, 98], Skilling and Brenner [79], Pollock and Cooper [99], Camilli and Chapman [80], Ganger [31],

Hochberg and Oksman [93]. Foord [100] has shown that for positive point discharges in pure nitrogen, the sudden fall in dielectric strength does not occur, as shown in Fig. 7.25, and points out that this supports the suggestion that the formation of a negative-ion space-charge cloud,

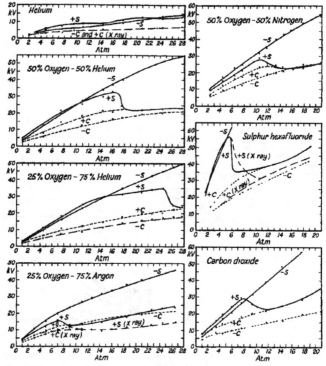

Fig. 7.27   Voltage characteristics for various gases between a point and a plane spaced at 0 3 cm. Curves S give the spark breakdown voltage   Curves C give the corona onset voltage

which might be particularly dense in CCl_2F_2 or SF_6, accounts for the above-mentioned peak.

Pollock and Cooper [99] have also studied the breakdown of point-plane gaps in a number of other gases, including O_2, CO_2, SO_2, SF_6, CCl_2F_2, A, He, and H_2, and mixtures of these gases. The results obtained are shown in Figs. 7.26 and 7.27. The point electrode used consisted of a tungsten wire with a hemispherical end of radius 0·025 cm. The gap was 0·3 cm. Besides measurement of the voltages for corona onset and for breakdown, oscillographic records were obtained of the current pulses occurring during the corona discharges (see p 173). Pollock and Cooper's results show that the peak in the breakdown voltage curve is most pronounced in those gases which form negative ions, such as SF_6 or CCl_2F_2, but is barely detectable in gases such as argon, helium, or hydrogen. The addition of a certain amount of oxygen, an electron-attaching gas, to the helium causes the peak to appear.

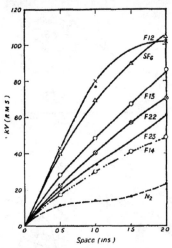

FIG. 7.28   Breakdown voltage curves for various gases (see Fig. 7.10) between a hemispherically-ended rod, of 0 1 m diameter, and a sphere of 1 0 m diameter   The gas pressure is 1 atm.

Other breakdown measurements for point-plane gaps in CCl_2F_2 and in mixtures of CCl_2F_2 with air or nitrogen have been made by Weber [94], Camilli and Chapman [80], Gossens [77], and Foord [100], with results generally similar to those given by Pollock and Cooper [99]. Results in air-CCl_4 mixtures have been obtained by Gossens [77]. Results in CO_2 are given by Ganger [31] and by Hochberg and Oksman [93].

Curves obtained by Camilli and Chapman [80] for the A.C. breakdown of several halogenated gases between a rod of 0·1 in. diameter with a hemispherical tip, and a sphere are given in Fig 7.28. Impulse breakdown voltages for these gases are also given by the same investigators   Fig. 7.29 shows the A.C. and impulse breakdown curves of SF_6 at several pressures for the same form of gap.

Point-point gaps and rod gaps

The breakdown characteristics of gaps between points and between rods have been widely studied [69, 71, 82, 85, 87, 89, 101 to 108], largely because of the frequent occurrence of such gaps in electrical apparatus and the consequent need for experimental data.

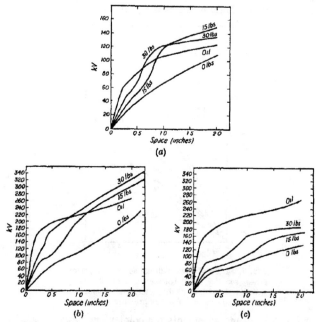

FIG 7 29. Breakdown voltage curves for SF$_6$ at various pressures as a function of the spacing between a homispherically-ended rod of 0 1 in diameter and a sphere of 1 0 in. diameter. (a) 60 c./s (b) Negative rod-plane (impulse) (c) Positive rod plane (impulse)

A curve given by Strigel [87] for the D.C. breakdown of gaps up to 1 cm. in air between points of 30° conical section is given in Fig. 7.19. Other results, for gaps up to 10 cm between points, have been obtained by Marx [85].

With increasing gap length the difference between the breakdown voltages for gaps between points and between rods decreases to a negligible value. Most of the results for longer gaps refer to gaps between rods of $\frac{1}{2}$ in. or $\frac{3}{8}$ in. square section with square-cut ends. Curves for the 60 c./s. A.C. breakdown voltage, and for the positive and negative impulse

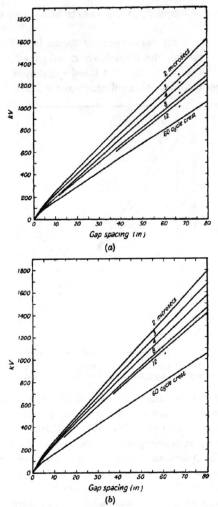

Fig. 7.30. 60 c./s. and impulse breakdown voltage
curves for a rod-gap in air at N.T.P. with one rod
earthed (absolute humidity of 6 5 grams per cubic
foot). (a) Positive impulse. (b) Negative impulse.

breakdown voltages, of gaps between rods of $\frac{1}{2}$ in. square section with
square-cut ends are given in Fig. 7.30 [82]. The impulse breakdown
values were obtained with a 1·5/40 μsec. wave, and the corresponding

curves are labelled according to the times to breakdown on the wavetail (see p. 214).

A comparison of the measurements made in different European laboratories for the impulse breakdown of rod gaps has been made by Allibone [104], who points out that the dispersion of the results is as large as ± 8 per cent. though good agreement exists between the mean

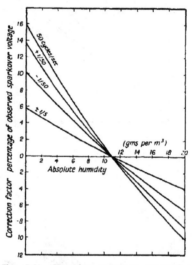

Fig. 7 31. Humidity correction curves for rod-gaps ($\pm 1/50$, $\pm 1/5$, and 50-cycle waves)

values and the mean values recorded in similar tests in the U.S.A. [41, 101] It is possible that the wide dispersion in the results may be caused partly by differences in the circuit used [71]. Errors may also have been introduced by the sphere-gap techniques used in the voltage measurements [64]. The American values for the breakdown voltages of gaps between $\frac{1}{2}$ in. square rods are given in Table 7 15. With the 1 5/40 μsec. wave there appears to be a pronounced scatter in the breakdown voltages measured for certain gaps but no reason for this increased scatter has been given.

The breakdown characteristics of rod gaps are influenced by humidity, the effect depending on the wave-shape, polarity, and amount of over-voltage in excess of the minimum breakdown value [41, 101, 103, 104, 106, 110, 111]. To correlate breakdown voltage data the measured values are corrected to a standard humidity of 11 gm./m.³, corresponding

TABLE 7.15

Minimum Breakdown Voltages in kV for 60 c./s. A.C. and Impulse Voltages applied to Gaps between ½ in. Square Rods in Air (Barometer, 30 inches; Temperature, 77° F.; Humidity, 0·6085 in. vapour pressure) [101]

Gap in inches	A C.	1/5 μsec. wave		1 5/40 μsec wave	
		+	−	+	−
0 5	16 5	22	23	22	23
0 75	24·5	30		30	
1 0	32	38	38	38	38
1 5	45	51	51	51	51
2	57	60	62	60	62
3	71	76	83	75	82
4	82	97	103	91–95	102
5	93	120	124	106–114	123
6	105	143	146	128–141	143
7	.	.	167	141–155	163
8	125	187	188	159–166	183
9	.	.	209	175–178	203
10	150	233	231	190	224
12	.		273		259–271
14	.	.	315		293–318
15	210	340	335	275	309–342
16		.	360	.	323–359
18			400	.	360–386
20	280	440	445	350	395–415
25			550		490
30	420	640	660	505	575
40	545	835	875	650	740
50	680	1,035	1,085	800	910
60	815	1,230	1,300	945	1,070
70	940	1,425	1,515	1,095	1,235
80	1,070	1,620	..	1,240	1,405
90	1,180	1,815		1,385	1,570
100	.	2,010		1,530	

to a relative humidity of 64 per cent. at 20° C. Humidity correction curves for rod gaps [41, 104], based largely on results given by Fielder [103, 110], are shown in Fig. 7.31. In a more recent investigation by Lebacqz [111] the breakdown voltages of gaps ranging from 6 in. to 30 in. between sharp-ended and square-cut rods have been measured as a function of the humidity. The breakdown voltage at all gaps up to 30 in. increases with increasing humidity, but the percentage increase per mm. Hg vapour pressure is not constant and attains a maximum of 1 2 per cent./mm. Hg for spacings between 18 in. and 24 in.

The breakdown of gaps between points and between rods in compressed air has been studied by Skilling [75] and by Gänger [31] Curves given by Gänger are shown in Fig. 7.32.

The breakdown of commercial-grade nitrogen in gaps between two rods, each of $\frac{1}{4}$ in. square section and with square-cut ends, has been studied by Nonken [78] for gaps up to 10 cm. and gas pressures up to 200

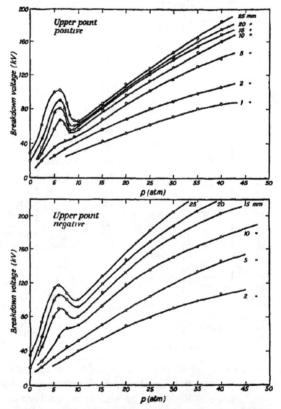

Fig 7 32. Breakdown voltage curves in compressed air between two points. The points are arranged vertically, the lower point being earthed The figures against the curves give the gap lengths

lb./in.[2] Some curves showing the A C. and the impulse breakdown voltages and the A.C. corona onset voltage are given in Fig. 7.33. The curves show that a peak in the breakdown voltage is reached at a particular gas pressure. A feature of the results is that for the 6-cm. and the 10-cm. gap the impulse breakdown voltage is lower than the A.C. breakdown voltage. This effect is more pronounced in the case of break-

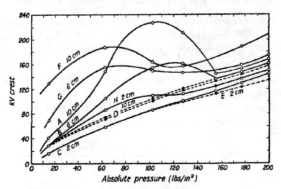

Fig. 7.33. 60 c./s. and impulse breakdown voltage curves between 0 5 in. square-cut rods in compressed nitrogen. The figures against the curves give the gap lengths.

Curves A, B, and C	60 c./s. breakdown.
Curves D and E	60 c./s. corona onset.
Curves F, G, and H	impulse breakdown

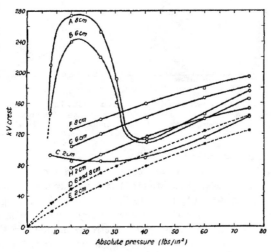

Fig 7 34. 60 c./s. and impulse breakdown voltage curves between 0 5 in. square-cut rods in CCl₂F₂ The figures against the curves give the gap lengths

Curves A B, and C	60 c /s. breakdown
Curves D and E	60 c./s. corona onset.
Curves F, G, and H	impulse breakdown.

down of freon, as shown in Fig. 7.34. It is possible that the reason for the effect is that with impulse voltages the rate of rise of voltage across the gap is so rapid that space charges have insufficient time to accumulate appreciably in the gap to modify the field distribution before the voltage reaches a value sufficient to cause propagation of a streamer across the gap (see pp. 186–205).

Other studies of the breakdown in freon, or in mixtures of freon with air or nitrogen, have been made by Skilling and Brenner [79] and by Weber [33].

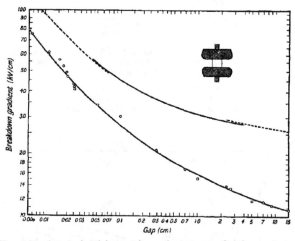

FIG. 7 35   60 c /s breakdown voltage characteristic for the sparkover across the surface of a glass or porcelain cylinder in air   The upper curve gives the breakdown characteristic of a uniform field in the absence of the cylinder.

Sparkover of insulating surfaces

The sparkover voltages in air for a variety of forms of insulators including glass, porcelain, bakelite, etc., have been measured by numerous investigators [83, 91, 101, 113, 116 to 120, 164]   A curve given by Maxstadt [117] for the A.C sparkover of right cylinders of glass or porcelain when placed between plane electrodes is shown in Fig. 7.35, which also includes the curve obtained for the breakdown of air when the cylinder is absent   The plane electrodes are of the Rogowski form [2, 3] so that the field in the gap, in the absence of the insulating cylinder, may be considered to be uniform.   An increase in humidity of the air is observed to cause a marked reduction in the sparkover voltage [116, 117,

159]. Curves obtained by Inge and Walther [116] for the A.C., D.C., and impulse sparkover voltage of a glass cylinder, 2 cm. in length, between parallel planes in air at pressures below atmospheric are given in Fig. 7.36.

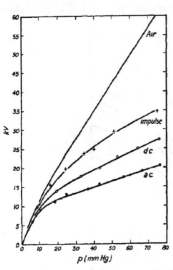

FIG. 7.36. Breakdown voltage curves for the sparkover across the surface of a glass cylinder, of length 2 cm , in air The curve labelled 'air' gives the breakdown voltage of the gap when the glass cylinder is absent.

FIG 7 37. 50 c./s breakdown voltage curves for sparkover across porcelain surfaces in compressed air. The figures against the curves give the gap lengths.

The A.C. sparkover of insulating surfaces in compressed air has been studied by Reher [29], Weber [121], Gossens [77], and Böcker [159]. Some results given by Weber for the A.C. sparkover of porcelain surfaces, the gaps ranging from 4·2 to 35·5 cm., are shown in Fig. 7.37. The breakdown voltage increases steadily with increasing air pressure for gaps up to 10·6 cm. For longer gaps peaks appear in the curve relating breakdown voltage with pressure, and for a given gap a limiting breakdown voltage is reached which cannot be surpassed by an increase in the air pressure. Sparkover of long gaps may then occur at a given gas pressure for which sparkover of a shorter gap does not occur [95]. The results are explained by considerations of space charge on the dielectric surface and the consequent distortion of the field The influence of dust particles and electrode roughness is also discussed by Weber [121].

Measurements of the A.C. sparkover of surfaces in compressed nitrogen have been made by Goldman and Wul [122], Palm [32], and Nonken [78]. Some results are also given by Palm [32] for carbon dioxide, by Nonken [78] for freon, and by Gossens [77] for air-CCl_4 and air-CCl_2F_2 mixtures.

The D.C. breakdown over the surfaces of solid insulators in compressed nitrogen has been studied by Trump and Andrias [123]. The insulators

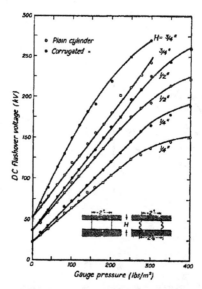

FIG 7 38.   D.C. breakdown voltage curves of
Textolite cylinders in compressed nitrogen.

were in the form of cylinders up to 1 in. in length and 2 in. in diameter, and were mounted between parallel plane electrodes. Materials of three types were investigated, namely Lucite, Textolite, and Isolantite. Lucite is a polymerized methylmethacrylate; Textolite is a laminated plastic with a paper base bonded with a phenolic compound; Isolantite is a porcelain ceramic. The approximate dielectric constants of these materials are 4·2, 4·6, and 6·1, respectively. Cylinders with straight smooth surfaces and also with corrugated smooth surfaces were studied. Some curves obtained for the Textolite cylinders are given in Fig. 7 38. Curves for ½-in. corrugated cylinders of the three materials are shown in Fig. 7.39, which also includes the breakdown curve of a ½-in. gap in compressed nitrogen alone. In all cases breakdown occurred over the surface of the cylinders and the breakdown voltage is increased by the use of corrugations. The influence of surface defects, non-uniform surface leakage, surface-bound space charges, and end conditions is discussed by Trump and Andrias The low breakdown voltage obtained with Isolantite is ascribed largely to imperfect contact between the

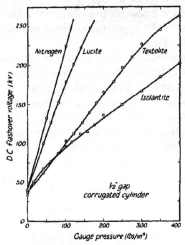

FIG 7.39   D C. breakdown voltage curves for a uniform field gap in nitrogen and for the surface sparkover of corrugated cylinders of various materials in nitrogen Gap length = 0 5 in.

insulator and the metal electrodes. Ionization then occurs readily in the small gas-filled gaps between the insulator and the electrodes and produces surface-bound space charges which distort the electric field. To confirm this hypothesis the ends of one of the Isolantite cylinders were ground to improve the contact with the electrodes, and the breakdown voltage was found to increase by about 20 per cent This difficulty is not encountered with Lucite or Textolite as both these materials could be machined accurately, and the plastic nature ensured close contact at the electrode-insulator boundaries.

Gases as insulators

The external design and to some extent the internal design of most high-voltage equipment is governed either by the breakdown strength of gaps in air or by the sparkover of solid insulating surfaces in air. During recent years the possibilities of compressed air or other gases for the insulation of electrical apparatus have been widely studied, and already practical application has been made of this technique in the design of certain types of condensers [32, 91, 114, 124, 125, 126], cables [127–31, 158], transformers [132, 133], electrostatic generators [134 to 149], and electrostatic voltmeters [150, 156]. By the use of increased gas pressures, and of such gases as CCl_2F_2 or SF_6 with particularly high breakdown strengths, it is possible to make appreciable reductions in the physical dimensions of the apparatus as compared with those required for operation in open air.

Besides the requirement of a high electric breakdown strength there are other properties which have to be considered in the choice of a suitable gas for insulating purposes. Among such properties are the following:

1. The gas should be chemically inert with respect to the solid material used in the construction of the apparatus.
2. A low temperature of liquefaction is required so that the gas can be used at high pressures at normal temperatures.
3 There should be negligible dissociation of the gas.
4. A high heat conductivity is preferable for cooling purposes.
5. The gas should be readily available and inexpensive.

Compressed air has the advantage of availability and has been used in condensers [123, 125, 126] and in electrostatic generators [134, 135, 136]. In the latter application the corona discharges present cause the decomposition of the air into nitric and nitrous oxides, and corrosion of the apparatus occurs. The fire risk in apparatus using compressed air is appreciable and special precautions are taken in electrostatic generators to cover this liability. The fire risk is minimized by the use of compressed nitrogen, which is chemically inert and does not decompose. Nitrogen has been used in preference to air for electrostatic generators [137, 138, 139] and for cables [127]   Carbon dioxide has a slightly higher breakdown strength than nitrogen [28] and has been used mixed with nitrogen for electrostatic generators [139]. However, because of its decomposition into carbon monoxide and oxygen it has not been widely applied. Sulphur dioxide, which has a high breakdown strength [143] rapidly dissociates when corona discharges are present with consequent corrosive effects. Hydrogen has a lower breakdown strength than the above gases but has the advantage of a much greater heat conductivity, and has been usefully applied in cases where it is necessary to remove heat losses in large rotating machinery [151]. Consideration has been given to the use of helium in cables [129].

Wide attention is being given to the use of gases containing chlorine

or fluorine because of their much higher breakdown strengths compared
with air or nitrogen at the same pressure. However, many of these gases
are liquid at normal room temperature. For example CCl_4, which has a
breakdown strength 6·3 times that of nitrogen [35], is a liquid at 20° C.
when its vapour pressure is only about 64 mm. Hg. Nevertheless, it is

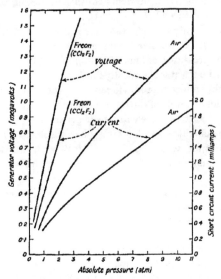

FIG. 7·40. Curves giving the variation of maximum
terminal voltage and of short-circuit current with gas
pressure of air and freon in an electrostatic generator.

only necessary for a small proportion of CCl_4 to be present in air or
nitrogen to cause a large increase in the breakdown strength, and such
mixtures have been used with success in certain electrostatic generators
[136, 140, 141, 142]. A disadvantage of CCl_4 is that it decomposes in
an electrical discharge, the carbon forming a conducting deposit on solid
insulating surfaces and the chlorine causing corrosion.

CCl_2F_2, known as 'freon' or 'arcton', has a breakdown strength about
2·5 times that of air at the same pressure and temperature and has been
widely applied in electrical apparatus, including electrostatic generators
[136, 137, 142, 145 to 148] and transformers [132]. Its vapour pressure
at room temperature is 85 lb./in.²; its temperature of liquefaction is
−28° C. It can be used at pressures up to about 70 lb./in.² at room
temperature, but may be used at higher total gas pressures in mixtures

with air or nitrogen. As a consequence of the improved breakdown strength obtained by raising the gas pressure or by the use of freon it is possible to cause a considerable increase in the voltage and current rating of a given electrostatic generator, as shown in Fig. 7.40. A disadvantage of CCl_2F_2 is that it dissociates in the presence of corona discharges, the dissociation products causing corrosion, particularly in the presence of moisture [147, 152, 153].

Sulphur hexafluoride [80, 99, 154] has a breakdown strength about equal to that of CCl_2F_2 but has the advantage that it may be used up to much higher gas pressures, of about 350 lb./in ^{2} at 20° C., before liquefaction occurs. Although it is chemically inert, it dissociates in an electrical discharge and the resultant products can cause corrosion of the apparatus [154, 155]. As an example of the improvement to be obtained by the use of sulphur hexafluoride in an electrostatic generator, the output voltage of a particular generator was 5·6 MeV when insulated with SF_6 at 200 lb./in.^{2} This output was roughly 4·2 MeV with a nitrogen-CCl_2F_2 mixture at the same pressure, and was only about 3 MeV with nitrogen or air [147].

While many gases containing chlorine or fluorine have been investigated it appears at present that CCl_2F_2 and SF_6 are the most satisfactory for insulating purposes.

REFERENCES QUOTED IN CHAPTER VII

1. H Ritz, *Arch. Elektrotech.* **26** (1932), 219.
2. W. Rogowski and H. Rengier, ibid **16** (1926), 73.
3. C. Stoerk, *Elektrotech. Z.* **52** (1931), 43.
4. W. O Schumann, *Arch. Elektrotech.* **11** (1923), 1.
5. W. Holzer, ibid **26** (1932), 865.
6. F. M Bruce, *J. Instn. Elect Engrs.* **94** (II) (1947), 138.
7. *Rules for the Measurement of Voltage with Sphere Gaps*, Brit. Stand. No. 358, 1939.
8. W. O. Schumann, *Arch. Elektrotech.* **12** (1923), 594.
9. W. Spath, ibid. 331.
10. A. Klemm, ibid. 553.
11. F. Muller, ibid. **13** (1924), 478.
12. W. Sahland, ibid **19** (1927), 145.
13. R. Cooper, D. E. M. Garfitt, and J. M. Meek, *J. Instn. Elect. Engrs.* **95** (II) (1948), 309.
14. J. G. Trump, F. J. Safford, and R. W. Cloud, *Trans. Amer. Inst. Elect. Engrs.* **60** (1941), 132.
15. F. S. Edwards and J. F. Smee, *J. Instn. Elect. Engrs.* **82** (1938), 655.
16. W. O. Schumann, *Elektrische Durchbruchfeldstärke von Gasen*, J. Springer, Berlin, 1923.
17. W. O. Schumann, *Z. Phys.* **60** (1930), 201.

18. J. D. STEPHENSON, *J. Instn. Elect. Engrs.* **73** (1933), 69.

19. D. Q. POSIN, *Phys. Rev.* **50** (1936), 650

20. K. S FITZSIMMONS, ibid. **58** (1940), 187.

21 W. R. HASELTINE, ibid. 188.

22. L. H FISHER, ibid **72** (1947), 423

23 B. L. GOODLET, F. S. EDWARDS, and F R. PERRY, *J. Instn. Elect. Engrs.* **69** (1931), 695

24. S. FRANCK, *Arch. Elektrotech.* **21** (1928), 318.

25. A. KOHLER, ibid. **30** (1936), 528.

26. R. COOPER, *J. Instn. Elect. Engrs.* **95** (II) (1948), 309.

27. A. H. HOWELL, *Trans Amer. Inst Elect Engrs.* **58** (1939), 193.

28. E. FINKELMANN, *Arch. Elektrotech.* **31** (1937), 282.

29. C. REHER, ibid. **25** (1931), 277.

30 M N. FELICI and Y. MARCHAL, *Rev. Gen. Elec.* **57** (1948), 155.

31. B. GANGER, *Arch. Elektrotech.* **34** (1940), 633; ibid. 701.

32. A. PALM, ibid **28** (1934), 296.

33. W. WEBER, ibid **36** (1942), 166

34. E E. CHARLTON and F. S. COOPER, *Gen Elect. Rev.* **40** (1937), 438

35. B. HOCHBERG, *Elektrichestvo*, No. 3 (1947), p. 15.

36. R. GEBALLE and F. S LINN, *J. Appl. Phys.* **21** (1950), 592.

37. B. HOCHBERG and E SANDBERG, *J. Tech. Phys. U S S.R* **12** (1942), 65.

38. E. SANDBERG, ibid. **17** (1947), 299.

39. J. W. WARREN, W. HOPWOOD, and J. D. CRAGGS, *Proc. Phys. Soc. B,* **63** (1950), 180

40 J. W. WARREN and J D. CRAGGS, *Inst. Petrol. Symp. Mass Spectrometry,* 1952.

41. *Measurements of Test Voltage in Dielectric Tests A.I.E.E. Standards,* No. 4, 1940.

42. R. DAVIS and G. W. BOWDLER, *J. Instn. Elect. Engrs.* **82** (1938), 645.

43. E. HUETER, *Elektrotech. Z.* **57** (1936), 621.

44. R. ELSNER, ibid. **56** (1936), 1405.

45. S. WHITEHEAD and A. P. CASTELLAIN, *J. Instn. Elect Engrs* **69** (1931), 898.

46. P. L. BELLASCHI and P. H McAULEY, *Elect J* **31** (1934), 228.

47. W. DATTAN, *Elektrotech. Z* **57** (1936), 377, 412.

48. W. DATTAN, *Arch. Elektrotech.* **31** (1937), 343.

49. W. WEICKER, *Elektrotech Z* **60** (1939), 97.

50. W. WEICKER and W HORCHER, ibid **59** (1938), 1029, 1064

51. S. WHITEHEAD, *J. Instn. Elect. Engrs.* **84** (1939), 408.

52. W. JACOBI, *Z. tech. Phys.* **5** (1938), 159.

53. L. BINDER and W. HORCHER, *Elektrotech. Z.* **59** (1938), 161.

54. R. W SORENSON and S. RAMO, *Elect. Engng* **55** (1936), 444.

55. C. S. SPRAGUE and G. GOLD, ibid. **56** (1937), 594.

56. J. R. MEADOR, ibid. **53** (1934), 942, 1652.

57. D. W. VER PLANCK, *Trans Amer. Inst. Elect. Engrs* **60** (1941), 99.

58. S. FRANCK, *Arch. Elektrotech.* **23** (1929), 226.

59. A. B. LEWIS, *J. Appl. Phys.* **10** (1939), 573.

60. F. O. McMILLAN and E. G. STARR, *Trans. Amer. Inst Elect. Engrs.* **49** (1930), 859.

61. F. O. McMILLAN, ibid. **58** (1939), 56.

62. F. D FIELDER. *Elect. World*, **102** (1933), 433.
63. J. M. MEEK, *J. Instn. Elect. Engrs.* **89** (1942), 335.
64 J M. MEEK, ibid. **93** (II) (1946), 97.
65 J M. MEEK, *J. Franklin Inst* **230** (1940), 229.
66 L B. LOEB and J. M. MEEK, *The Mechanism of the Electric Spark*, Stanford University Press, 1941.
67. M. TOEPLER, *Elektrotech Z.* **53** (1932), 1219.
68. J. CLAUSSNITZER, ibid **57** (1936), 177.
69. P L. BELLASCHI and W. L. TEAGUE, *Elect J.* **32** (1935), 120.
70. J. H. HAGENGUTH, *Trans. Amer. Inst. Elect. Engrs.* **56** (1937), 67.
71. J. H. HAGENGUTH, ibid. **60** (1941), 803.
72. R. COOPER, *J Instn. Elect Engrs.* **95** (II) (1948), 378.
73. O. ZEIER, *Ann. Phys. Lpz.* **14** (1932), 415.
74. J. G. TRUMP, R. W CLOUD, J. G. MANN, and E. P. HANSON, *Elect Engng.* **69** (1950), 961.
75. H. H. SKILLING, *Trans Amer Inst Elect. Engrs.* **58** (1939), 161.
76. H H. SKILLING and W. C BRENNER, ibid. **60** (1941), 112
77. R. F. GOSSENS, *Conf Int Gr. Res. Elect.*, Paper No. 117, 1948.
78. G. C NONKEN, *Trans. Amer. Inst. Elect. Engrs.* **60** (1941), 1017.
79 H. H. SKILLING and W. C BRENNER, ibid. **61** (1942), 191.
80. G. CAMILLI and J. J. CHAPMAN, ibid **66** (1947), 1463.
81 G. KOWALENKO, *J Tech Phys U.S.S R.* **3** (1940), 455.
82 P L. BELLASCHI and W. L. TEAGUE, *Elect Engng.* **53** (1934), 1638.
83. T. E. ALLIBONE, W. G HAWLEY, and F R PERRY, *Journal I.E E.* **75** (1934), 670.
84 R. R. NEWTON, *Phys. Rev.* **73** (1948), 1122.
85. E. MARX, *Arch Electrotech* **20** (1928), 589.
86. E. UHLMANN, ibid. **23** (1929), 323.
87. R STRIGEL, ibid **27** (1933), 377.
88. R STRIGEL, *Elektrische Stoßfestigkeit*, J. Springer, Berlin, 1939.
89. W. HOLZER, *Z. Phys* **77** (1932), 676.
90. A. A. GOREV, A. M. ZALESKY, and B. M. RIABOV, *Conf. Int. Gr. Res. Elect*, Paper No. 142, 1948.
91. M SCHULZE, *Brown Boveri Rev* **30** (1943), 244.
92. W. BAER, *Arch. Elektrotech.* **32** (1938), 684.
93 B. M. HOCHBERG and J A OKSMAN, *J. Phys. U.S.S.R.* **5** (1941), 39.
94 W. WEBER, *Arch. Elektrotech* **36** (1942), 166.
95. T. E. ALLIBONE and J. M. MEEK, *J Sci. Instrum Phys Ind.* **21** (1944), 21.
96. I. GOLDMAN and B. WUL, *J Tech. Phys. U.S.S R.* **1** (1935), 497
97 I. GOLDMAN and B WUL, ibid. **3** (1936), 16.
98. I GOLDMAN and B. WUL, ibid. **5** (1938), 355
99. H. C. POLLOCK and F S. COOPER, *Phys. Rev.* **56** (1939), 170.
100. T. R. FOORD, *Nature*, **166** (1950), 688.
101. *Electrical Engineering*, **56** (1937), 712.
102. T. E ALLIBONE and F. R. PERRY, *J. Instn Elect. Engrs.* **78** (1936), 257.
103. F. D. FIELDER, *Elect J.* **29** (1932), 459.
104 T. E. ALLIBONE, *J. Instn. Elect Engrs.* **81** (1937), 741.
105. P. JACOTTET, *Elektrotech. Z* **58** (1937), 628.

106. P. L BELLASCHI and P. H McAULEY, *Trans. Amer. Inst. Elect. Engrs* **59** (1940), 669.
107. W. WANGER, *Bull. Ass. Suisse Elect* **34** (1943), 193
108. P. L. BELLASCHI and P. EVANS, *Elect. Engng.* **63** (1944), 236.
109. P. L. BELLASCHI and W. L. TEAGUE, *Elect. J.* **32** (1935), 56.
110. F. D. FIELDER, ibid. 543.
111. J. V. LEBACQZ, *Trans. Amer. Inst. Elect. Engrs.* **60** (1941), 44
112 S. WHITEHEAD, *Dielectric Phenomena*, E. Benn, London, 1927.
113. F. W. PEEK, *Dielectric Phenomena in High Voltage Engineering*, McGraw-Hill, 1929.
114. A A BOLSTERLI, *Schweiz. Elek. Bull.* **22** (1931), 245.
115. W. M. THORNTON, *Phil. Mag.* **28** (1939), 266.
116. L. INGE and A. WALTHER, *Arch. Elektrotech.* **26** (1932), 409.
117. F. W. MAXSTADT, *Elect. Engng* **53** (1934), 1062.
118. W. L. LLOYD, *Trans. Amer. Inst. Elect. Engrs.* **51** (1932), 510.
119. G. D. HEYE, *Gen. Elect Rev.* **37** (1934), 548.
120. A. SCHWAIGER, *Theory of Dielectrics* (translated by R. W. Sorenson), J Wiley, 1930.
121. W. WEBER, *Arch. Elektrotech.* **35** (1941), 756
122. I. GOLDMAN and B. WUL, *J. Tech. Phys U.S.S.R* **3** (1936), 519.
123. J. G. TRUMP and J. ANDRIAS, *Trans Amer. Inst. Elect. Engrs.* **60** (1941), 986.
124. B. HOCHBERG, M. GLIKINA, and N. REINOV, *Journ Tekhn. Fiz.* **8** (1942), 8.
125. B. HOCHBERG and N. REINOV, ibid. **15** (1945), 713
126. H. SCHERING and R. VIEWEG, *Z. tech. Phys.* **9** (1928), 442.
127. C. J. BEAVER and E. L. DAVEY, *J. Instn. Elect. Engrs* **90** (I) (1943), 452.
128. L. G. BRAZIER, ibid. **93** (II) (1946), 415.
129. I. T. FAWCETT, L. I. KOMIVES, H. W. COLLINS, and R. W. ATKINSON, *Trans. Amer. Inst. Elect Engrs.* **61** (1942), 658
130. G. B SHANKLIN, ibid. 719.
131. B. HOCHBERG and M. GLIKINA, *Journ. Tekhn. Fiz.* **12** (1942), 3.
132. E. E. CHARLTON, W. F. WESTENDORP, L E. DEMPSTER, and G HOTALING, *J. Appl. Phys.* **10** (1939), 374.
133. H. HARTMANN, *Brown Boveri Rev.* **28** (1941), 84.
134 R. G. HERB and E. J. BENNET, *Phys Rev.* **52** (1937), 379.
135. R. G HERB, D. B. PARKINSON, and D. W. KERST, ibid. **51** (1937), 75.
136. D. B. PARKINSON, R. G. HERB, E. J. BENNET, and J. L. McKIBBEN, ibid. **53** (1938), 642.
137. R. L. FORTESCUE and P. D. HALL, *J. Instn. Elect. Engrs.* **96** (I) (1949), 77.
138 J. G. TRUMP, *Elect Engng.* **66** (1948), 525
139. R. J. VAN DE GRAAFF, J. G. TRUMP, and W. W BUECHNER, *Rep. Phys. Soc. Progr. Phys.* **11** (1946–7).
140. M T. RODINE and R. G. HERB, *Phys. Rev.* **51** (1937), 508.
141. F. JOLIOT, M FELDENKRAIS, and A. LAZARD, *C. R. Acad Sci Paris*, **202** (1936), 291.
142. J. C. TRUMP, F. H. MERRILL, and F. J. SAFFORD, *Rev. Sci Instrum* **8** (1938), 398.
143. C. M HUDSON, L. E. HOISINGTON, and L. E. HOYT, *Phys. Rev.* **52** (1937), 664.
144 H. A. BARTON, D. W. MUELLER, and L. C. VAN ATTA, ibid. **42** (1932), 901.

145. J. G. TRUMP and R. J. VAN DE GRAAFF, ibid. **55** (1939), 1160.
146. W. B. MANN and L. G. GRIMMETT, *Proc. Phys. Soc.* **59** (1947), 699.
147. W. W. BUECHNER, R. J VAN DE GRAAFF, A. SPERDUTO, L. R. McINTOSH, and E. A. BURRILL, *Rev. Sci. Instrum.* **18** (1947), 754.
148. R. G HERB, C. M TURNER, C. M HUDSON, and R. E. WARNER, *Phys. Rev.* **58** (1940), 579.
149. W. W. BUECHNER, R. J. VAN DE GRAAFF, A. SPERDUTO, E. A BURRILL, L. R. McINTOSH, and R C. URQUHART, ibid. **69** (1946), 692.
150. H. BOOKER, *Arch. Elektrotech.* **33** (1939), 801.
151. D. S. SNELL, R. H. NORRIS, and B. O. BUCKLAND, *Trans. Amer. Inst. Elect. Engrs.* **69** (1950), 174.
152. N, V. THORNTON, A. B. BURG, and H. I. SCHLESINGER, *J. Amer. Chem. Soc.* **55** (1933), 3177.
153. F. T. DE WOLF, *Elect. Engng.* **60** (1941), 435.
154. B. HOCHBERG, *Elektrichestvo*, No. 3 (1947), 15.
155. W. C. SCHUMB, J. G. TRUMP, and G. L. PRIEST, *Industr. Engng. Chem.* **41** (1949), 1348.
156. W. ROGOWSKI and H. BOCKER, *Arch. Elektrotech.* **32** (1938), 44.
157. M. O. JORGENSEN, *Elektrische Funkenspannungen*, E. Munksgaard, Copenhagen, 1943.
158. B. M. HOCHBERG and N. M. REINOV, *Doklady Akad. Nauk*, **70** (1950), 837.
159. H. BOCKER, *Arch. Elektrotech.* **40** (1950), 37.
160. W. A. WILSON, J H SIMONS, and T. J. BRICE, *J. Appl. Phys.* **21** (1950), 203.
161. A. BOUWERS and A KUNTKE, *Z. tech. Phys.* **18** (1937), 209.
162. J. M. MEEK, *Proc. Instn. Elect. Engrs.* **98** (II) (1951), 362
163. F. M. BRUCE, ibid. 363.
164. H. RITZ, *Arch. Elektrotech.* **32** (1932), 58.

VIII

IRRADIATION AND TIME LAGS

THE time lag between the application of an impulse voltage and the consequent breakdown of a gap may be separated into two components, (1) the statistical time lag, caused by the need for an electron to appear in the gap during the period of application of the voltage in order to initiate the discharge, and (2) the formative time lag, corresponding to the time required for the spark discharge, once initiated, to develop across the gap.

The statistical time lag depends on the amount of pre-ionization, or irradiation, of the discharge gap. A gap subjected to an impulse voltage may break down if the peak voltage reaches the D.C. breakdown value provided that the gap is sufficiently irradiated so that an electron is present in the gap to initiate the spark process when the peak voltage is reached. With a lower amount of irradiation the voltage must be maintained above the D C. breakdown value for a longer period before an electron appears. Therefore, to cause breakdown of a gap with an impulse voltage of defined wave shape, increasing values of peak voltage are required as the amount of irradiation is decreased, so that the voltage is in excess of the D.C. breakdown value for increasingly long time intervals.

The fact that irradiation reduces the statistical time lag has long been known, and it is now usual to supply initiating electrons in the gap by the use of radioactive materials or by ultraviolet illumination. Reference to the effect is made in the rules which have been issued for the measurement of high voltages by means of sphere gaps. In the British Standards [1] it is stated that 'in localities where the random ionization is likely to be low, irradiation of the gap by radioactive or other ionizing media should be used when voltages of less than 50 kV peak are being measured'. The use of a mercury-arc lamp is recommended in the American specifications [2] as being necessary for sphere gaps of up to 6·25 cm. diameter when the gap is less than one-third the sphere diameter. Although these specifications both imply that the impulse breakdown voltage of a sphere gap of length greater than about 2 cm. is not affected by irradiation, recent measurements [3, 4] show that the influence of irradiation is appreciable for much longer gaps, even up to 20 cm. However, the presence of irradiation has a decreasing effect on the statistical time lag as the gap length is increased because of the greater gap volume and the consequent increase in probability of the appearance of a suitably placed electron to initiate the discharge.

Most of the measurements of statistical time lags have been made for gaps in air at atmospheric pressure. The technique generally used is to observe oscillographically the waveform of the voltage applied to the gap, and to record the time between the instant when the voltage exceeds the D.C. breakdown value and that when breakdown occurs as shown by the sudden collapse of voltage across the gap. Some measurements of statistical time lags have been made by the Kerr cell technique [5, 6, 7]. Another method involving a device known as a 'time-transformer' has been used by Strigel [51].

The various sources of irradiation include radioactive materials, X-rays, and ultraviolet illumination provided by a suitable arc-lamp or by a spark or corona discharge. Some of the results obtained with several of these sources will now be described.

Irradiation with the mercury-arc lamp

Early observations concerning the influence of the illumination provided by an arc-lamp on the breakdown of a discharge gap subjected to intermittent voltages were made in 1887 by Warburg [10], who was able to deduce that the effect of the illumination is to reduce the time lag to breakdown. In these experiments an open-air arc between carbon electrodes was used, but the type of lamp which has been most widely employed during recent years by different investigators is the quartz-enclosed mercury-arc lamp.

In the irradiation of the normal forms of discharge gaps the lamp is arranged to the side of the gap, so that the direction of illumination is oblique to the cathode, and is sufficiently far away from the gap so that the field distribution in the gap is not affected. Under such conditions it is unlikely that the photo-current density in the gap is greater than about 10^{-11} A/cm.^{2} for the normal types of cathode material used in air, and in many cases, particularly for the longer gaps, and for contaminated surfaces it is probable that the density will be appreciably lower than this value. In certain special forms of gaps where the anode is in the form of a wire gauze the lamp may be arranged behind the cathode to illuminate the cathode in the direction normal to its surface, and photo-currents as high as 10^{-10} A/cm.^{2} may then be obtained with aluminium cathodes [11].

It is of interest to consider the photo-current in terms of the rate of electron emission. A current density of 10^{-12} A/cm.^{2} is equivalent to an

emission of 1 electron/cm.^{2}/0·16 μsec. In a gap where the sparking area is
1 cm.^{2} the consistent breakdown of the gap may then be expected when
impulses of duration 1 μsec. or more are applied to the gap. But in a
0·1-cm. gap between spheres of 1 cm. diameter, for which the electron
initiating the spark must appear within a cathode area of about 0·2 cm.

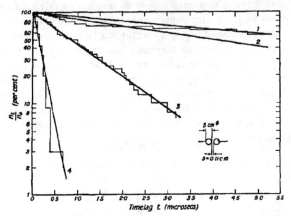

FIG. 8 1   Influence of the nature of the electrode surfaces on the
time-lag distribution curves for a gap of 1 1 mm between 5-cm.
diameter spheres. The gap is subjected to an impulse voltage of 6·5
kV, the D.C. breakdown voltage being 5 0 kV.

1. Oxidized copper electrodes, weakly illuminated, cleaned with alcohol
2 Copper electrodes, unilluminated, cleaned with alcohol.
3. Copper electrodes, weakly illuminated, not cleaned with alcohol.
4. Copper electrodes, weakly illuminated, cleaned with alcohol.

diameter, a photo-current density of 10^{-12} A/cm.^{2} gives an average
electron liberation of 1 electron/5·1 μsec. in the gap. In this case irregular
behaviour of the gap may be expected to occur for impulse voltages of
duration as short as 1 μsec.

Measurements showing the influence of illumination from a mercury-
arc lamp on the behaviour of short gaps when subjected to impulse
voltages have been made by Strigel [8, 9, 12, 52 to 55]. Some results are
given in Fig. 8 1 for the statistical time lag of a 0·11-cm. gap between
spheres of 5 cm. diameter when subjected to an impulse voltage of 6·5 kV
peak [9]. The D.C. breakdown voltage for the gap is 5 kV. The curves
give the variation of the ratio n_t/n_0 with the time lag t between the
application of the impulse and the breakdown of the gap, where n_0 is
the total number of impulses applied to the gap and n_t is the number
for which a time lag greater than t is recorded. The results show that

$\log(n_t/n_0)$ varies in an approximately linear manner with t, at a rate which depends on the condition of the electrodes and the amount of illumination from a mercury-arc lamp. Curve 1 refers to oxidized copper electrodes, from which grease has been removed by alcohol, and which are weakly illuminated, curve 2 refers to grease-free copper electrodes, unilluminated; curve 3 refers to greasy copper electrodes, weakly illuminated; curve 4 refers to grease-free copper electrodes, weakly illuminated.

The linear variation of $\log(n_t/n_0)$ with t can be explained in a simple manner following the analysis originally put forward by von Laue [13] and Zuber [14]. Let β be the rate at which electrons are produced in the gap by the ionizing agent, p_1 the probability of an electron appearing in a region of the gap where it can lead to a spark, and p_2 the probability that such an electron in a given field will lead to a spark. Then the number dn of sparks which occur in a time interval between t and $t+dt$ is given by

$$dn = -p_1 p_2 \beta n_t \, dt,$$

where n_t is the number of time lags of duration greater than t. On integration

$$\log(n_t/n_0) = -\int_0^t p_1 p_2 \beta \, dt, \qquad (8.1)$$

where n_0 is the total number of time lags observed. For constant values of p_1, p_2, and β, equation (8.1) becomes

$$n_t = n_0 e^{-p_1 p_2 \beta t} = n_0 e^{-t/\sigma}, \qquad (8.2)$$

where σ is the mean statistical lag. This expression (8.2) gives a linear relation between $\log(n_t/n_0)$ and t in accordance with the experimental observations such as in Fig. 8.1.

The results of Fig. 8.1 show the large effect of the electrode surface on the statistical time lag. With surfaces which have a high work function fewer electrons are emitted for a given amount of illumination and the statistical time lag is thereby decreased [9]. For this reason clean copper electrodes yield considerably lower time lags than grease-covered copper or oxidized copper electrodes. This is emphasized again in the curves of Fig. 8.2, which show the variation of statistical time lag with overvoltage for a 0·11-cm. gap between 5 cm. diameter spheres of several materials when illuminated weakly by ultraviolet light (the material Elektron is an alloy consisting mainly of magnesium) [8] The distribution of the curves corresponds with that expected from consideration of the work function of the electrodes and therefore of the probable rate of electron liberation in the gap. The work functions quoted by Strigel

[9] for these electrodes are. copper oxide, 5·3 eV, copper, 3·9 eV; silver, 3·0 eV, aluminium, 1·8 eV; magnesium, 1·8 eV. With an 80 per cent. overvoltage applied to the gap the statistical time lags for the copper and the copper oxide electrodes of Fig 8.2 are 0·45 and 2·2 μsec. respectively. These results were obtained with gaps which are only weakly illuminated. By the use of more intense illumination the corresponding time lags fall to 0·042 and 0·18 μsec.

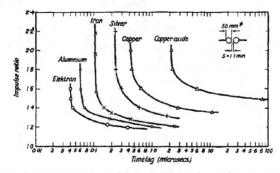

FIG. 8 2. Dependence of the mean statistical time lag on over-voltage for different electrode materials. Gap of 1 1 mm. in air at atmospheric pressure; D C. breakdown voltage of 5 0 kV

As the statistical time lag depends largely on the photo-emission of electrons from the cathode the nature of the electrode material is impor-tant, mainly with regard to the cathode of the gap. This is shown in experiments made with the above gap, with 30 per cent. overvoltage applied to the gap [12]. An average time lag of about 560 μsec. is measured for a cathode surface of copper oxide, when the anode consists of either Elektron or copper oxide, whereas, with an Elektron cathode, the same gap breaks down with an average time lag of 0·16 μsec. for an anode surface of either material.

By consideration of the mean photo-current flowing in the gap Strigel [9] has estimated the number of electrons released from the cathode during the period of the statistical time lag. The number of electrons required to initiate a spark decreases with increasing overvoltage applied to the gap. For a 0·11-cm. gap in air subjected to an impulse voltage of 9 kV, corresponding to an 80 per cent. overvoltage, the measurements show that the average statistical time lag corresponds closely to the estimated time interval between the release of successive individual electrons from the cathode of the gap

The distribution of the statistical time lags changes with repeated sparking between the electrodes. This is particularly noticeable for copper or silver electrodes, which gradually become oxidized and give rise to increasingly longer time lags [8, 9] In the case of copper spheres of 5 cm. diameter spaced at 0·11 cm. when subjected to an impulse voltage of 7 kV, the average statistical time lag for the first 50 sparks is

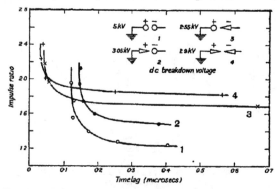

Fig. 8.3. Dependence of the mean statistical time lag on over-voltage in non-uniform fields for a gap of 1 1 mm.

2·6 μsec. After a further 150 sparks the average statistical time lag for the next 50 sparks is 480 μsec.

The statistical time lag for short non-uniform gaps of about 1 mm. between points, or between a point and a sphere, has also been investigated by Strigel [12], who shows that it varies in a similar manner to that for a uniform gap, and can be expressed in the form given by expression (8.2) The time lag is also affected, though to a lesser extent, by the amount of illumination from a mercury-arc lamp. Some results for a 0·11-cm. gap are given in Fig. 8.3. Experiments with longer gaps, of several centimetres or more, show that irradiation has little effect on the breakdown of such gaps The lack of influence is explained by the fact that corona discharge develops readily in the intense localized fields surrounding a point electrode, at voltages well below that required to cause breakdown, and consequently when the voltage reaches the breakdown value a considerable number of electrons are already present in the gap.

Studies of the effect of illumination by a quartz mercury-arc lamp on the impulse breakdown of gaps between spheres of 6·25 cm. diameter have been made by Nord [15], who records marked differences in the breakdown characteristics of short gaps, up to several millimetres in length. Oscillographic measurements show time lags up to 100 μsec. or more, when the gap is unirradiated. The effect decreases with increasing gap length and is not detectable, according to Nord, for gaps requiring above about 70 kV to cause breakdown. On the other hand, experiments by Meek [4] show a distinct difference between the irradiated and unirradiated cases for the breakdown of a 3-cm. gap between 12·5 cm. diameter spheres, as shown in Fig. 8.4.

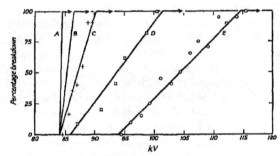

FIG. 8 4. 3 0-cm. gap between 12 5-cm. diameter spheres.

A. Positive 1/5 wave. Gap irradiated by light from impulse-generator spark gaps or by radium.
B. Positive 1/5 wave. Gap irradiated by light from mercury-arc lamp, at a distance of 30 cm.
C. Positive 1/5 wave. Gap irradiated by light from mercury-arc lamp, at a distance of 100 cm.
D. Positive 1/500 wave. Gap unirradiated.
E. Positive 1/5 wave. Gap unirradiated.

The influence of illumination on the breakdown of gaps between 50 cm. diameter spheres, when subjected to a 60 c./s. A.C. voltage, has been studied by Sprague and Gold [16], who show that illumination causes a material improvement in the consistency of the sphere gap.

Measurements by Hardy and Craggs [17, 50] on the photo-currents emitted from different cathode surfaces in air when illuminated by a mercury-arc lamp, show that the currents decrease with time, and vary over a wide range for different cathode materials. Also the method of cleaning the cathode surface has a large influence on the current. Consequently large variations can occur in the breakdown characteristics of discharge gaps, for a given ultraviolet source of irradiation, unless the cathode surface conditions are precisely defined, and it would appear that more satisfactory results can be obtained, in a simpler manner, by the use of radioactive materials, which give irradiation effects independent of the cathode material.

Irradiation with radioactive materials

The breakdown of short gaps irradiated by means of radioactive materials has been examined by various workers [3, 4, 18, 19, 21, 50, 56, 59, 60, 61]. In experiments by van Cauwenberghe [18] a capsule containing up to 10 mg. of radium is inserted inside one of the electrodes, close behind the sparking surface, and this convenient arrangement has been used in the later experiments by Garfitt [19] and Meek [3, 4]. The latter measurements show that the use of 0·5 mg. of radium is insufficient to facilitate the reliable breakdown of gaps of the order of 0·1 cm. when subjected to impulse voltages of short duration. Oscillographic records [3] of the breakdown of a 0·08-cm. gap between spheres of 1·3 cm. diameter for various conditions of irradiation give values of time lags as shown in Table 8.1, which refers to impulse voltages of peak values 4·1 kV and 6·1 kV respectively as compared with the D.C. breakdown voltage of 3 8 kV. The wave-shape of the impulse voltage in this case is such that it rises to a peak value in 1 μsec. and then declines to half-value in about 1,000 μsec. (The time lags as given in Table 8.1 are measured from the instant of application of the voltage to the gap and, therefore, include the wavefront time of 1 μsec.)

TABLE 8.1

		Kilovolts applied to gap	4 1	6 1
Time-lags in μsec		Gap screened from irradiation	No breakdown	68, 118, 148
		Gap with 0 5 mg. of radium	No breakdown	1, 4, 6, 11, 19, 20, 45, 118
		Gap with polonium-coated cathode	5, 5, 6, 7, 10, 14, 18, 30, 34	
		Gap illuminated by a spark	1, 1, 1	

With the unirradiated gap long time lags to breakdown are observed, as shown in Table 8.1, even though the voltage is considerably in excess of the minimum D.C. breakdown value. The introduction of the 0·5 mg. of radium causes some reduction in the time lag but there is still an appreciable scatter. The effect of the radium is seen to be negligible in comparison with that of ultraviolet illumination provided by a nearby spark, when the time lag is consistently at a low value of about 1 μsec., even at a voltage little greater than the D.C. breakdown value. A coating on the cathode surface of radioactive polonium, with an activity such that about one α-ray is produced in the gap per μsec , gives considerably improved results.

Although the use of 0·5 mg. of radium is not as satisfactory as other means of irradiation for short gaps it gives good results for longer gaps of 1 cm. or more [4]. This is because of the greater volume of the gap, so that a greater amount of ionization is produced in the effective portion of the gap, and there is, therefore, a greater chance of a suitably placed electron appearing in the gap. Measurements of the ion pairs produced by 10 mg. of radium inserted in one of the electrodes, for a gap of length 0·14 cm. and with a sparking area of 0·2 cm diameter, have been made by van Cauwenberghe [18] The saturation current under these conditions was found to be equivalent to $2·4 \times 10^6$ electrons/sec , corresponding to one electron/0·42 μsec. With an 0·5 mg. source of radium the rate of electron liberation in the same gap may then be expected to be one electron per 8·4 μsec. As these electrons are distributed over the gap only a fraction of them are in a position to initiate electron avalanches of sufficient intensity to lead to a spark, and therefore time lags in excess of the time intervals between the release of successive electrons in the gap may be expected, in accordance with the experimental observations [3, 4]. With increasing gap length and sparking area many more electrons are produced in the gap by a given amount of radium, and greatly reduced time lags are observed [3, 4]

The pronounced effect of a small amount of radium, i e 0·5 mg., when placed behind the sparking surface of one of the electrodes in a 3-cm. gap between spheres of 12·5 cm. diameter is shown in Fig. 8.4, which records the percentage breakdown of the gap as a function of the peak value of the applied impulse voltage waves for different conditions of irradiation [4] In order to determine each of the points given in Fig 8.4 a series of ten or more impulse voltages of the same peak value is applied to the gap, and the number of impulses for which breakdown occurs is counted. At low voltages no breakdown is observed, and at high voltages breakdown takes place on each application of the impulse. However, there is an intermediate region in which the breakdown of the gap occurs at random, and where the characteristics of the gap can be conveniently represented by a quantity representing the probability of breakdown. In Fig. 8.4 the percentage breakdown of the gap is used to express this probability and defines for a given peak voltage the ratio of the number of impulses for which breakdown occurs to the number of applied impulses.

The results of Fig. 8 4 refer to the breakdown of the gap when subjected to an impulse voltage rising to its peak value in 1 μsec. and

declining to half-value in 5 μsec. For the unirradiated gap a peak voltage
of about 96 kV causes breakdown of the gap for only one out of ten of
the applied impulses, and the magnitude of the voltage has to be raised
to 115 kV before breakdown occurs on every application of the impulse
waves to the gap. When the gap is irradiated by the light from a mercury-
arc lamp placed at a distance of 100 cm. from the gap, the transition
from no breakdown to 100 per cent. breakdown occurs at considerably

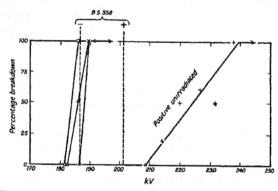

Fig. 8 5   9·0-cm. gap between 12·5-cm diameter spheres· 1/50 wave

+ + + Positive impulse   Gap unirradiated.
× × × Positive impulse. 0 5 mg. radium in high voltage sphere.
○ ○ ○ Negative impulse   Gap unirradiated.
□ □ □ Negative impulse   0 5 mg radium in high-voltage sphere.

lower voltages, between 84 kV and 91 kV. A still sharper transition is
observed when 0 5 mg. of radium is used, when breakdown is not
observed for impulses of 84 kV peak, and is recorded for each application
of impulses of 85 kV peak. Clearly occasional impulses will cause the
unirradiated gap to break down at voltages lower than 90 kV, but the
chance of such an occurrence is remote.

As the gap length between spheres is increased differences are observed
between the effects of irradiation on the behaviour of the gap according
to the polarity of the voltage applied to the gap, when one sphere is
earthed. Fig. 8.5 shows the characteristics of a 9-cm. gap between
spheres of 12·5 cm. diameter when subjected to an impulse wave rising
to peak value in 1 μsec. and declining to half-value in 50 μsec. [4]. The
addition of irradiation by 0·5 mg. of radium has little effect on the transition from no breakdown to 100 per cent. breakdown for a negative applied
voltage, but causes a marked change in the transition for a positive

applied voltage, for which the 50 per cent. breakdown voltage is reduced from about 224 kV to 188 kV.

The difference in the behaviour of the sphere gap when subjected to positive and negative polarities, for the longer gaps with one sphere earthed, has been explained by Meek [4]. For short gaps the initial electron avalanche crosses the full gap between the spheres whereas for long gaps the avalanche leading to breakdown develops in the region of the high-voltage sphere. In the latter case, for a negative impulse voltage the avalanche leading to breakdown is initiated at the surface of the high-voltage cathode, and for a positive impulse voltage the avalanche is initiated by an electron in the gas surrounding the anode. The chance of this initiating electron appearing in the gas is probably small compared with that at the cathode surface, and a larger scatter of results may therefore be expected for the unirradiated gap when the high-voltage sphere is of positive polarity. With adequate irradiation the scatter is reduced and the positive impulse breakdown becomes as consistent as the negative impulse breakdown, with an equally sharply defined breakdown voltage.

Because of the wide variations in the measurements for unirradiated gaps, both with regard to scatter and to the mean observed breakdown voltage, as exemplified in the results given in Figs. 8.4 and 8.5, it is clearly important that adequate means of irradiation should be provided for discharge gaps such as sphere gaps which are widely used in the measurement of impulse voltages With the longer gaps, because of the high voltages involved, it is difficult to arrange a mercury-arc lamp sufficiently close to the gap to produce adequate irradiation, and in such cases the use of radium is preferred.

Some experiments on the effect of 0·5 mg. of radium on the breakdown at 50 c./s. voltages of short gaps between spheres of 1·3 cm. diameter have been made by Edwards and Smee [20]. In all the gaps studied, up to about 0·9 cm., the voltage required to cause breakdown of the unionized gap is appreciably higher than that for the ionized gap. With longer gaps and larger spheres the difference between the breakdown voltages decreases.

The use of radioactive materials other than radium has been investigated by various workers. A coating of radioactive polonium, as used by Garfitt [19], is found to give a marked improvement in the consistency of performance of short gaps, but is not recommended as the coating wears away rapidly with repeated sparking. The use of radioactive cobalt, either as the cathode material, or as an insert in the cathode, has

been suggested by Hardy and Craggs [17, 50]. The possible application
of a radioactive polonium alloy for spark-plug electrodes has been
considered by Dillon [21] and Evans [56].

Illumination by spark discharges

The influence of the ultraviolet illumination from a nearby spark in
facilitating the spark breakdown of a discharge gap was observed by
Hertz in 1887 [22]. During more recent years spark sources of irradiation
have been examined by Street and Beams [23], White [5, 6], Wilson [7],
Garfitt [19], Meek [3, 4], and others [24, 25]. In the experiments by
Street and Beams [23] two small sphere gaps G_1 and G_2 are connected
in parallel and, under normal conditions in air, when an impulse voltage
of a given peak value is applied to the gaps, sparking occurs equally
frequently on either gap when the spacings are the same If, however,
the gap G_1 is enclosed in a screened vessel containing carefully dried and
filtered air, and ions are removed from the gap by a transverse low
electric field before the impulse voltage is applied, then it is found that
the external gap G_2 has to be increased to about 16 times greater than
that of G_1 before sparking occurs equally frequently on G_1 and G_2. The
breakdown of the gaps G_1 and G_2 can be caused to take place again
with equal regularity, for the same spacing, if the enclosed gap is illuminated by the light from a spark discharge which is arranged to occur
simultaneously with the application of the impulse voltage. These and
similar experiments by other workers confirm the importance of irradiation in securing consistent behaviour of discharge gaps, particularly in
enclosed forms of gaps [26].

In many of the experiments where impulse voltages are used the
impulses are derived from a generator in which a number of condensers
are charged in parallel and discharged in series through spark gaps
The latter, therefore, provide a source of ultraviolet illumination
simultaneously with the development of the impulse wave and for
various purposes it is convenient to use this method of irradiation.
Examination of the influence of this source of ultraviolet light on the
behaviour of discharge gaps has been made by Garfitt [19] and Meek
[3, 4]. As shown in Table 8.1 the time lags to breakdown are considerably
lower than those obtained in the presence of 0·5 mg. of radium, and therefore the transition from no breakdown to 100 per cent. breakdown is
more sharply defined. With spark illumination of this gap no breakdown
occurs at 3·8 kV and 100 per cent. breakdown does not occur until the
peak impulse voltage is raised to about 5·0 kV.

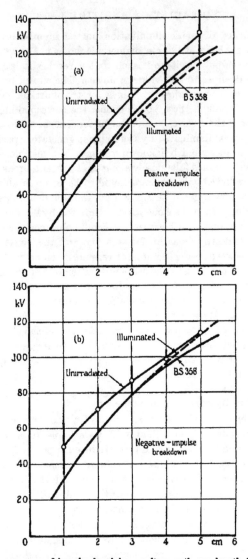

Fɪɢ. 8.6. The variation of impulse breakdown voltage with gap length for a 6 25-cm. diameter sphere gap The circled points denote the 50 per cent. breakdown voltages of an unirradiated gap, for a 1/5 wave, the vertical lines through the points denote the corresponding spreads between the 1 per cent and 99 per cent. breakdown voltages The broken-line curve gives the breakdown voltage of the same gap for the same conditions, but illuminated by the light from the impulse-generator spark gaps.

The influence of spark illumination on the impulse breakdown of gaps between spheres of 6·25 cm. diameter is shown in Fig. 8.6. The wave-shape of the impulse voltage is such that it reaches its peak value in 1 μsec and declines to half-value in 5 μsec. The circled points denote the 50 per cent. breakdown voltages for the unirradiated gaps; the vertical lines through these points denote the corresponding spread of values between the 1 per cent. and the 99 per cent. breakdown voltages. When the gap is illuminated by the impulse generator spark gaps, the spread of the measured breakdown voltages is negligible and the re-sultant curve obtained falls closely along the 1 per cent. breakdown value for the unirradiated gap. There is found to be no noticeable difference between the curves given in Fig. 8.6 for the illuminated gap and the curves obtained when the same gap is irradiated by 0·5 mg. of radium. Differences are observed between the results obtained for the positive impulse breakdown, shown in Fig. 8.6 (a), and the negative impulse breakdown, shown in Fig. 8.6 (b), for reasons given on p. 358.

For gaps of 1 cm. or more the use of irradiation by a spark or by 0·5 mg. of radium gives closely similar results, but with decreasing gap length the spark illumination is the more effective method of producing ade-quate ionization in the gap. As the gap length is increased, to values of 10 cm. and beyond, the use of radium inside one of the electrodes is to be preferred as it becomes increasingly difficult to arrange the gap to be illuminated satisfactorily at a close enough distance to the spark source to make such illumination effective [4].

Illumination by corona discharge

In a study of the impulse breakdown of a gap between spherically shaped electrodes Wynn-Williams [27] was able to show that the voltage required to cause breakdown is reduced appreciably by a nearby corona discharge and that the consistency of behaviour of the gap is greatly improved. It was also established that the influence of the corona discharge can be attributed to the light emitted and not to the passage of ions from the corona discharge into the main gap. The effect has been applied with advantage in the three-electrode gap [62] which is widely used in the measurement of impulse voltages up to several kV.

The impulse breakdown of a discharge gap in which the electrodes are separated by a solid insulating spacer is affected by the corona discharges occurring at the boundaries of the insulator and the electrodes. This has been clearly demonstrated in experiments by Berkey [26], who observed the breakdown characteristics of a form of gap as used in lightning arresters. In the absence of the insulating ring, with the electrodes supported externally and separated by a spacing of 0·165 cm., the impulse ratio between the average impulse breakdown voltage and the average 60 c./s. breakdown voltage was 1·87 when the gap was enclosed in a light-tight box. The insertion of a porcelain insulating ring between the electrodes caused the impulse ratio to fall to a value of 1·36. An insulating ring of rutile (titanium dioxide) gave a further improvement and the impulse ratio recorded for a 0·21-cm. gap was 1·07. The considerable difference caused by rutile can be attributed to its high dielectric constant, of about 80, as compared with a value of about 5 for porcelain. Greater electric gradients are, therefore, produced in the gas near the contacts between the insulating ring and the electrodes when rutile is used, and surface corona discharges are thereby enhanced, with a consequent improvement in the impulse characteristics of the main gap.

Influence of electrode surface

The materials forming the electrodes of a gap are clearly important when ultraviolet light is used as a means of irradiation because of their effect on the photo-current produced in the gap. However, in the absence of any external source of irradiation, changes in the condition of the electrode surface can lead to appreciable differences in the observed impulse breakdown characteristics of the gap.

Slepian and Berkey [28] have investigated the breakdown of gaps of several millimetres in length when small particles of various insulating materials such as porcelain, silicon carbide, rutile, or alumina are placed on the electrode surfaces. The results of the experiments show that the time lag of the gap is materially reduced by the presence of the particles, and that the impulse ratio of the gap can be lowered to a value close to unity by the suitable choice of average particle size. Single large particles are not desirable as they cause a lowering in the alternating breakdown voltage by a proportionately greater amount than the impulse breakdown voltage, so that the impulse ratio is not reduced. By the use of a large number of sufficiently small particles the impulse breakdown characteristics can be affected appreciably with only a small change in the alternating voltage breakdown. The most satisfactory results appear to be obtained when the average linear dimensions of the particles are between about 0·002 and 0·015 cm. Two groups of measured values for the breakdown voltages of a plain gap and for the same gap when the electrodes are coated with particles of average size 0·006 cm. are given in Table 8.2.

TABLE 8.2

Electrode surface	Breakdown kV		Impulse ratio
	60 c/s	Impulse	
Plain electrode	13 4	18 8	1 4
Rutile particles on electrode .	12 4	12 9	1·04
Plain electrode	14 2	21 3	1 5
Alumina particles on electrode	12 6	14 1	1 12

The presence of the particles is considered to cause localized intense field distortions at the electrode surfaces, so that the rate of electron emission from the cathode is increased, with a consequent reduction in the statistical time lag Alternatively the particle may break down under the increased stress and so provide the necessary electrons to initiate a spark across the main gap.

Similar results are observed for discharges between electrodes on which an oxide layer has been formed if the oxide is an insulator. This occurs for electrodes of aluminium or magnesium and is particularly noticeable in low-pressure discharges, as for instance in Geiger–Müller counters [63]. Pactow [29] has found that for cathodes on which insulating oxides are present a single discharge produces conditions which lead to a considerable reduction in the statistical time lag for subsequent discharges. In this case the passage of the first discharge may cause the particles of insulating material on the cathode to become charged by positive ion bombardment or by photo-electric processes, so that high localized electric fields then develop across the particles and enhance the electron emission [64].

The influence on statistical time lags of oxide films on tungsten electrodes has been investigated by Llewellyn Jones [30] for gaps ranging from 0·025 to 0·05 cm. Some typical results are given in Fig. 8.7, where the number of breakdowns is plotted against the time lags observed. Curve A refers to a series of 402 time lags recorded for heavily oxidized tungsten electrodes, curve B to a similar series for slightly oxidized tungsten electrodes, and curve C for clean smooth electrodes. While the shape of the distribution curves is affected by the waveform of the applied voltage, the curves clearly show that as the oxidation of the surface is increased there is a reduction of the mean time lag, from about 200 to 20 μsec. In these experiments no external form of irradiation was used other than that due to normal atmospheric ionization processes. From the results of measurements with anodes and cathodes of different surface conditions it is concluded that the anode surface has little influence, and that the minimum time lags are obtained for a heavily oxidized cathode.

In some of the experiments a pre-breakdown corona could be seen over an electrode which had been used for a large number of sparks, the current values ranging from some microamps to milliamps. Micro-photographs of these particularly active surfaces showed that parts

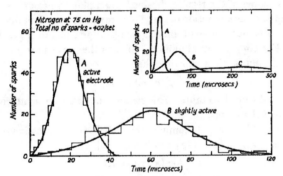

Fig 8 7. Distribution of time lags for a gap in nitrogen at about atmospheric pressure. Curve A for heavily oxidized tungsten electrodes, curve B for slightly oxidized tungsten electrodes, curve C for clean tungsten electrodes.

of the surface projected to distances of about 10^{-3} cm. On some parts spikes project through the oxidized layer, and on other parts spikes occur on the clean metal Such projections at the cathode could lead to field emission of electrons with a consequent reduction in statistical time lag. Also the radiation emitted from minute corona discharges at the projections could cause photoionization in the gas. While surface roughness can affect the ionization produced, Llewellyn Jones [30] has shown experimentally that even with smooth oxide layers the statistical time lags are reduced, and he concludes that there is an enhanced emission of electrons from the oxide layer because of positive charge on the surface.

Llewellyn Jones's experiments apply mainly to air at atmospheric pressure, but reference is made to results obtained at higher and lower pressures, and also to the breakdown of oxygen and nitrogen. In general the mean time lag increases with reduction in gas pressure and is greatest for oxygen and least for nitrogen. The long time lags in oxygen are attributed to the influence of electron attachment.

Statistical time lags at high gas pressures

Experiments have been conducted by Cobine and Easton [31] to determine the influence of gas pressure on the time lag of spark breakdown. The measurements were made for an unirradiated 0·0125-in. gap between spheres of 1·25 in diameter in nitrogen, with several sphere materials. The applied impulse voltage reached its peak value in 0·25 μsec. and decayed to 97 per cent. of its peak value in 5,000 μsec. Fig. 8.8 shows some of the results obtained when plotted as curves, relating $\log_{10} \log_e(n_0/n_t)$ as a function of $\log_{10} t$, where n_0 is the total number of time lags, and n_t is the number of time lags of duration equal to or greater than t μsec For copper and aluminium spheres the graphs are in the form of straight lines, thus showing that

$$\log_e(n_0/n_t) = bt^c, \tag{8.3}$$

where b and c are constants. This straight line relation is not obtained for the nickel spheres.

Cobine and Easton [31] compare the distribution expression (8.3) with that given by von Laue [13] and Zuber [14], namely

$$\log_e(n_0/n_t) = Kt \tag{8.4}$$

Expression (8 4) gives a straight line relation when $\log_{10} \log_e(n_0/n_t)$ is plotted as a function of $\log_{10} t$, the slope of the line being unity. However, in Cobine and Easton's measurements, the graph for copper has a slope of 0·496 in Fig. 8 8 (a) and of 0 422 in Fig. 8 8 (b), while that for aluminium has a slope of 0 200 in Fig. 8.8 (c) and of 0·162 in Fig. 8.8 (d).

The results obtained by Cobine and Easton are explained by consideration of the variation of voltage across the gap during the period of the time lag. Such variation, even though small, may cause appreciable changes in the quantities p_1, p_2, and β, defining the probability of initiation of a spark (see p. 351). In order for the experimental results to conform to the theoretical analysis it is necessary that

$$\int_0^t p_1 p_2 \beta \, dt = bt^c. \tag{8.5}$$

The fact that the exponent c is less than unity can be explained in a qualitative manner. The longer the time lag the lower is the voltage across the gap when the initiating electron arrives, and there is therefore a reduced probability that the electron will lead to breakdown. Consequently an electron arriving late in the surge is less likely to cause breakdown than one which arrives early, and hence a greater proportion of time lags fall in the long time range than would be the case with a constant gap voltage. This causes the slope of the graph relating $\log_{10}\log_e(n_0/n_t)$ with $\log_{10} t$ to be less than that with a constant applied voltage.

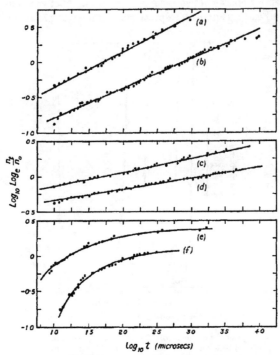

Fig 8 8. Time lag distribution curves in compressed nitrogen.

(a) Copper electrodes, 37 0 per cent overvoltage, 64 7 lb /in ^{2} abs pressure
(b)   „   „   82 6   „   „   24 7   „   „   „
(c) Aluminium „   37 0   „   „   49 7   „   „   „
(d)   „   „   82 6   „   „   19 7   „   „   „
(e) Nickel   „   37 0   „   „   24 7   „   „   „
(f)   „   „   14 1   „   „   29 7   „   „   „

In the light of this analysis Cobine and Easton [31] have replotted Tilles's earlier data [32] on the breakdown of a 0·0269-in. gap between copper spheres, and have shown that the results are better represented by the distribution

$$n_t/n_0 = \exp[-2\cdot52 \times 10^{-8}\, t^{1\cdot34}] \qquad (8.6)$$

than by the simpler relation (8.2). The fact that $c = 1\cdot34$ and is therefore greater than unity implies that the voltage rose during the period of the surge used by Tilles.

The most probable time lag T' is plotted as a function of gas pressure in nitrogen for sparks between copper spheres and between aluminium spheres in Figs. 8.9 and 8.10 respectively. At low gas pressures the lags

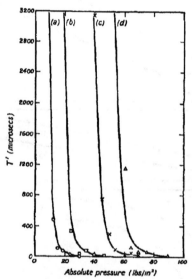

Fig. 8.9. Most probable time lag as a function of gas pressure between copper spheres in compressed nitrogen. (a) 128 per cent. overvoltage; (b) 82 6 per cent. overvoltage; (c) 37 0 per cent. overvoltage, (d) 14 1 per cent overvoltage (Points for $T' > 3,200$ μsec have been omitted to give more convenient scale)

for aluminium are shorter than those for copper, as noted previously by Strigel [9], but at high gas pressures the situation is reversed. Cobine and Easton explain this result by consideration of the mechanism of production of initiating electrons in the gap.

As the gap is not subjected to deliberate irradiation the free ions in the gap are produced by cosmic rays and radioactive emanation from the earth. Under such conditions the rate of electron production is small and the statistical time lag is large, of the order of 10^{-2} sec. With increasing values of electric field a spark may be initiated by electrons released in the gap by field emission, by positive ion bombardment of the cathode, or by negative ion disintegration. Because of the observed influence of cathode material it is probable that the first two mechanisms are more important than the third. While no precise analysis can be given because of the lack of experimental data concerning γ, Cobine and Easton [31]

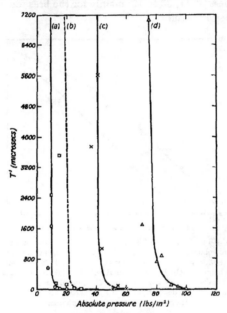

FIG. 8.10　Most probable time lag as a function of gas pressure between aluminium spheres in compressed nitrogen. (*a*) 128 per cent. overvoltage, (*b*) 82 6 per cent. overvoltage, (*c*) 37 0 per cent overvoltage; (*d*) 14 1 per cent. overvoltage.

consider that the observed trends for copper and aluminium electrodes are in general accordance with the limited data available for γ. The anomalous results obtained with nickel electrodes, as shown in Fig. 8.8, are accounted for in terms of field emission, attributable to the Paetow effect [29].

The lowering of breakdown voltage by intense irradiation

The magnitude of the breakdown voltage of a discharge gap is un-affected by ultraviolet illumination of the cathode, provided that the photo-current so produced is less than about 10^{-12} A/cm.^{2} However, if the gap is exposed to more intense illumination, such as that from a nearby spark, the breakdown voltage of the gap may be lowered appreciably, down to as much as 20 per cent. below the normal D C. breakdown value. The effect has been studied extensively by Rogowski, Wallraff, Fucks, and others [3, 11, 33 to 49] mainly for the breakdown of gaps in air at atmospheric pressure.

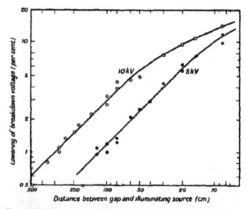

FIG 8 11 Lowering of sparking voltage between aluminium spheres as a function of distance from the illuminating spark gap The two curves correspond to different gap lengths such that the normal D.C breakdown voltages are 5 kV and 10 kV respectively The illuminating spark is caused by the discharge of a 0 1 μF capacitance at 10 kV through a gap between zinc spheres.

In most of the measurements the gap is illuminated by the light from a nearby spark produced by the undamped discharge of a large capacitance. Variation of the intensity of illumination is caused by alteration of the distance between the illuminating source and the gap, or by the use of absorbing screens   Two curves obtained by Brinkman [43] for the lowering of the breakdown voltage of gaps between aluminium spheres of 5 cm. diameter are given in Fig. 8.11. The gap lengths are such that the D.C. breakdown voltages are 10 kV and 5 kV respectively. The curves show the maximum lowering of the D.C. breakdown voltage as a function of the distance of the gap from the illuminating spark, which is formed by the discharge of 0·1 μF charged to 10 kV. Similar results are quoted by the other investigators.

It is of interest to relate the lowering to the value of the photo-current i_0 produced in the gap, but in the above experiments with spark illumination no exact measurements have been made of the magnitude of i_0 or of its variation with time. However, the total number of electrons released in the gap during the illuminating period has been measured with an electrometer technique by several investigators [11, 37], and the variation of this quantity has been studied as a function of the distance between the cathode and the illuminating source. As this distance is

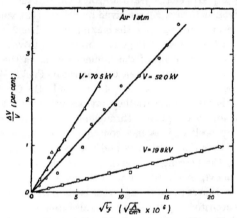

FIG 8 12. Dependence of the lowering of sparking voltage on the square root of the current density in gaps in air at atmospheric pressure

increased the number of electrons released declines more rapidly than the inverse square of the distance, because of absorption of the radiation from the source in atmospheric air[11]. The influence of cathode material and of different illuminating sources on electron liberation has also been studied [11]. From a knowledge of the total number of electrons released in the gap an approximate calculation can be made of the magnitude of i_0 if an assumption is made as to the duration of the illumination. Such a calculation is clearly inaccurate, particularly as i_0 will vary during the illuminating period, but it may be expected to give the order of magnitude. On the assumption that the duration of the illuminating spark is $\sim 10^{-4}$ sec. values of i_0 as high as 10^{-6} A/cm.^{2} may be attained.

If steady sources of illumination, such as that provided by a quartz mercury-arc lamp, are used the lowering observed is much smaller than that produced by the more intense spark sources, but the values of i_0 can be measured directly. Some curves obtained by Fucks and

Schumacher [38] for the breakdown of three gaps in air, for which the normal D.C. breakdown voltages are 19·8, 52·0, and 70·5 kV respectively, are given in Fig. 8.12, where the lowering is plotted as a function of $\sqrt{i_0}$ The curves show that the lowering is directly proportional to $\sqrt{i_0}$. In other measurements by Brinkman [11] a lowering of 5 per cent. for a 10-kV gap between aluminium spheres is recorded when the gap is irradiated by a quartz mercury-arc lamp at 5 cm. distance. Some results for gaps in argon at 10 mm. Hg are given by Schade [48], who also shows that for values of i_0 ranging up to the maximum used, of 2×10^{-8} A/cm.^{2}, the lowering of breakdown voltage is proportional to $\sqrt{i_0}$.

A theoretical explanation of the influence of i_0 on the lowering of breakdown voltage has been given by Rogowski, Fucks, Wallraff, and their colleagues [34, 36, 40, 41, 44, 45, 57, 58]. On the basis of the Townsend mechanism for a spark they have shown that the lowering may be accounted for by the growth throughout the gap of a space-charge field caused by positive ions as a consequence of the large difference in mobility between the electrons and the positive ions. From such considerations the percentage lowering is calculated, and is shown to vary directly as $\sqrt{i_0}$ in accordance with the experimental results.

An interpretation of the lowering on the basis of the streamer theory of the spark has been given by Meek [49], who also assumes that the lowering is attributable to the change in the field distribution in the gap caused by space charge. Because of the field distortion the integral of the Townsend coefficient across the gap is greater than that for the externally applied field alone. The breakdown criterion is then satisfied for a lower voltage gradient than when i_0 is negligible, and a lowering of about 10 per cent. is estimated for a 1-cm. gap when $i_0 = 10^{-8}$ A/cm.^{2}

REFERENCES QUOTED IN CHAPTER VIII

1. *Rules for the Measurement of Voltages with Sphere-gaps*, Brit. Stand., No. 358, 1939.
2 *Measurement of Test Voltage in Dielectric Tests*, A.I.E E Standards, No. 4, 1940.
3. J. M. MEEK, *J. Instn. Elect. Engrs.* **89** (1942), 335.
4. J. M. MEEK, ibid. **93** (1946), 97.
5. H. J. WHITE, *Phys. Rev.* **46** (1934), 99.
6. H. J. WHITE, ibid. **49** (1936), 507.
7. R. R. WILSON, *Phys. Rev.* **50** (1936), 1082.
8. R. STRIGEL, *Arch. Elektrotech.* **27** (1933), 137.
9. R. STRIGEL, *Elektrische Stossfestigkeit*, J. Springer, Berlin, 1939.

10. K. WARBURG, *Wied. Ann.* **62** (1897), 385.
11. C BRINKMAN, *Arch Elektrotech* **33** (1939), 1, 121.
12. R. STRIGEL, ibid. **27** (1933), 379.
13. M. VON LAUE, *Ann. Phys. Lpz.* **76** (1925), 261.
14. K. ZUBER, ibid. 231.
15 G. NORD, *Trans. Amer. Inst. Elect. Engrs.* **54** (1935), 955.
16. C. S. SPRAGUE and C. GOLD, ibid **56** (1937), 594.
17. D. R. HARDY and J D. CRAGGS, *Nature*, **164** (1949), 356.
18. R. VAN CAUWENBERGHE, *Bull. Soc. Franç. Élect.* **7** (1937), 1005.
19. D. E. M. GARFITT, *Proc. Phys. Soc.* **54** (1942), 109.
20. F. S. EDWARDS and J. F. SMEE, *J. Instn. Elect. Engrs.* **82** (1938), 645.
21 J H. DILLON, *J. Appl. Phys.* **11** (1940), 291.
22. H. HERTZ, *Wied. Ann* **31** (1887), 983.
23. J C. STREET and J. W. BEAMS, *Phys. Rev.* **38** (1931), 416.
24. L. B SNODDY, ibid. **40** (1932), 409.
25 J W. FLOWERS, ibid **48** (1935), 954.
26. W. E. BERKEY, *Trans. Amer. Inst. Elect. Engrs* **59** (1940), 429.
27. C. E. WYNN-WILLIAMS, *Phil. Mag.* **1** (1926), 353.
28. J. SLEPIAN and W. E. BERKEY, *J. Appl. Phys* **11** (1940), 765.
29. H. PAETOW, *Z. Phys* **111** (1939), 770.
30. F J. LLEWELLYN JONES, *Proc Phys. Soc. B*, **62** (1949), 366.
31 J. D COBINE and E. C. EASTON, *J. Appl Phys.* **14** (1943), 321.
32. A. TILLES, *Phys. Rev.* **46** (1934), 1015.
33 W. ROGOWSKI and A. WALLRAFF, *Z. Phys.* **97** (1935), 758.
34. W. ROGOWSKI and W. FUCKS, *Arch. Elektrotech.* **29** (1935), 362.
35. W. ROGOWSKI and A. WALLRAFF, *Z. Phys* **102** (1936), 183.
36. W. ROGOWSKI, ibid. **114** (1940), 1.
37. A. WALLRAFF and E HORST, *Arch Elektrotech* **31** (1937), 789.
38 W. FUCKS and G SCHUMACHER, *Z Phys* **112** (1939), 605.
39. W. FUCKS and H. BONGARTZ, *Z. tech. Phys* **20** (1939), 205.
40. W. FUCKS, *Z. Phys* **98** (1936), 11
41. W. FUCKS and W. SEITZ, *Z. tech. Phys.* **17** (1936), 387.
42. H. J. WHITE, *Phys. Rev.* **48** (1935), 113.
43. C. BRINKMAN, *Z. Phys* **111** (1939), 737.
44. W. ROGOWSKI and A. WALLRAFF, ibid. **106** (1937), 212.
45 W. ROGOWSKI and A. WALLRAFF, ibid **108** (1938), 1.
46. M TOEPLER, *Phys. Z.* **40** (1939), 206.
47 M. ARNOLD, ibid. 687.
48. R SCHADE, *Z. Phys.* **105** (1937), 595.
49 J. M. MEEK, *Proc. Phys. Soc.* **52** (1940), 547, 822.
50. D. R HARDY and J. D. CRAGGS, *Trans. Amer. Inst Elect. Engrs.* **69** (I) (1950), 584.
51. M. STEENBECK and R. STRIGEL, *Arch. Elektrotech.* **26** (1932), 831.
52. R STRIGEL, ibid. 803
53. R. STRIGEL, *Wiss. Veroff. Siemens-Werk*, **11** (1932), 52.
54. R. STRIGEL, ibid. **15** (1936), 15
55. R. STRIGEL, *Elektrotech. Z.* **59** (1938), 31.
56. R. D. EVANS, *J. Appl. Phys.* **11** (1940), 561.
57. W. FUCKS and F. KETTEL, *Z. Phys.* **116** (1940), 657.

58. W. Fucks, *Naturwiss.* 9 (1948), 282.
59 W. E. Berkey, *Elect. J.* 31 (1934), 101.
60. R. van Cauwenberghe and G. Marchal, *Rev Gén. Élect.* 27 (1930), 331.
61. K. Berger, *Assoc Suisse des Élect Bull.* 24 (1933), 17.
62. J. D. Morgan, *Phil. Mag.* 4 (1927), 91.
63. S C. Curran and J D. Craggs, *Counting Tubes*, Butterworth, London, 1950.
64. L. Malter, *Phys Rev.* 50 (1936), 48.

HIGH-FREQUENCY BREAKDOWN OF GASES

THE breakdown voltage of a gap for an alternating voltage at 50 c./s. is substantially the same as that for D.C. conditions. However, if the frequency is raised to a value at which positive ions have insufficient time to cross the gap in half a cycle a positive space charge is gradually built up in the gap leading to field distortion and a lowering of the breakdown voltage below the D.C. value. At very much higher frequencies, the breakdown mechanism is further complicated as a consequence of the amplitude of oscillations of electrons in the gap becoming comparable with the gap length so that cumulative ionization can be produced in the gap by an electron travelling many times the gap length in the direction of the alternating field.

In studies of high-frequency discharges in air Reukema [1] has shown that for gaps up to 2 5 cm between spheres of 6·25 cm. diameter there is no appreciable change in breakdown voltages for frequencies up to 20 kc./s. From 20 to 60 kc./s. there is a progressive lowering of the breakdown voltage of a given gap as the frequency is raised but for higher frequencies, up to the maximum of 425 kc /s. used by Reukema, the breakdown voltage of a given gap is constant at about 15 per cent. below the value at 60 c./s.

Similar results have subsequently been obtained by other investigators [2 to 10], who have extended the measurements to different types of gaps for higher frequencies and voltages. In several of the German investigations [2 to 6] voltages up to about 150 kV at frequencies up to 1 Mc./s have been used. The results show that the lowering of the breakdown voltage at the higher frequencies is appreciably greater for gaps in which point electrodes are present than for uniform or nearly uniform fields, as shown by the curves of Fig. 9.1 given by Misere [3]. Luft [4] quotes breakdown voltages similar to those of Misere but also includes values for point-plane gaps, the lowering at 370 kc./s. being 46 per cent. for a 3-cm gap and 70 per cent for a 25-cm. gap, as compared with the 50 c /s. values.

In the case of breakdown between spheres or plates, several of the investigators including Lassen [5], Misere [3], and Müller [6] record a critical gap length below which the breakdown voltage is independent of the frequency, the critical gap length decreasing with increasing frequency. This gap is stated by Muller to be 0·45 cm. at 110 kc./s and 0·09 cm. at 995 kc./s. Consideration of positive ion mobilities shows that these critical gaps correspond roughly to those for which accumulation of positive ions may be expected to occur in the gap, with a consequent space-charge distortion of the field and a lowering of the breakdown voltage.

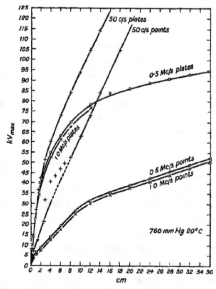

Fɪɢ 9 1. Breakdown voltage in air at atmospheric pressure as a function of gap length between plates and between points for frequencies of 50 c /s , 0 5 Mc./s. and 1·0 Mc./s.

The breakdown of several compressed gases, including air, nitrogen, and freon, has been studied by Ganger [7] for various electrode arrangements at frequencies mainly in the region of 105 to 125 kc./s. Some results for air in a uniform field at 105 kc /s. are shown in Fig. 9.2 where they are compared with the D.C. breakdown values. The results for nitrogen are similar to those for air. Measurements in nitrogen for a gap between a 5-cm. diameter sphere and a plate show that there is only a small difference between the D.C. breakdown value and that at 115 kc./s , the observed values being 67 kV and 65 kV respectively for a 4-mm. gap at 6 atm. With a smaller sphere, of 5 mm. diameter, the difference is more noticeable. Curves are also given by Gänger for the breakdown of gaps in freon, between a 5-cm. diameter sphere and a plate, and no difference is recorded between the 50 c./s. and 110 kc./s. breakdown voltage for gaps ranging up to 8 mm. and pressures up to 6 atm. Break-

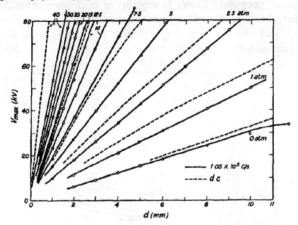

Fig. 9.2. Breakdown voltage as a function of gap length between plates in compressed air for D.C and $1\ 05 \times 10^5$ c /s. voltages.

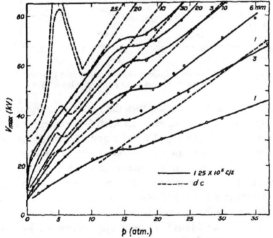

Fig 9 3 Breakdown voltage curves in nitrogen, for different gaps between a point and a plane, as a function of gas pressure in atmospheres.

down curves in compressed nitrogen for gaps between a point and a plane are shown in Fig 9.3 for D.C. and 125 kc /s. voltages.

Experiments have been made by Hale [11] to determine the break-

down voltages in xenon and argon for the range of frequencies from
5 to 50 Mc./s. and for gas pressures of approximately 20 to 50 microns Hg.
Hale has also proposed a theory to explain the mechanism of breakdown
under such conditions, and suggests that breakdown occurs when the
electrical field and the frequency are such that an electron acquires
the ionizing energy at the end of one mean free path. On this basis

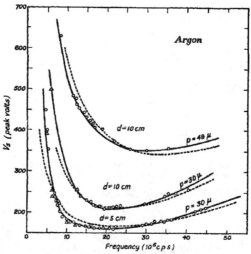

Fig 9 4. Breakdown voltage curves in argon as a function of
frequency. The dotted curves give the calculated values.

the breakdown voltage characteristics for xenon and argon have been
calculated and are in close agreement with the experimentally determined values, as shown in Fig. 9.4. for argon. Calculations, using this
theory, have also been made to determine the minimum breakdown
voltage in hydrogen, and also the gas pressure at which the minimum
occurs. Comparison between these calculated values and those measured
by Thomson [12] over a range of frequencies, gives reasonable agreement
for the minimum breakdown voltage at the higher frequencies, of about
100 Mc./s., but considerable differences in the corresponding gas pressures
are found. However, as Hale [11] points out, the theoretical expressions
which he has used in his calculations are over-simplified. For instance,
the electron mean free path used is the kinetic theory value, whereas
account should be taken of the variation of free path with electron
energy during the sinusoidal cycle. It is also assumed in Hale's calcula-

tions that an electron acquires the ionizing energy at the end of a mean free path, whereas it is known that the ionization probability increases rapidly with electron energy for energies slightly above the ionizing potential for the gas.

Following earlier work by several investigators [13, 14, 15] Gill and von Engel [16] have investigated the electric field strengths required

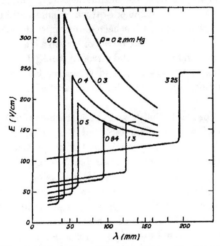

FIG. 9 5 Breakdown field strength as a function of wave-length for a flat-ended cylindrical tube of 3 55 cm. internal length containing hydrogen at different pressures.

to initiate high-frequency discharges in gases at low pressures. The measurements in gases at pressures about 10^{-3} mm. Hg and for wave-lengths between about 4 and 80 metres show that the starting fields are independent of the gas and only slightly dependent on the pressure. With increasing wave-length the starting field strength varies directly with the length of the discharge chamber in the direction of the field. At such low gas pressures, electron collisions with gas molecules occur too infrequently to produce cumulative ionization in the gas, and Gill and von Engel have put forward a theoretical explanation of their results on the basis that secondary emission of electrons from the walls of the discharge chamber is the operative factor.

Later work by Gill and von Engel [17] extends the results to various gases at higher pressures, between 0·2 and 350 mm. Hg, for wave-lengths ranging from 5 to 2,000 m. At these pressures the nature of the gas becomes important. The starting field rises gradually with wave-length until a discontinuous cut-off occurs, as shown in Fig. 9.5 for hydrogen. This discontinuity persists in nitrogen, hydrogen, and deuterium up to the highest pressures used, but is observed in neon and helium only at pressures of less than 0·2 and 0·5 mm. Hg respectively. The cut-off occurs when the amplitude of oscillation becomes equal to the length of the tube, and from values of the cut-off wave-length and the associated field the electron drift velocity has been determined in general agreement with known data. Above the cut-off wave-length a high field is necessary to start a discharge in hydrogen and nitrogen. In hydrogen, as the wave-length is increased, the starting field decreases because of secondary electron production caused by the bombardment of the walls by positive ions.

The onset of corona discharge and the spark breakdown of gaps in air between wires of various diameters and a coaxial cylinder of 5 cm. diameter have been studied by Bright [18] for frequencies up to 12 Mc./s. The peak voltages (kV) required to cause visual corona discharge at several frequencies are given in Table 9.1. With increasing voltage the corona streamers extend from the wire until, when they reach the outer cylinder, breakdown occurs. From photographic and oscillographic records Bright has measured the time taken by the streamer to cross the gap, and gives values ranging from 40 μsec. for a 0·457-mm. diameter wire to 170 μsec. for a 0·234-mm. diameter wire when the frequency is 10·3 Mc./s.

TABLE 9.1

Wire diameter in mm.	Frequency					
	50 c/s	1 6 Mc/s.	3 0 Mc./s.	3 5 Mc./s.	4 0 Mc./s.	9 4 Mc./s.
0 316	7 95	7 0	6 85	6 40	6 15	6 01
0 274	7 30	6 55	6 40	5 95	5 60	5 62
0 234	6 75	5 90	5·85	5 50	5 00	5 10

Bright [18] has also investigated the breakdown of short gaps between spheres in air, nitrogen, oxygen, and CCl_2F_2. Curves obtained for nitrogen are given in Fig. 9.6 and show the existence of critical gap lengths above which the high-frequency breakdown voltage falls below the D.C. value. The curves for air exhibit a series of ripples at gaps above the critical value and this is attributed by Bright to the presence of complex ions [19]. In oxygen the critical gap is more sharply defined

than in air or nitrogen and the high-frequency breakdown voltage falls
to 30–35 per cent. below the 50 c./s. value. No difference is detectable
between the curves at 50 c./s. and at 8 Mc./s. in freon for gaps up to
0·035 in. Calculations of the critical gap have been made by Bright on
the basis that positive ions are trapped and oscillate in the gap when the

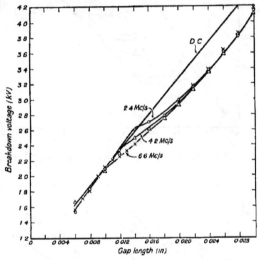

Fig. 9.6. Breakdown voltage curves in nitrogen as a function
of gap length between spheres of 2 0 cm. diameter.

frequency exceeds a critical value. From consideration of positive-ion
mobilities Bright has deduced a theoretical relation between the critical
frequency f_c and the critical gap length d_c in nitrogen as follows:

$$f_c = \frac{3\ 6E}{\pi d_c},\qquad (9.1)$$

where E is the peak voltage gradient. Reasonable agreement is obtained
between the experimental results and those predicted by equation (9.1).

Experiments on the high-frequency breakdown characteristics of
parallel-plate gaps in air when subjected to alternating voltages at
yet higher frequencies, i e. between 100 and 300 Mc./s., are described
by Pim [20]. Gaps of length up to 1 mm. in air at pressures ranging
from 50 mm. Hg to 1,000 mm. Hg have been investigated. The breakdown voltage is found to follow a smooth curve until a certain gap length

is reached when, for a further increase in the gap length, there is a decrease in breakdown voltage, as shown in Fig. 9.7.

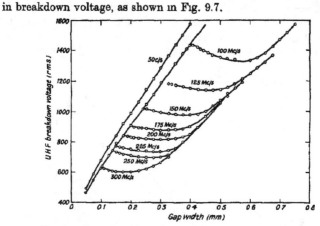

FIG. 9 7   Variation of breakdown voltage with gap length between
plates in air at different frequencies

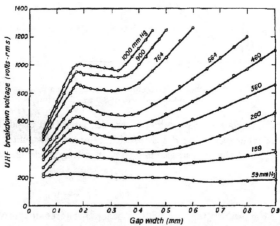

FIG. 9 8   Variation of breakdown voltage with gap length between
plates in air at 200 Mc /s for different gas pressures

The critical gap length, where the discontinuity occurs, depends on the frequency. At gaps less than the critical value the breakdown voltage is about 10 per cent. below the 50 c./s. value for gaps at atmospheric pressure. At appreciably longer gaps the breakdown voltage is about 40 per cent. below the 50 c./s. value, and under such conditions the breakdown gradient is observed to tend to a substantially constant value of about 29 kV/cm.

Further curves obtained by Pim [20] for the breakdown voltage in air at various gas pressures, as a function of gap length, are shown in Fig. 9.8. The critical gap length remains approximately constant as the gas pressure is reduced. For gaps much greater than the critical value, the breakdown gradient tends to become directly proportional to the gas pressure.

Pim [20] has explained his results by consideration of the growth of individual electron avalanches in the gap and on the assumption that the breakdown is governed by the concentration of positive ions in an avalanche. In the normal case at low frequencies a spark is initiated in a gap at the minimum breakdown voltage by an electron avalanche which travels across the gap from cathode to anode If the frequency is raised sufficiently a condition is reached where electrons oscillate in the field, the amplitude of the oscillation becoming comparable with the gap length. At a particular frequency for a given gap, Pim proposes that an electron avalanche which has travelled across the gap may reach the opposite electrode just as the electric field is reversing its direction. The e^n electrons formed in the passage of the avalanche will now return back across the gap, so that a total of e^{2n} electrons is produced before electrons are removed at the electrode from which the original single electron started. A discontinuity in the breakdown characteristics of the gap is therefore to be anticipated when the condition of the gap length and frequency are such as to produce the reverse avalanche which effectively grows across twice the gap. Further, because of the greater length of growth of the 'double' avalanche, the ion density produced is greater, and, therefore, the external electric field required to create an ion density sufficient to initiate a spark is reduced, in accordance with Pim's results for gaps above the critical value.

The first published results for breakdown at ultra-high frequencies appear to be those of Cooper [21], who has made measurements of the breakdown of air in coaxial lines and wave-guides for gaps between 0·1 and 0 3 cm at gas pressures ranging from 20 to 760 mm. Hg. The voltage was applied in the form of 1 μsec. pulses repeated at 400 p./s. At the two wave-lengths used, of 10·7 cm. and 3·1 cm., the breakdown gradient was found to be approximately 70 per cent. of the D.C. breakdown value. The precision of the measurements was increased appreciably by the use of irradiation, by radioactive materials, to provide adequate electrons in the gap to initiate the spark on the application of a

voltage pulse, as shown by Fig. 9.9. The variation of breakdown voltage
with air pressure for a coaxial line, at a wave-length of 10·7 cm., is shown

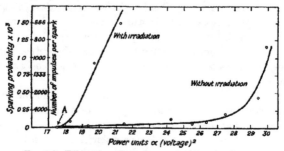

FIG. 9 9. Effect of irradiation on coaxial-line spark-gap.
Diameter of outer conductor = 1 04 cm.
Diameter of inner conductor = 0 535 cm.
A, zero sparking probability; onset of sparking

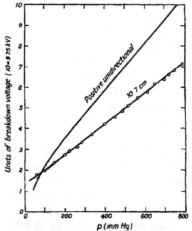

FIG 9 10   Variation of breakdown voltage with
pressure for coaxial-line spark-gap.
Diameter of outer conductor = 1 04 cm
Diameter of inner conductor = 0 404 cm.

in Fig. 9.10, which includes also the curve obtained for the positive D.C.
breakdown of the same line.

Similar measurements at wave-lengths of 10, 3, and 1·25 cm. have
been made by Posin [22] for the breakdown of gaps up to 0·15 cm. in
air at pressures up to 2 atm. for pulse voltages of durations ranging
between 0·3 and 5 μsec. Posin finds that for 3-cm. waves the breakdown

voltage of a 0·043-cm. gap in air under atmospheric conditions is substantially independent of the pulse duration, provided that the duration exceeds 4 μsec.; with decreasing pulse duration the breakdown voltage rises, the increase being about 30 per cent for a 0·3-μsec. pulse. The influence of irradiation on the breakdown voltage is also discussed.

The nature of the spark mechanism in a cavity resonator at these wave-lengths has been studied by Prowse and Cooper [23] and by Prowse and Jasinski [24], using photographic and spectroscopic methods. The results for air at atmospheric pressure show at breakdown a bright spot on one electrode with a luminous pencil of discharge extending into the gap. If the gas is illuminated by a collimated beam of ultra-violet light of sufficiently short wave-length from an external spark, numbers of mid-gap streamers are observed for each voltage pulse. The formation of these mid-gap streamers is attributed to photoionization of the gas.

A series of investigations on the microwave breakdown of gases in cylindrical cavities and between coaxial cylinders, at a wave-length of 9·6 cm., has been made by S. C. Brown and his colleagues [25 to 30]. The gaps studied range from 0·06 to 7·6 cm. in air at pressures from 0·1 to 100 mm. Hg. The results are discussed in terms of a new theory for ultra-high-frequency breakdown, which is based on the criterion that the ionization rate equals the diffusion rate for electrons   Other processes of removal of electrons, such as attachment and recombination, are considered to be negligible for the type of discharges studied, where the gap length is small compared with the wave-length, and the electron mean free path and amplitude of oscillation of an electron are small compared with the gap length. The breakdown condition is obtained from consideration of the continuity equation for electrons

$$\frac{\partial n}{\partial t} = \nu n - \nabla . \Gamma, \tag{9 2}$$

where n is the electron density ν is the net production rate of electrons per electron and denotes the difference between the ionization rate and the attachment rate. Γ is expressed in electrons per second per unit area and represents the electron current density lost to the walls by diffusion. As the electric field is raised, ν increases until the time derivative $\partial n/\partial t$ becomes positive. The threshold for breakdown is considered to occur when $\partial n/\partial t$ goes through zero. The breakdown is then the characteristic value of the electric field obtained from the solution of the equation

$$\nu n - \nabla \ \Gamma = 0 \tag{9 3}$$

and the boundary condition that the electron density vanishes at the cavity surfaces.

By the introduction of the term $\psi = Dn$, where D is the diffusion coefficient for electrons, the breakdown criterion may be rewritten as

$$\nabla^2\psi + \zeta E^2\psi = 0, \qquad (9.4)$$

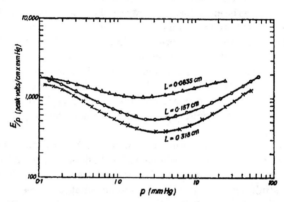

FIG 9.11. Breakdown voltage gradient between plates in air as a function of pressure for three gap lengths (cavity resonant wavelength ∼ 9 6 cm)

where E is the r.m s value of the high-frequency electric field ζ is known as the ionization coefficient, and may be expressed as $\zeta = \nu/DE^2$. Values for ζ have been computed by Brown *et al.* from their breakdown measurements under parallel plate conditions in cylindrical cavities, and are expressed as functions of E/p and $p\lambda$, where p is the gas pressure and λ the wave-length The data are then used to calculate the breakdown voltages in air between coaxial cylinders and the results are found to be in close agreement with the experimentally determined breakdown voltages for a range of gas pressures from 0·2 to 100 mm. Hg. A comparison [25] between the calculated curves and the measured points at various gas pressures in air, for three gap lengths, is given in Fig. 9.11.

In a further extension of the above theory [29], the ionization coefficient ζ has been computed theoretically on the basis of kinetic theory without using any gas discharge data other than experimental values of the ionization potential and collision cross-section of helium. The breakdown voltages for gaps of 0·15 to 2·54 cm. in helium at pressures from 0·3 to 300 mm. Hg are calculated from equation (9.4). The agreement

between the theoretically calculated breakdown voltages and the
measured values is within the experimental error of 6 per cent.

If the applied frequency is greater than the frequency of inelastic
collisions and less than the frequency of elastic collisions, Holstein [31]
considers that the energy distribution of electrons in a high-frequency

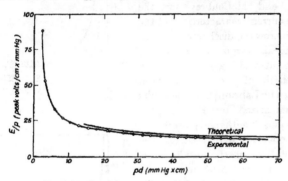

FIG. 9.12. Breakdown voltage gradients in argon.

field is closely the same as that of electrons in a static field equal in
magnitude to the r.m s value of the high-frequency field [32]. The
growth of ionization may then be calculated on the assumption that
the ionization coefficient α has the value corresponding to the r.m.s.
value of the field. Holstein deduces the breakdown condition for a
non-attaching gas, the criterion for breakdown being that the rate of
production of electrons by ionization must exceed the rate of loss of electrons by diffusion. In the case of a uniform field between parallel plates
the calculated relation between the breakdown gradient E, the gap
length d, and the gas pressure p is

$$(pd)^2 = \frac{\pi^2 k T_e}{e(E/p)(\alpha/p)},\qquad(9.5)$$

where α is the Townsend first ionization coefficient, T_e is the electron
temperature, and k is Boltzmann's constant.

Krasik, Alpert, and McCoubrey [33] have carried out an experimental
investigation on the breakdown of argon at pressures between 3 and 100
mm. Hg and for a frequency of 3,000 Mc./s. The discharge gap used was
constructed in a resonant cavity and gave an approximately uniform
field. A curve showing the experimental results for E/p as a function
of pd is given in Fig. 9.12 where it is compared with the theoretical curve
as calculated from equation (9.5) using Kruithof and Penning's data for

α [40] and Townsend and Bailey's data for T_e [41]. The experimental curve lies from 4 to 8 per cent. below the theoretical curve and the investigators consider the agreement is such as to give satisfactory verification of Holstein's breakdown theory over the region of applicability.

Curves are given also by Krasik, Alpert, and McCoubrey for the voltage-current characteristics of the argon microwave discharge.

The breakdown of air and neon at low pressures in wave guides for a wave-length of 3 2 cm. has been examined by Labrum [34]. The results are summarized by Prowse [35]. Various pulse lengths and pulse repetition frequencies were used. In the interpretation of the results it is assumed that the ionization increases exponentially with time so that the number of electrons increases from n_0 to n in a time t according to the relation

$$n = n_0 e^{(G-L)t}, \qquad (9.6)$$

where G is the number of ion pairs produced by an electron per sec. and L is the number of electrons which disappear per sec. If the criterion for breakdown is that there should be at least n_D electrons in the gap then $(G-L)$ must exceed $\dfrac{1}{T}\log_e\!\left(\dfrac{n_D}{n_0}\right)$ for breakdown to occur, where T is the

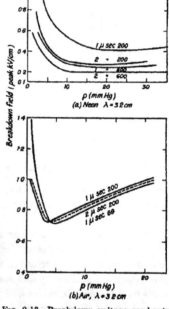

(a) Neon $\lambda = 3.2$ cm

(b) Air, $\lambda = 3.2$ cm

Fig 9 13 Breakdown voltage gradients in neon and in air. (The figures on the curves give the pulse duration and the pulse repetition frequency in pulses per sec.)

pulse duration. Because of the logarithmic form of this criterion no great precision is needed for its specification provided n_D is large. A probable value for n_D is 10^{13}. G is of the same nature as the Townsend ionization coefficient α. In a non-attaching gas, where diffusion is the chief agent in removing electrons, L is small and it is assumed that no electrons are lost during the first few microseconds. The breakdown voltage is then determined by the pulse duration and breakdown occurs when

$$G = \frac{1}{T}\log_e\frac{n_D}{n_0}. \qquad (9.7)$$

Fig. 9.13 (a) shows the variation of the breakdown field in neon with gas pressure for two pulse lengths and two pulse repetition frequencies. Similar curves are given in Fig. 9.13 (b) for air. The pulse repetition rate affects the results in so far as ionization remains in the gap from pulse to pulse.

In an explanation of the curves attention is drawn by Labrum [34, 35] to the calculations by Townsend and Gill [32] of the average energy V_e gained by an electron, per mean free path, in a field E of angular frequency ω, namely

$$V_e = \frac{e}{2m}\frac{1}{\nu^2+\omega^2}E^2, \qquad (9.8)$$

where ν is the collision frequency for an electron with the gas molecules. It is assumed that the electron accumulates energy at the rate V_e per collision until it acquires an energy V_i, where V_i is the ionization potential of the gas. The electron then loses all its energy in one ionizing collision and the process starts again. This picture of the ionizing mechanism leads to an expression for G and thence to the breakdown field X_c which is given by

$$X_c^2 > \frac{4V_i}{e/m}\frac{\omega^2+\frac{1}{3}\nu^2}{\nu}\frac{1}{T}\log_e\!\left(\frac{n_D}{n_0}\right). \qquad (9.9)$$

A series of theoretical papers on high-frequency gas discharges has been written by Margenau and Hartman [36, 37, 38], who have discussed methods for determining electron energy distributions and have shown how such functions can be used in the calculation of breakdown fields on the assumption that the only mechanism for electron removal is recombination with positive ions. The calculated values are appreciably lower than the measured values, in the case of helium and neon at about 1 mm. Hg for a frequency of 3,000 Mc./s., and the discrepancy is explained by the consideration that electrons must also be removed by other mechanisms.

REFERENCES QUOTED IN CHAPTER IX

1 L. E. Reukema, Trans. Amer. Inst. Elect. Engrs. 47 (1929), 38.
2. J. Kampschulte, Arch. Elektrotech. 24 (1930), 525.
3. F. Misere, ibid. 26 (1932), 123.
4. H. Luft, ibid. 31 (1937), 93
5. H. Lassen, ibid. 25 (1931), 322.
6. F. Muller, ibid. 28 (1934), 341.
7. B. Gänger, ibid 37 (1943), 267.
8. E. W. Seward, J. Inst. Elect. Engrs. 84 (1939), 288.
9. P. Jacottet, Elektrotech. Z. 92 (1939), 60.
10. W. Fucks, Z. Phys. 103 (1936), 709.
11. D. H. Hale, Phys. Rev. 73 (1948), 1046.

12. J. Thomson, *Phil Mag* **23** (1937), 1.
13 E W. B. Gill and R H. Donaldson, ibid. **12** (1931), 719.
14. C. and H. Gutton, *Comptes Rendus*, **186** (1928), 303.
15. M. Chenot, *Ann. Phys. Paris*, **3** (1948), 277.
16. E. W. B. Gill and A. von Engel, *Proc Roy. Soc.* A, **192** (1948), 446.
17. E. W. B. Gill and A. von Engel, ibid. **197** (1949), 107.
18. A. W. Bright, *Brit. Elect. and Allied Ind. Res. Assoc. Report*, Ref. L/T 229 (1950)
19. O. Tuxen, *Z. Phys.* **103** (1936), 463.
20. J. A. Pim, *J. Instn. Elect. Engrs.*, Part III, **96** (1949), 117.
21. R Cooper, ibid. **94** (1947), 315.
22. D. Q. Posin, *Phys Rev* **73** (1948), 496.
23. W. A. Prowse and R. Cooper, *Nature*, Lond., **161** (1948), 310.
24 W. A. Prowse and W. Jasinski, ibid. **163** (1949), 103.
25. M. A. Herlin and S. C Brown, *Phys. Rev.* **74** (1948), 291.
26. M. A. Herlin and S. C. Brown, ibid. 910.
27. M. A. Herlin and S. C. Brown, ibid. 1650
28. S. C. Brown and A. D. MacDonald, ibid. **76** (1949), 1629.
29. A. D. MacDonald and S. C. Brown, ibid. **75** (1949), 411
30 A. D. MacDonald and S. C Brown, ibid. **76** (1949), 1634.
31. T. Holstein, ibid **70** (1946), 367
32. J. S. Townsend and E. W. B. Gill, *Phil. Mag*, **26** (1938), 290.
33 S Krasik, D. Alpert, and A O McCoubrey, *Phys. Rev.* **76** (1949), 722
34 N. R Labrum, *C S.I R. Australia R P R* **85**, December 1947
35 W. A Prowse, *J Brit I.R E.* **10** (1950), 333.
36 H. Margenau, *Phys. Rev.* **73** (1948), 297, 326.
37. H Margenau and L. M. Hartman, ibid. **73** (1948), 309.
38. L. M. Hartman, ibid 316.
39 T. Holstein, ibid **69** (1946), 50 (A).
40. A. A. Kruithof and F. M. Penning, *Physica*, **4** (1937), 430.
41. J S. Townsend and V. A. Bailey, *Phil. Mag.* **44** (1922), 1033.

X

THE SPARK CHANNEL

The formation and initial expansion of spark channels

THE manner of growth of the spark channel, i e. the bright narrow discharge of low impedance and high current density which passes across a discharge gap after breakdown (see Chapter IV), is not clear. Loeb and Meek [1] refer to a potential wave which passes from the cathode to the anode after a streamer bridges the gap, but the physics of this mechanism is not explained. Formation of a cathode spot (see Chapter XI) is presumably necessary in order that the electron emission, needed to give the transition from the breakdown discharge to the lower impedance spark channel, may be produced, but experimental and theoretical work on this phase of the discharge is difficult and little has been done. Although various workers disagree as to the mechanism of channel expansion [2 to 6] it seems likely that the channels expand rapidly at the commencement of current flow (see Fig. 10.1, Pl. 9) at a rate of the order of the thermal velocity of the gas atoms. It appears that, during the earliest parts of the channel régime, there are pressure discontinuities amenable to treatment on shock wave theory. Various authors [7, 8, 9] have also pointed out that the self-field of the discharge, giving inwardly directed momentum to the charged particles in the channel, may lead to the establishment of radial pressure gradients (see Flowers [6] for an estimate of the energy involved) The channels expand more slowly in the later stages (Fig. 10.1) because of diffusion of charged, neutral, and excited particles Recombination processes are probably also important The variation of spark channel gas density, or gas temperature and pressure, with time is not yet known, but the slow rate of expansion after ~ 1 μsec. suggests [4, 5] that a constant pressure state of low density is reached more quickly in the channel than many early workers supposed [2, 3, 6, 10]. The properties of the channel have been studied by many workers following the pioneer work of Lawrence and Dunnington [11], but the experimental data are incomplete, and a full detailed account of the spark channel has not yet been given. We shall proceed to describe some of the recent experiments in this field and, later, to summarize the conclusions to be drawn from them

The propagation of shock waves from sparks has been studied by Funfer [12], Toepler [13], Foley [14], Craig [4], and others. McFarlane [15] has also carried out experiments to measure the sound radiation from spark discharges. Anderson and Smith [16] published in 1926 the details of their method for determining gas temperatures in spark channels by finding the velocity of propagation of a pressure wave (whose transit shows as a local change in the brightness of the channel) through the spark. Various refinements and extensions of this method are

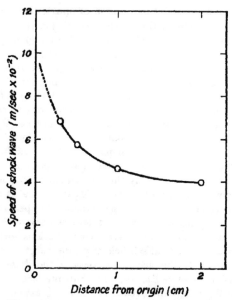

Fig 10 2   Speed of propagation of the shock wave
from a spark.

described in Suits's work on arcs [17, 18, 19]. However, the detailed relationship between the shock-wave movement and the diffusion processes remains to be worked out   Foley's results have been re-analysed by Craggs, using Davies's [21] recent theoretical work on shock wave propagation. Fig. 10.2 shows the data given by Foley for the speed of propagation of the shock wave as a function of the distance from the centre of the spark channel. The results give an initial gas temperature in the air spark of 4,500° K., corresponding to a pressure of 15 atm , but the large extrapolation required in the calculations make the result highly inaccurate. Fig. 10.3, Pl. 9, gives photographs of shock waves from sparks as photographed by Adams.

If the initial rate of expansion of a spark channel is equal to the velocity of propagation of the shock wave, then the expansion rate should follow a similar law to that deduced by Taylor [22] for the expansion of the hemispherical luminous envelope of the Los Alamos atom bomb explosion, when $r^{5/2}$ varied as t, but with the alteration, necessary for the case of the cylindrical shock wave from a spark, that

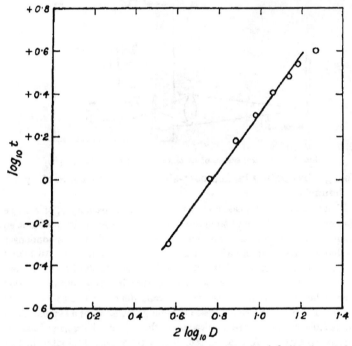

FIG. 10.4. Application of Taylor's analysis [22] of an atom bomb flash to the initial stage of spark channel expansion. Experimental data from Higham and Meek [3] for a spark in air

$D =$ visible spark diameter in mm.

$t =$ time in μsec.

r^2 varies as t. Fig 10.4 shows a plot of data obtained from the records of Higham and Meek [3].

For the study of luminous channel expansion rates Flowers [6] employed a technique previously used on arcs by Kesselring [23] and Suits [17–19]. It consists (see Fig 10 5) in one form of an optical system in which the spark image is focused on to a fine slit, which is perpendicular to the discharge axis. The image of the slit is photographed by some form of rotating mirror or drum camera. The latter thus records, for that part of the discharge gap seen by the slit, the luminous diameter of the spark channel as a function of time. A typical record obtained by Craig is shown in Fig 10.1. Komelkov [24] has used this technique for work on pre-channel parts of the spark discharge (see Chapter IV).

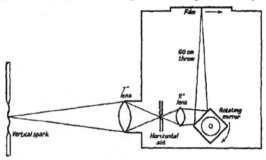

FIG 10 5   Optical system of streak camera for spark photography

Energy dissipation in, and post-shock wave expansion of, spark channels

Flowers also made measurements, on the same discharges, of voltage gradients as a function of time although his measurements were not so accurate as those of later workers. The data for Flowers's condenser discharge currents (damped or undamped) are given in Table 10.1, and in Fig. 10.6 and Fig. 10.7. Current densities of nearly 30,000 A/cm.^{2} were measured; Norinder (see below) has reported initial values of 170,000 A/cm^2 but after a few μsec. the current densities fell to $\sim$ 30,000 A/cm.^{2} Flowers has also discussed the energy balance for a typical spark where 2,000 watt sec. were required to produce the discharge, in which the current was 94,000 A after 5 μsec. The total energy dissipated in this time was calculated to be 2,040 watt sec., comprising 990 watt sec. of circuit loss, radiation and diffusion losses, etc., in the channel, 130 watt sec. of sound energy (associated with the initial rapid radial expansion of the channel and the accompanying shock wave), and 920 watt sec. lost in producing the ionization in the channel. However, it must be emphasized that the assumptions made in these calculations, in the absence of complete experimental data, are probably not entirely justified. Flowers assumes that the channel grows radially from a streamer and, discussing this process in some detail, indicates certain inconsistencies.

Flowers's optical technique has been used by various subsequent

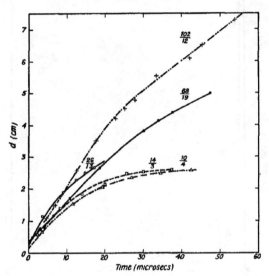

FIG. 10.8 a   Temporal variations of spark channel diameter in the experiments of Norinder and Karsten.

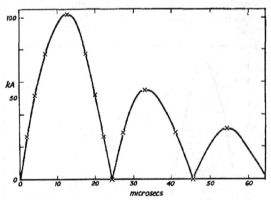

FIG. 10.8 b.   Current oscillogram of a spark discharge studied by Norinder and Karsten

$\sim 10^5$ cm./sec. at 1μsec after channel formation. This rate of expansion, probably due to ambipolar diffusion, decreased with increasing time, and detailed results are given for a wide range of variables. The gases studied include air, nitrogen, oxygen, and hydrogen. Branched channels,

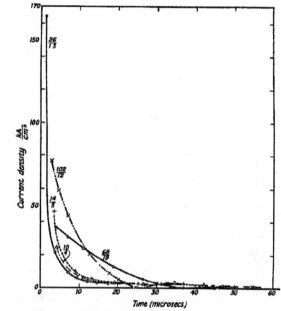

FIG 10 9 a Temporal variations of current densities in spark discharges studied by Norinder and Karsten The fractions x/y marked on the curves refer to $\dfrac{\text{peak current}}{\text{quarter period in }\mu\text{sec}}$ values.

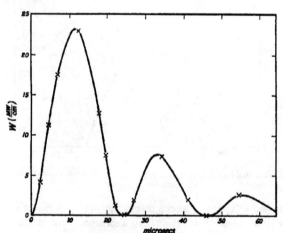

FIG. 10.9 b Temporal variation of power dissipation in spark discharge of Fig. 10.8 b.

observed previously by Craggs, Hopwood, and Meek [36], were also photographed, Fig. 10.10, Pl 10, shows an example in which the apparent outward movement of the bright parts of the channels is clear. This may be due to the shock wave emitted by each channel affecting the other channel [4]; such effects have been previously observed for single channel discharges by Suits [17, 18, 19] for arcs and Anderson and Smith [16] for exploded wire discharges Measurements of channel cross-section were also made with the method of Saxe and Higham [37], in which the channel is scanned diametrically with a slit and multiplier tube system (see Fig. 10 11, Pl. 10, for an example of the record obtained, this method clearly has great potentialities).

Rough calculations of electron density can be made [3, 29] from the mobility equation (used previously by Craggs and Meek [10], Norinder and Karsten [2], and Flowers [6] for sparks), namely $i = Nev$, where i is the channel current density, e is the electron charge, and v is the electron drift velocity. In such calculations the value of i is determined from the measured visible channel diameters, the procedure is an approximate one which may be improved later as further data on the radial variations of electron concentrations [33, 37] become available. The drift current of positive ions may usually be neglected, but the effect of positive ion scattering on electron velocity is analysed by Gvosdover [38] and others [39] and shown to be appreciable for large electron concentrations [5]. Results are given in Table 10.1 for nitrogen at 1 atm. [3, 29]. The effect of the positive ion scattering for hydrogen sparks has been discussed by Craig and Craggs [5].

TABLE 10 1

Current wave† .	500 (10)			250 (10)			250 (28)		
Time from breakdown (μsec) .	0 5	2	8	0 5	2	8	0 5	2	8
Channel diameter (cm)	0 14	0 27	0 41	0 11	0 19	0 28	0 11	0 19	0 27
Current density(kA/cm²)	32	7·9	2·2	29	7·8	2 4	29	8 9	3·7
Voltage gradient (V/cm)	125	60	30	125	60	30	100	59	36
Electron drift velocity (km./sec.) .	4	3 3	3 8	4	3 3	2 8	3 8	3 1	2 8
Electron density (×10⁻¹⁷/cm³)	5	1·5	0·5	4·5	1·5	0 5	4 8	1·8	0 8

† Figures give peak in amps , decaying to half-value in (   ) μsec

Further errors, impossible to assess, are caused by a lack of knowledge of v for *atomic* nitrogen (the gas used for the data of Table 10.1) and by a general ignorance of density conditions in the channel. Discussions of these matters are given in more detail for the case of hydrogen sparks by Craig and Craggs [5] and for hydrogen arcs by Edels and Craggs [40].

Care must also be taken that the correct allowance is made for pressure and gas temperature (see Loeb [41], p. 187) when deducing v as $f(E/p)$ from, say, the Healey and Reed tables [40, 42].

Higham and Meek's second paper [29] describes detailed measurements, with a non-linear potential divider, of the voltage drop across spark gaps, during the channel process. Point-point and point-plane gaps were used in air, nitrogen, oxygen, and hydrogen at pressures lying between 100 and 700 mm. Hg. The gradients fell from several kV/cm. to $\sim$ 20–40 V/cm. for currents $\sim$ 200–300 A (reaching their peaks in 0·25 μsec. and falling to half-value in either 10 or 28 μsec.). The spark lengths, corrected for deviations from the linear inter-electrode distance, varied from 10 to 35 cm. The sparks tended to be shorter for higher overvoltages, and a typical set of channels for nitrogen at 1 atm. in a 20-cm. gap at 200 kV applied voltage is given in Fig. 10.12, Pl. 11. Records for hydrogen under the same conditions are shown in Fig. 10.13, Pl. 11, where the channels are seen to be more tortuous, a fact recorded previously and discussed by Craggs, Hopwood, and Meek [36].

The voltage gradients, in general, follow the changes in visible channel size, being greater when the latter is small in the early stages of the discharge. Spark channels artificially constricted by a tube [29], and also those obtained in liquid breakdown, show increased gradients, relative to those obtained with discharges in gases at atmospheric pressure. No quantitative interpretation of these results has yet been made, and the accumulation of further experimental data seems necessary.

Additional information on the voltage gradients, and structure of spark channels, in hydrogen [5] and in argon, neon, and helium [32] is given in papers by Craig and Craggs, who also discuss expansion mechanisms in more detail.

Experimental work on the expansion of luminous spark channels, using a Kerr cell technique described by Holtham and Prime [30], has been carried out by Holtham, Prime, and Meek [31] for discharges carrying approximately rectangular current pulses. Details of the Kerr effect, and its use in the photography of transient phenomena, are given by von Hámos [43] and other workers [44, 45] (see p. 183).

Holtham's results are exemplified by the photograph of Figs. 10.14 (a) to (d), Pl. 11, showing 2,100 A spark channels about 6 mm. long, in air, between tungsten electrodes Figs. 10.14 (a) to (c) were taken at times of 0, 4, and 5·5 μsec after the commencement of current flow. Figs. 10.15 (a) to (e), Pl. 12, show similar discharges in hydrogen at times 0, 4, 5·5, 8·5, and 11 μsec. in which cathode spots (lower electrode) are visible, together with a bright central pencil of light from the anode. A microphotometer plot of a channel photograph in hydrogen, showing peculiar central features is given in Fig. 10.16. The radial variation in brightness (obtained as described below by the method of Brinkman [46] and of Grimley and Saxe [35]) is shown in Fig 10.17. The central

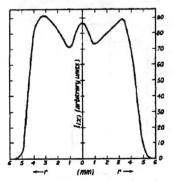

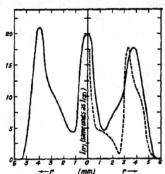

Fig 10 16 Microphotometer plot, cali-
brated for intensities, of a spark channel
(Holtham).

Fig. 10 17. Radial variation of light
intensity in the discharge of Fig. 10.16.

trough, which is observed only in the early stages of the channel growth, is discussed by Craig and Craggs [5] (see pp. 489–91). The central pencil of light (Fig. 10 16) is, as yet, not explained but it may possibly be akin to the discharge form observed by Bruce [47] (see pp. 462–4), or it may be attributable to the erosion processes occurring at the anode [48].

The structure of spark channels

In view of the interest attached to the structure of spark channels, it is worth describing in detail how the relation between the observed channel profile and the true luminous structure [35], shown for example in Fig. 10.17, can be found. This was done over 15 years ago by Hörmann [49] and later by Brinkman [46] whose work is summarized in a paper by van den Bold and Smit [50]. The same analysis has been made by Huldt [51] and by Grimley and Saxe [35]. The following summary is taken very closely from Brinkman's work. It is first assumed that, by some suitable optical technique using a fine slit, the luminosity curve taken in one direction across the discharge has been obtained.

Take the origin of rectangular coordinates at a point in the slit, which may conveniently be the collimator slit of a spectrograph, the latter being parallel to the z axis. The y axis is perpendicular to the slit and in its plane, and the x axis is then perpendicular to the arc axis and the slit (Fig. 10.18). The observed intensity variation $I(x)$ is then obtained by a suitable quantitative micro-photometer technique. $I(r)$ is the required radial variation of light intensity. We have

$$I(x) = \int\limits_{-\infty}^{+\infty} I(r)\, dy = 2 \int\limits_{0}^{\infty} I(r)\, dy.$$

(10.1)

The solution of this equation is, according to Abel [46],

$$I(r) = -\frac{1}{\pi} \int\limits_{r}^{\infty} \frac{I'(x)}{\sqrt{(x^2 - r^2)}}\, dx$$

$$= -\frac{1}{\pi} \int\limits_{x=r}^{\infty} \frac{I'(x)}{x}\, du,$$

(10.2)

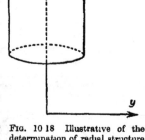

where $u^2 = x^2 - y^2$.

The graphical determination of $I(r)$ now proceeds as follows. $I(x)$ is differentiated and so $I'(x)$ is obtained. The value of $I(r)$ for a given r is determined by plotting $I'(x)/x$, for all values of $x \geqslant r$, against u and measuring the area under the curve so obtained.

Fig. 10.18   Illustrative of the determination of radial structure of cylindrical discharge columns.

$I(r)$ is readily found by carrying out this procedure for various values of $r \geqslant 0$.

Examples of these graphical determinations of $I(r)$ from $I(x)$ are given by Brinkman, and some of his results for a carbon arc, impregnated with a barium salt, are given in Fig. 10.19. Brinkman states that the graphical analysis of the measured function $I(x)$ can be performed with such accuracy that differences between the measured $I(r)$ and that obtained from $I(r)$ by integrating back are < 1 per cent.

The energy balance in spark channels

Rompe and Weizel [52] have used the energy balance equations to study the later phases of spark discharges, i.e. the conditions in the spark channels. According to an equation derived by Toepler the conductivity of a spark at any given time should be equal to the product

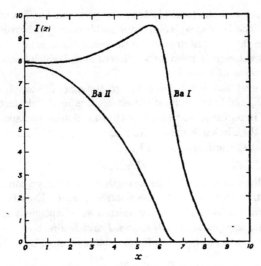

Fig. 10.19 a. $I(x)$–x plots for Ba impregnated arc (Brinkman)

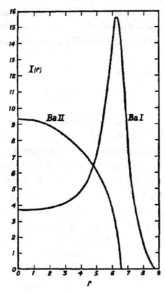

Fig. 10.19 b. $I(r)$–r plots for Ba impregnated arc (Brinkman).

of a constant and the total electron charge which has flowed through the spark. The Toepler equation is applicable usually [52] over a limited range, since the constant depends on the spark duration and length. The following treatment is given for a different model which leads to results similar to those of Toepler.

It is assumed that, in the channel stage, the spark fills evenly a channel of radius R, and that the localized effects due to the electrodes are of negligible importance compared with the charge transport in the regions of the discharge remote from the electrodes. The current and electric field strength are related by

$$I = \pi R^2 n_e k_e E e, \tag{10.3}$$

where I is the current, E the field strength, n_e the electron concentration. k_e the electron mobility, and e the electron charge. Especially at high pressures the mobility may be considered as independent of the field strength and so equation (10 3) relates I and E directly.

The energy balance equation for the spark channel is

$$IE = S + W + \frac{dU}{dt}, \tag{10.4}$$

where W is the energy loss/sec. by heat conduction, S is the radiation term, and U is the internal energy of the spark channel. S, W, and U depend in some way upon N_e, and if there is, for example, thermal equilibrium in the channel, a temperature T can be invoked and S, W, and U will be functions of it.

The conductivity F is given by

$$F = \frac{I}{E} = \pi R^2 n_e k_e e \tag{10.5}$$

and can be related to the internal energy (relevant data are given by Davies [21] for air and its constituents), which comprises the translational energy of atoms, ions, and electrons, the energy of ionization, and also, for molecular gases, the excitation energies for rotational and vibrational states and the dissociation energies. It is assumed here that, as the electron gas takes energy from the field, it is passed only very slowly to the heavy particle constituents of the plasma, so that in the time considered neither the kinetic energy of the ions, atoms, or molecules changes nor does any appreciable degree of rotational and vibrational excitation occur. Molecular dissociation is also assumed to be negligible. Ionization, however, is assumed to take place and therefore this series of assumptions refers only to a short-lived discharge in a high electric field. As is shown elsewhere, conditions similar to the above are not easily produced since they refer to the earliest stages of spark production, i.e. to the processes taking place when the channel is formed. It is difficult indeed to arrange conditions in which negligible energy is passed to atoms by excitation or by elastic collisions. With these limitations, however, it is interesting to continue the treatment as given by Rompe and Weizel. The internal energy is then (k = Boltzmann's constant)

$$U = \pi R^2 n_e \tfrac{3}{2} kT + \pi R^2 n_e eV_i, \qquad (10.6)$$

where the first term gives the kinetic energy of the electrons and the second term the energy given up by the electrons in ionization. The internal energy and conductivity are both proportional to electron concentration, and the constants of proportionality are dependent upon temperature to a slight extent. We shall ignore this dependence, and so, approximately, the conductivity can be expressed as

$$F = aU, \quad \text{where } a = k_e e/(\tfrac{3}{2} kT + eV_i). \qquad (10.7)$$

If a state of thermal equilibrium does exist in the channel (this can hardly be reconciled with the assumptions made above as to the limited degree to which electron/atom interactions occur) then F and a can be expressed in terms of temperature.

In order to derive an expression which would connect the conductivity of the spark channel with the electrical characteristics of the circuit it will be assumed that the discharge occurs in so short a time that radiation and heat loss by conduction are negligible. Rompe and Weizel put this time as less than 10^{-7} sec , certainly not more than 10^{-6} sec., and it is necessary to repeat that assumptions as to the extent to which thermal equilibrium could be established in such conditions must be made with caution. We then have the simple equation

$$IE = I^2/F = \frac{dU}{dt}$$

and, since $U = F/a$,

$$I^2 = (F/a)dF/dt = \frac{1}{2a}\frac{dF^2}{dt},$$

so that

$$F^2 = 2a \int I^2 \, dt.$$

The measurement of ion concentrations in hydrogen spark channels

The possibility of determining ion concentrations in spark channels in hydrogen, as a development from the early qualitative work of Merton [54], Lawrence and Dunnington [11], Hulburt [55, 56], Finkelnburg [182], and others, is clearly worth investigation [10] as a check on the value of N derived from the drift current equation (see p. 398), and measurements have been made in rigidly defined electrical conditions by Craggs and Hopwood [57]. The circuit was essentially that of a simple radar modulator [58]; later improvements in technique are described by Craig and Craggs [5]. It enabled sparks, with the current waveform shown in Fig. 10.20, Pl. 13, to be passed between pointed tungsten electrodes, some 5 mm. apart in hydrogen at about 86 cm. Hg. The pulse

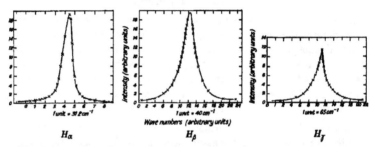

H_α H_β H_γ

Fig. 10 21. Stark profiles of hydrogen lines H_α, H_β, and H_γ with 110 A, 1 μsec. spark discharges.

length was varied from about 1 to about 10 μsec. for peak currents of about 110 A.

The spectra were observed either with a photomultiplier tube attached to a small Hilger constant-deviation spectrometer or by direct photography. Line profiles for the first three members of the Balmer series were obtained respectively by traversing the lines across a telescope slit mounted before the photomultiplier or by taking microphotometer plots of the lines on the photographic plates. It was necessary, of course, to calibrate the latter, and a standard step wedge technique [57] was adopted. Fully corrected profiles for H_α, H_β, and H_γ are given in Fig. 10.21 where the current pulse durations are given. Further experimental details are given in the paper [57].

The Holtsmark statistical theory for Stark broadening of spectral lines in inhomogeneous fields was developed to take full account of the fine structure of the lines in the above paper [57] and independently by Verweij [59] whose work was later discussed by Unsöld [60]. This theory enables a resultant profile for, say, H_α to be determined for a given value of the interatomic field F_n, or (ion+electron) concentration N, and Fig. 10.22 gives an example of such a curve fitted at one point to the experimentally measured profile for the whole spark channel. Craggs and Hopwood [57] arbitrarily took the ion concentration appropriate to the

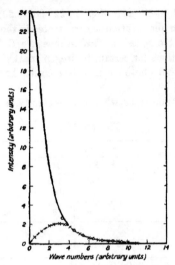

FIG 10 22   Fitting of theoretical Stark profile (for inhomogeneous field of $F_n = 124$ kV/cm) to observed profile from spark discharge for H_α

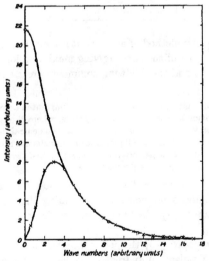

FIG. 10 23   As for Fig 10 22 but for H_β and with $F_n = 133$ kV/cm.

profile giving the best fit (as in Figs. 10.22 and 10.23) to be that for the spark channel, with qualifications clearly stated in the paper (see below).

It appears possible *by this method* to measure F_n to ± 10 per cent., and ion concentrations for certain hydrogen sparks, using respectively photographically and photo-electrically recorded line widths, are given in Tables 10.2 and 10 3.

From Figs. 10 22 and 10.23, which are the results, as the paper indicates

TABLE 10.2

Plate no	Pulse length (μsec)	F_n (kV/cm)			Mean F_n	N ($ions/cm^3$)
		H_α	H_β	H_γ		
1 5 4	1	124	133	130	130	2 1 × 10^{17}
30 4 3	4	as with 1 μsec pulse to within about $\pm 5\%$			130	2 1 × 10^{17}
1 0 3	10	85	79	76	80	1 0 × 10^{17}

TABLE 10.3

Pulse length (μsec)	F_n ($kV/cm.$)		Mean F_n	N ($ions/cm^3$)
	H_β	H_γ		
1	165	163	164	2 9 × 10^{17}
4	As with 1 μsec pulse to within about $\pm 5\%$		164	2 9 × 10^{17}

[57], of an arbitrary method of analysis, it is seen that matching is only obtained at the skirt of the curve (giving maximum ion concentrations in the discharges) and the following comments are made on this point in the paper.

Neither the Holtsmark theory, nor that of Hulburt cited above, applies to the undisturbed radiation at the centre of the profile, where the experimental and calculated intensities are widely different, due to the central undisplaced Stark components and the fact that $F_n \to 0$ for some of the radiating atoms. The form of the line near the centre of the pattern is probably governed partly by other effects, e g. a Doppler broadening $F_n \to 0$ where $N \to 0$.

It is also stated [57] that radial variations in ion concentration exist in spark channels, and a method for determining the rate of change of channel structure for sparks having rectangular current wave shapes is outlined.

In an earlier paper [10] the ion concentration for similar sparks (120 A, 1–4 μsec. pulses) in hydrogen was deduced by several methods as being approximately 10^{17} ions/cm.^{3}, in agreement with the later results. In particular, attention is here directed to the method of equations (2) and (3) of the early paper, in which the ion concentrations are found from a knowledge of the channel radius, taken as 0·75 mm., and the voltage drop. The latter, about 90 V/cm, corresponded to $N \sim 6 \times 10^{17}$ ions/cm.^{3} Stress is particularly placed on the order of the ion concentrations, although the Stark profiles obtained can be used to give ion concentrations more accurately.

Grimley and Saxe [35] have analysed the above data in a different fashion and extended its usefulness considerably by determining from the line profiles (Fig. 10.21) the variation of ion concentration in the spark channel with radius. This treatment, and similar but graphical methods (Saxe, unpublished, 1948, and elsewhere [61]), replaces the earlier interpretations of Craggs and Hopwood [57] made on the basis of accurate Stark profiles for such discharges but without the necessary knowledge of radial variations of luminous intensity (see above).

The following extracts from the paper [57] indicate the need felt at that time for data on channel structure, later to be provided theoretically by Grimley and Saxe [35] and experimentally by various authors [3, 5, 30, 31, 32, 33, 34, 35, 37].

In order to obtain preliminary data on channel structure, photographs of 1, 2, and 4 μsec. sparks (constant peak current) were taken and the channel images traced on the microphotometer The figures (Figs 21 and 22 of the paper [57]) show that the channels were non-uniform in luminosity and are brighter at the centre. Although the relation between intensity of light emission and ion concentration is not yet accurately known, it is reasonable to suggest a correspondingly high radial change in ion-concentration The channel profiles could owe their shapes to the mechanism of growth of a channel from a streamer, in the sense that the photographic record integrates an infinite number of succeeding stages of the spark from its beginning as a streamer, thin compared with the spark channel which succeeds it, in which case the postulation of radial changes in ion concentration would require verification. It is considered, since the growth of spark luminosity is so slow with visible light, that this alternative explanation is not likely, and that the channel profiles indicate strong radial changes in N at the times 1–4 μsec.

Grimley and Saxe [35] point out that it is possible to derive the radial variation of ion concentration in spark channels, or indeed in similar discharges where the necessary data may be obtained, if besides a knowledge of the Stark profiles (see Fig. 10.21) the radial variation of light intensity radiated from the channel is known. It is first necessary, therefore, for the purpose of this full analysis to obtain by experiment the variation in light emission across the discharge channel (as viewed, for example, through a transverse slit), and to correct this for the varying finite depth of discharge. The latter is, reasonably, assumed to

be circular in cross-section. The theoretical treatment is developed on
the assumption that the nearest neighbour approximation can be made,
i.e. that a radiating atom experiences only the field of the nearest ion,
since the use of the Holtsmark treatment, involving as it does [57] two
series of terms for the calculation of the effective total field acting on the
radiating atom due to all the ions and electrons in the discharge, would be
excessively cumbersome. Chandrasekhar [62] suggests that the proba-

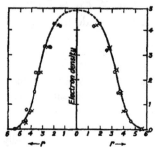

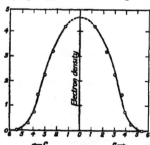

FIG. 10.24 Radial variation of electron
density in a 110-A rectangular current
pulse hydrogen spark 2 μsec after the
initiation of the discharge. Units of r
are 0 12 mm Electron density units are
10¹⁷/c.c.

FIG. 10.25 Radial variation of electron
density in a 110-A rectangular current
pulse hydrogen spark 3 9 μsec. after the
initiation of the discharge Units of r
are 0 12 mm. Electron density units are
10¹⁷/c.c.

bility distribution of inter-atomic fields varies little from the Holtsmark
statistical distribution, but this seems to be in error (a fuller discussion is
given by Hopwood, Craig, and Craggs [33]) and thus the use of an analysis
based on the nearest neighbour approximation does not seem justified.

Graphical methods of analysis, therefore, would appear to be the
simplest techniques available [33, 61], and radial variations of ion concentration for a 110 A 4 μsec. spark in hydrogen for times of 2 and 3·9
μsec. after the start of current flow are given in Figs. 10 24 and 10.25.
Such data are required for the development of spark channel theory,
and the study of hydrogen afterglows at high pressures.

It may be stated, as a summary to the above discussion, that electron
concentrations determined from Stark broadening or from the drift
current equation agree as well as may be expected in view of the uncertain knowledge of channel temperatures Relevant discussions are
given by Craggs and Meek [10], Grimley and Saxe [35], and Craig and
Craggs [5].

If spark channels may be considered as discharges in thermal equilibrium (and theoretical, but not final, evidence for this is given on

pp. 51 and 52), then Saha's equation [63] may be used for the determination of the degree of ionization if the temperature is known. Usually, and still not with complete justification, spark temperatures are calculated from Saha's equation and a knowledge of electron concentrations derived either from Stark effect data or from use of the drift current equation. Except in the earliest part of a spark discharge, a state of constant density cannot be assumed to exist and, as no data on density are available, the use of Saha's equation cannot be fully justified.

The latter is conveniently expressed as

$$\log_{10} \frac{N_i N_e}{N_a} = -5040 \frac{V_i}{T} + \tfrac{3}{2} \log_{10} T + 15 \cdot 38, \qquad (10.8)$$

where N_i, N_e, and N_a are the positive ion, electron, and normal atom concentrations, V_i is the ionization potential, and T the discharge temperature. For a plasma, such as a spark channel, $N_i = N_e$. Useful nomograms are given by Andrade [64] and Taylor and Glasstone [65] and a table of dissociation and ionization data for varying gases is given by Davies [21]. For example, if $V_i = 15$ V, then at 10,000° K. and a density corresponding to $p = 1$ atm. at room temperature, the degree of ionization ~ 1 per cent. or $N_e = 2 \cdot 7 \times 10^{17}/\text{cm}^3$ A fuller discussion of Saha's work is given by Edels [63].

It should be noted that the excitation temperature measurements of Craggs et al. [66] do not suggest that simple thermal equilibrium holds for the hydrogen sparks studied. Reference should be made to the work of von Engel and Steenbeck [67] and Ornstein and Brinkman [68], who discuss the relative contributions from electron-atom and atom-atom collisions in cases where thermal ionization is taking place.

The luminosity of spark channels and of high-current transient discharges

The most luminous part of the spark discharge occurs during the channel process, especially at high pressures, and since sparks can be easily controlled in time they have naturally been used as light sources in high-speed photography. Various authors have measured the light emission from sparks, and typical early work is by Kornetzki et al. [69], who used an integrating circuit with a photo-cell to study the radiations from sparks in air, using condenser discharges ($V = 4$–50 kV, $C = 10^{-3}$ to 1 μF). The radiation H varied from 0·01 to 200 candle-sec. and it was found that

$$H = aCV^n, \qquad (10.9)$$

where a is a constant, $\sim 0 \cdot 02$, and n varied from 2·5 to 4. The temporal variation of the total visible radiation emitted during the discharges was also measured. About 200 candle-sec. were emitted from a 45 kV, 85×10^{-3} μF discharge in about 0·5 μsec. Other authors, Laporte [70], Craggs and Meek and their collaborators [10, 57, 66, 76], have measured total visible radiations from sparks as a function of time with photo-multiplier tubes, rotating mirrors, and other devices. Similar measurements have been made by van Calker and Tache [71], Standring and Looms [72], and Beams *et al.* [73]. The temporal variations of intensity for single spectral lines have been studied by Raiskij [74], Wolfson [75], and others [76], and further reference is made on p. 414 to this work.

More recently the development of flash tubes (containing rare gases, often Kr or Xe or mixtures thereof) of various types, with a greatly increased luminous emission, has largely rendered sparks obsolete in many applications, although [72] when a relatively feeble source is to be used at repetition rates much greater than, say, 10 c./s. a spark may be preferable. Representative papers on the new sources are by Edgerton [77, 78] and Aldington and others [79] (see also references [80–85]). The development of these sources in Britain originally owed much to the work carried out at the Armaments Research Establishment by Mitchell and others [80] and also in the B.T.H. and G.E.C. Research Laboratories.

The radiations from high-pressure inert gas arcs, bearing some resemblance to those from flash tubes, have been studied by Schulz and others [81, 82]. It is often important to consider the afterglows found with flash sources and spark discharges and the production of short light flashes ($\sim 10^{-7}$ sec. duration) is considered by various authors [175, 176, 177, 179].

Dushman [86] has summarized the luminous efficiency of various light sources, mainly of the high-pressure type, ~ 100 atm., where perhaps 50 per cent. of the energy input may be radiated, and even at ~ 1 atm. the efficiency may be ~ 40 lumens/watt (27 per cent. efficiency). Mohler [87], discussing the luminous efficiency of discharges in highly ionized Cs vapour, calculates an efficiency of 140 lumens/watt for a 1 atm discharge at 7,000° K. The above data refer, of course, to stable arcs. The new flash sources may have efficiencies of ~ 40 lumens/watt [79].

Observations of emission spectra of spark channels have been made by many workers; recent typical work is by Craggs *et al.* [66, 76, 88]. Similar data on lightning have been accumulated by Israel and Wurm [92] and others [93, 94, 95].

Spark ignition of gases

The study of sparks used for the initiation of explosions or combustion appears to resolve, at least for certain cases, into a study of the channel since it is during this period of the spark that most of the heat available for chemical reactions is developed. The lowest temperature at which chemical reaction in a given gaseous medium becomes self-sustained is usually termed the ignition temperature [96]. The monograph by Ellis and Kirkby [96] gives a clear but condensed account of ignition mechanisms and the following paragraphs are based on pp. 57-67 of their book. The establishment of an electric spark in a given gaseous atmosphere is not, of course, the only means by which an explosion may be produced but it is the only mechanism which will be considered here.

As a somewhat over-simplified generalization, it may be stated that two main theories of ignition are current. They are, firstly, the purely thermal theory [97, 98] and, secondly, the chain reaction theory, and the following brief treatment makes reference to both. An excellent review of some of the earlier work is given by Bradford and Finch [99].

It appears from the work of Thornton, Wheeler, Morgan, Finch, and others that when a condenser discharge is used the amount of heat required is less than when a spark in an inductive circuit is used. If a hot wire is used for ignition the amount of heat required is greater than with an inductive circuit, but this fact could mean either that ionization processes might be important or that the rate of production of energy in the discharge is the controlling factor. Thus Jones [98] concluded that ignition depends upon heating a sufficient volume of gas to a sufficient temperature and that the most effective spark is that which heats the greatest volume of gas to the ignition temperature Finch *et al.* [100] performed experiments with H_2 and CO in D.C. discharges at low gas pressures. They concluded that combustion is dependent upon a suitable excitation of the reacting molecules, and that ignition ensued when the concentration of these molecules is sufficiently great.

Guénault and Wheeler [101] showed that a temperature rise of several hundred degrees could be produced by the type of discharge used by Finch *et al*, and it seemed therefore that purely thermal effects may have been operative at the same time as the electrical effects (excitation of molecules, etc.). Finch and collaborators [102, 103] studied in great detail the effect of the various kinds of ignition coil discharges that they produced, but it is difficult to distinguish clearly from these and other experiments the desirable electrical characteristics that a spark should possess in order to give efficient ignition. Before chemical reactions can occur at any speed, some of the participating molecules must have their internal energy increased. This may be done by thermal means, so that the molecular energy of translation is increased, or by collisions with electrons, etc., so that energy of excitation is produced. It seems likely that the latter process will be preponderant in the early part of ignition by a spark and that heat will then be liberated by virtue of chemical reaction between these excited or ionized molecules. The production of heat leads then to a spreading reaction and finally to combustion, although whilst this process is proceeding the later stages of the spark will have developed as an additional source of heat. It appears necessary to distinguish in future work between the relative importance of the spark channel heating and the heat of reaction developed as the excited or ionized molecules react.

Slater [104] discusses the effect of increased gas temperatures on the frequency of collisions which lead to chemical reactions. Thus the probability of reaction equals $A \exp(-Q_1/kT)$, where A and Q_1 are constants. This expression suggests that only those collisions in which the reacting molecules taken together have an energy (translational, rotational, and vibrational) of Q_1 or more lead to a reaction since, by the Maxwell distribution law, the fraction of molecules having this energy will contain the factor $\exp(-Q_1/kT)$. Q_1 is connected with the inter-atomic or inter-molecular forces that must be overcome in order to produce a new stable molecule.

The thermal and electrical mechanisms of spark ignition have also been reviewed in some detail by Jost [105] The thermal case is first considered [105, chapter 2], in which ignition proceeds outwards by virtue of thermal conduction of heat from the primary spark zone where a great increase in gas temperature is locally produced by virtue of the dissipation of electrical energy in the spark. In a later part of his book (chapter 10) Jost proceeds to discuss electrical mechanisms, i.e. the importance of excited or ionized atoms and molecules [96]. It was shown, for example, by various workers [106, 107 ; Jost, p. 359] that the introduction of atoms (from dissociated molecules) into the explosion space often reduced the minimum ignition temperature, and it is not unreasonable to suppose that electrical ignition proceeds, at least some-times, partly in this way. Landau [108] derived a general theory for ignition processes in which the concentration of active particles (species unspecified) at a point governs the ignition process, and made the sim-plified assumption that the reaction velocity is proportional to the con-centration of active particles. The propagation of ignition conditions is then set by an equation for the diffusion of the active particles (or the heat conduction) away from the spark region. Relevant recent work is by Linnett, Raynor, and Frost [109] and by Frost and Linnett [110, 111], by Blanc, Guest, von Elbe, and Lewis [112] and by Lewis and von Elbe [113]. In one of these papers [112] the method of measuring spark energy is discussed in detail, together with the experimental results. It was found that, for any particular experimental arrangement, a certain minimum energy is necessary for ignition of an explosive mixture (mixtures of CH_4, O_2, and N_2 were used) and a theory which explains this result is given in a second paper [113]. The energy distribution in a spark gap, as a function of time or position in the discharge, is clearly of great importance in this work and the detailed arguments given are not always in agreement with the latest measurements of energy distribution in short sparks (see, for example, Craig [115] and the work of Boyle and Llewellyn [114]).

Many recent contributions to the study of spark ignition have not been considered in detail above in the interests of brevity [116–21].

Afterglows in sparks and related discharges

Afterglows following spark discharges have been observed and studied by Laporte and Pierrejean [122] in Ne, A, etc., at pressures of 1–10 mm. Hg, by Craggs, Meek, and Hopwood [10, 32, 36] using pressures ∼ 1 atm. in various gases, and by Bogdanov and Wolfson [123]. The latter authors used a rotating-mirror technique and showed for Kr discharges that spark lines reached their maximum strength before arc lines and that the latter persisted much longer. They suggested that, whilst this could be due to the high electric fields existing in the early stages of the discharge, a more likely explanation is that the degree of ionization is then higher than the degree of excitation, so that arc lines are only emitted feebly at the commencement of the channel process.

Craggs et al. [10, 32] studied afterglows following sparks with rectangular current pulses in order to avoid the difficulty of interpreting afterglows in the presence of small residual currents, which renders the earlier work uncertain. Examples of the records are shown in Fig. 10.26, Pl. 13, where it is seen that afterglows in argon are much longer than those in hydrogen [124] Various explanations are possible but not proved. Thus the long rare gas afterglows could perhaps be due to metastable atoms or to delayed recombination due to a slowly falling electron temperature caused by the small Ramsauer cross-sections in these gases [125]. The record of Fig. 10.27, taken by Craig [32], shows how in neon at

1 atm there is a sudden rise in luminosity at the end of the current pulse which was 10 μsec long. This is consistent with the suggestion that the gas temperature in this case is still much less than the electron temperature even after several μsec. and that the latter fell suddenly to the former at that time with a subsequent sudden increase in the probability of the electron-ion radiative recombination process.

Emission spectra of afterglows have been observed with low pressure discharges in helium, and in caesium and mercury vapour, by Mohler and his collaborators [126, 127], by Kenty [128] working with argon, and by Webb and Sinclair [129]. The afterglows have been attributed to radiative electron-ion recombination (see Chapter I) and recombination cross-sections have been deduced. Zanstra [130] has thus explained afterglows observed in hydrogen by Rayleigh [131]. More recent work on various gases has been performed by Holt and his collaborators [132, 178, 181] (see p. 21 for a discussion of the relevant work of Biondi and Brown [133, 134], who used a multiplier tube technique, previously developed by Rawcliffe [135], Craggs and Meek [10, 136], Dieke and his collaborators [137, 138], and others).

Non-isothermal plasmas: discharge temperatures

A detailed theory of the disintegration of a discharge plasma has been given by Granovsky [139] This paper also refers to experimental work with probes on decaying low pressure plasmas [140]. A simplified theoretical discussion of accommodation times (see p. 49) in relation to decaying plasmas will now be given.

Plasma decay is of importance in T.R. switches, used in radar practice, and Margenau et al. [141] have discussed the mechanism of energy dissipation in electron swarms for conditions applicable in such cases. The approach to thermal electron velocities takes place through the agency of collisions (largely elastic) between electrons and gas atoms in the switch, which occur at a rate v/λ per sec., where v = electron velocity, λ = electron mean free path. Cravath (see p. 4) puts the mean energy loss at an electron temperature T_e and gas atom temperature T_g as

$$\Delta\epsilon = \frac{4m}{M} k(T_e - T_g). \tag{10.10}$$

(A full discussion of discharge temperatures is given by Edels [63].) The rate of loss of mean energy $\frac{3}{2}kT_e$ is then

$$\Delta\epsilon\left(\frac{v}{\lambda}\right) = -\frac{\partial}{\partial t}(\tfrac{3}{2}kT) = \frac{4m}{M\lambda} k(T_e - T_g)\left(\frac{3kT_e}{m}\right)^{\frac{1}{2}}. \tag{10.11}$$

The solution of this equation is

$$\frac{\sqrt{T_e}-\sqrt{T_g}}{\sqrt{T_e}+\sqrt{T_g}} = \frac{\sqrt{T_{e_1}}-\sqrt{T_g}}{\sqrt{T_{e_1}}+\sqrt{T_g}} \exp(-g\sqrt{T_g}\,t), \qquad (10.12)$$

where the quantity T_{e_1} is the initial value of T_e and the constant g is given by

$$g = \frac{8}{M\lambda}\left(\frac{km}{3}\right)^{\frac{1}{2}}. \qquad (10.13)$$

The time in which the electron velocity falls from its initial value to α times the final velocity is then given by

$$t_\alpha = \frac{1}{g\sqrt{T_g}} \log_e\left(\frac{\alpha+1}{\alpha-1}\right), \qquad (10.14)$$

provided that $\sqrt{T_e} \gg \sqrt{T_g}$. For $\alpha = 2$, the relaxation time t_α is a few μsec. for electrons in argon at 10 mm Hg pressure. Beyond that time the approach to thermal velocities is slow.

A similar result is obtained by a simple argument used by Craggs, Hopwood, and Meek [36]. Since in one collision an electron loses on the average the fraction $2m/M$ of its excess energy, it must make $M/2m$ impacts to lose an appreciable part of its energy. The time between impacts is $\lambda/v \sim 10^{-10}$ sec. Hence the time in question for the argon discharge mentioned above is

$$\frac{M}{2m}\frac{\lambda}{v} \sim 4\times10^{-6} \text{ sec.}$$

(as T_g is always much less than the initial T_e).

In the analysis, therefore, we may assume with reasonable safety that after some 10 μsec. the electron speeds are approximately thermal in T.R. switch conditions of low pressure, etc. For the case of spark discharges at atmospheric pressure the relaxation time will be $\ll 10\,\mu$sec. if the gas temperature during the discharge is always very much less than the electron temperature (see p. 417). For other spark discharges, e.g. in hydrogen, it is possible that the gas temperature rapidly rises to a value close to the electron temperatures and so, at the end of a discharge lasting some μsec., or even less, $T_e \sim T_g$.

Data on (T_e-T_g) calculated from the Townsend method [142] are given by Norinder and Karsten [2]. They show that, for a Maxwellian distribution of electron velocities, on the assumption that η (the average fraction of its energy lost by an electron at an elastic collision) and λ

(the electron mean free path for elastic collisions) are independent of electron energy,

$$T_e - T_g = \frac{2 \cdot 65 \times 10^7 \lambda^2 E^2}{T_e \eta}, \tag{10.15}$$

where E is the electric field.

For $E = 100$ V/cm. in nitrogen at $T_g = 5,000°$ K. and $10,000°$ K. we have respectively that $(T_e - T_g)$ is about $190°$ K. and $40°$ K., and for $E = 200$ V/cm. it is about $550°$ K. and $100°$ K.

The above calculations are based on the assumption of equilibrium and can only apply to transient discharges if the relevant accommodation times (see pp. 49 and 418) are such that equilibrium between the electrons and gas atoms can be considered as virtually complete. Further, it is doubtful whether constant-density conditions persist for as long in a spark discharge as Norinder and Karsten suppose (see Craig and Craggs [32] and Craig [4]), so that the arc conditions of constant pressure and high temperature, and therefore of low density, may be reached in a few μsec. In that case the values of $(T_e - T_g)$ given above need correction, as λ would then be greater. Townsend mobility data, which refer to low T_g, must be used with extreme caution, therefore, in high pressure arcs and sparks.

As the above problems are of some interest, a further investigation on relaxation, or accommodation, times in discharges will now be described. It should be clear, however, that whilst excitation temperatures have been measured in spark discharges by several authors [63, 66, 76] measurements of gas temperatures and more especially of electron temperatures in such discharges are few [63] although an extensive programme is in progress in the authors' laboratory.

The conception of the electron swarm in a gaseous discharge as an electron gas has led [143, 144] to simple, though apparently approximate, evaluations of the relaxation time for the energy transfer from electrons to gas atoms. For the present, we shall follow Jagodzinski's analysis [144] and assume that Newton's law of cooling (valid for small temperature differences) applies. The heat dw transferred during time dt from the electron gas at temperature T_e to the true gas at temperature T_g is given by

$$dw = K(T_e - T_g)\, dt. \tag{10.16}$$

If the electrical power supplied is L, then

$$T_e - T_g = L/K. \tag{10.17}$$

If the current is suddenly switched off, then during the following transient period the electron temperature T at time t is given by

$$T - T_g = (T_e - T_g)\exp(-Kt/mc), \qquad (10.18)$$

where mc is the heat capacity of the electron gas. Now if the time at which the temperature difference is one-half of its original value is τ, then

$$T - T_g = \tfrac{1}{2}(T_e - T_g), \qquad (10.19)$$

therefore we have $K = (mc/\tau)\log_e 2$ or $(T_e - T_g) = \dfrac{L\tau}{mc\log 2}$. τ is governed by the number of electron collisions per unit time and, considering pressure variations only,

$$\frac{\tau}{\tau_0} = \frac{p_0}{p},$$

where p_0 is a standard pressure at which $\tau = \tau_0$. Also [144] for a monatomic gas

$$(mc)_{gas} = \frac{C_v + 2}{22414} \frac{T_0}{T_g} \frac{p}{p_0} \text{ cal./cm.}^3/°\text{K.}, \qquad (10.20)$$

where $p_0 = 760$ mm. Hg, T_g and T_0 are the gas and room temperatures, and C_v is the specific heat at constant volume

If x is the degree of ionization

$$(mc)_{electrons} = x(mc)_{gas} = \frac{5}{22414} \frac{T_0}{T} \frac{p}{p_0} x. \qquad (10.21)$$

If x_0 refers to p_0, then $x = x_0 \sqrt{(p_0/p)}$ from Saha's equation, when $x \ll 1$, so that finally

$$T_e - T_g = \frac{22414}{5\log_e 2} \frac{L\tau_0}{x_0} \frac{T}{T_0} \left(\frac{p_0}{p}\right)^{\frac{3}{2}}. \qquad (10.22)$$

It is seen that $(T_e - T_g)$ varies as $p^{-\frac{1}{2}}$, other parameters considered constant.

Experiments were conducted by Jagodzinski on discharges between carbon electrodes in air at pressures between 10 and 50 mm. Hg when the water vapour content was sufficient to enable measurements of the intensity of H_α (6,563 Å) to be made. The aim of the experiments was to compare the experimentally determined variation of intensity of H_α, as a function of pressure, with that predicted from theory. To link the above equations with the temperature variation of intensity of H_α it is necessary to use the expression

$$I = Nh\nu A_{nm} g_n \exp\left(-\frac{E_\alpha}{kT}\right) \qquad (10.23)$$

for the intensity I of a spectral line in terms of N, the number of atoms in the volume v under consideration, E_a the excitation energy of the upper level of the line, A_{nm} the transition probability, and g_n the statistical weight. We have $N = pN_0 v/p_0$, where N_0 is the particle density at the pressure p_0. Jagodzinski then assumes that conditions are such that the value of T in equation (10.23) is T_e and so, substituting for T_e from equation (10.22) and putting

$$R = \frac{22414}{5 \log_e 2} \frac{L\tau_0}{x_0} \frac{T}{T_0}$$

we obtain

$$\log I = \log(N_0 A_{nm} g_n v h\nu/p_0) + \log p - \frac{E_a}{kT_g + kR(p_0/p)^{\frac{1}{2}}} \quad (10.24)$$

or

$$\log I = A + \log p - \frac{B}{C + D(p_0/p)^{\frac{1}{2}}}. \quad (10.25)$$

It is unnecessary to describe the experiments in detail since they apply to a steady discharge, and the main interest of the work for the present purpose attaches to τ_0. By trial and error $\tau_0 \sim 10^{-11}$ sec. fitted best to the experimental results for the special conditions of the experiment. Using the value for τ_0 of 2×10^{-11} sec. with equation (10.22) it is found that at 1 atm., when $x_0 = 2 \times 10^{-3}$,

$$T_e - T_g < 1^\circ \text{K.}$$

In view of the high pressure and the short mean free path, and the consequent ease of interaction between electrons and gas atoms, this result may not be surprising [141, 143].

Mannkopff and Paetz [145] later used the result of Jagodzinski to deduce electron temperatures from a study of the relative intensities of arc and spark lines in a discharge and the use of Saha's equation. We have

$$T_e = T_g + \frac{\tau_0}{x_0} \frac{L}{p^{\frac{1}{2}}(mc)_g}. \quad (10.26)$$

Mannkopff and Paetz then take T_e (more properly T_g in the case of thermal equilibrium) from the electron density which is given in turn by I^+/I, the ratio of spark to arc line intensities, invoking Saha's equation, and so

$$\frac{I^+}{I} = F\left(p, \frac{\tau_0}{x_0}\right). \quad (10.27)$$

Thus if x_0 is found, I^+/I (measured) can be compared with the calculated values, using various values of τ_0, which will now be the relaxation or accommodation time of the electron temperature in an ionized gas. Mannkopff and Paetz studied various iron lines over a pressure range from 25 to 700 mm. Hg. They found that the intensity of the spark lines decreased with decreasing pressure from 760 mm. Hg but increased again after passing through a minimum at 130–40 mm. Hg, the intensity ratio of a pair of lines was the same at 50 and 760 mm. Hg. They attributed the initial decrease to a falling gas temperature at the lower pressures.

The intensity ratio I^+/I was then plotted as a function of p and compared with theoretical curves which were derived by putting τ_0/x_0 equal to various arbitrary values. It is found, using the approximate value of 3×10^{-3} for x_0, that $\tau_0 = 3 \times 10^{-10}$ sec. The value of τ_0 for rare gases should be greater since the interchange of energy between electrons and atoms would then be less efficient. Mannkopff and Paetz used an ionization potential $V_i = 7 \cdot 6$ V for iron in the Saha equation and therefore assumed that the discharge is through iron vapour. They state that the iron vapour content of the beam emitted by the cathode and observed spectroscopically was practically 100 per cent.

Measurements of excitation temperatures in spark discharges are scanty and also probably inaccurate owing to the uncertainty with which the various physical processes involved (self-absorption, presence of Boltzmann distribution for the populations of the excited states, etc.) can be applied to spark channel conditions.

The excitation temperature T_{ex} may be defined by the relation

$$\frac{N_n}{N_0} = \frac{g_n}{g_0} \exp\left(-\frac{E_n}{kT_{ex}}\right), \tag{10.28}$$

where N_0 and N_n are the concentrations of atoms in the ground state and the excited state n of excitation energy E_n, g_0 and g_n are the respective statistical weights, and k is Boltzmann's constant. T_{ex} may be determined more directly from a measurement of the relative intensities of two spectral lines of frequencies $\nu_{n,r}$ and $\nu_{m,r}$ given by transitions from the states n and m to r:

$$\frac{J_{n,r}}{J_{m,r}} = \frac{\nu_{n,r} \, g_n \, A_{n,r}}{\nu_{m,r} \, g_m \, A_{m,r}} \exp\left[\frac{E_m - E_n}{kT}\right], \tag{10.29}$$

where the A's are the Einstein coefficients. The application of this theory to arc temperatures was originally made largely by Ornstein and his collaborators [68] but a full bibliography is given by Edels [63].

Measurements have been made by Langstroth and McRae [146, 147], Williams et al. [76], Levintov [148], and Braudo et al. [88] on excitation temperatures derived from studies of metal lines, using electrodes of Cd, Ba, Mg, etc., and by Craggs et al. [66] on excitation temperatures in hydrogen sparks. Qualitative work with band spectra, on low pressure sparks in various molecular gases, has been carried out by Singh and Ramulu [149] and in nitrogen by Tawde [150].

The excitation temperatures (which, presumably [5, 52, 63, 76, 151] should be within $< 100°$ of the electron temperature) are found by the above-mentioned workers to vary from 4,000–20,000° K. This temperature range covers a wide variety of experimental conditions but it does not yet seem possible to compare excitation temperatures with electron temperatures, Saha temperatures (from measurements of ion concentrations, assuming thermal equilibrium), or gas temperatures. Efforts are being made in this direction for sparks [5, 32], and also for arcs where the conditions of stability render observation easier and where at least a close approach to thermal equilibrium is usually accepted [63, 68]. Excitation temperatures, measured for a series of excitation levels in the same spark discharge, show wide variations [66, 76, 88] and it appears that a Boltzmann distribution for the excited state populations in these cases does not exist. Similar results have recently been obtained [152] with a hydrogen arc at $\sim$ 1 atm. but a full interpretation of the data has not yet been satisfactorily propounded (see also the work of Barnes and Adams [153, 154] on Hg arcs and the comprehensive review of gas discharge temperatures by Edels [63] for further details).

Gas temperatures in exploded wire discharges have been measured by Anderson and Smith [16], and experiments on gaseous spark discharges, using the latter authors' technique, have also been made by Suits [17] and Craig [4]. The work of Suits on arc discharge gas temperatures, using a variety of ingenious techniques, is well known. This work is also reviewed by Edels [63].

Reignition of spark gaps, following arc and spark discharges

The study of afterglow conditions is of interest in relation to certain technical uses of discharge gaps, such as lightning arrestor gaps [155]. In this work direct measurements of the rate of re-establishment of the dielectric strength of a spark gap, after the passage of a discharge, are made.

McCann and Clark [156] studied the re-striking characteristics of 6 and 11 in. rod gaps in cases where the initial breakdown was caused by the application of a high voltage surge. For the 6-in. gap the peak currents varied from 1,390 to 23,000 A and, by the use of either of two arrangements, could be maintained either for a fraction of a μsec. or for about 5–10 μsec. The recovery curves are of a similar shape to those of Fig. 10.31. In 1 m.sec the dielectric strengths of the 6-in. and 11-in.

gaps recovered to respectively 71 and 82 per cent. of the initial values and there was little variation for the two types of current wave Studies were also made with longer current pulses of smaller amplitude. The authors suggest that the heating of the discharge path by the initial

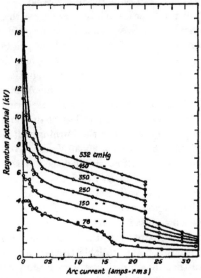

FIG 10 28 Arc reignition characteristics for constant pressures Gap = 1 mm., gas = air, parameter = pressure in cm. of Hg

discharge, and its subsequent finite rate of cooling during the restriking period can account for the finite time of recovery.

On the assumption that the spark channel is of uniform temperature (clearly only an approximation) and constant pressure so that the dielectric strength of the air is inversely proportional to temperature, McCann and Clark worked out a recovery voltage-time wave. A satisfactory fit with experiment was found in a particular case by taking a temperature of 4,000° K. and a channel radius of 0·13 in., assumed to apply at a time of 250 μsec. after the start of the first breakdown. The variation of channel radius with time was found by photographing successive discharges for varied time intervals between discharges and measuring the maximum deviation, from the first channel, of the second channel on the assumption that the second discharge takes place anywhere within the hot channel of the first discharge.

The investigators stress the relative crudity of many of their arguments but the evidence given appears to support, quite strongly, the suggestion that the decreased dielectric strength of a spark channel, after current flow has ceased, may be due entirely to the increased gas temperature. This suggests that de-ionization is relatively rapid.

The reignition characteristics of discharge gaps through which arc currents have passed have been determined by Cobine, Power, and Winsor [157] by observation of the arc striking points in the 60 c./s. current and voltage waves used for arc excitation (see also Garbuny and Matthias [180]). Fig. 10.28 shows that the reignition potential falls rapidly for arc currents up to about 0·25 A (r.m s) and then less slowly for currents up to 3 A, with transition points occurring in two regions (0·1–0·2 and 1·5–2·25 A). For currents less than about 0·2 A the discharge was of the glow type. Since the burning voltage for the intermediate region of 0·2–1·5 A was ∼ 80–30 V, for a 1-mm. gap, it was suggested that over this current range the discharge was an arc and not another form of high-pressure glow, and that the operative cathode mechanism was connected with field emission as distinct from the thermionic emission region above 1·5 A. The investigators suggest that the relatively slow voltage recovery rate for the gap following the passage of an arc, as compared with that following the glow, is because in the latter case there is no appreciable thermal ionization in the discharge channel so that de-ionization is very rapid, with a necessarily high reignition voltage. The difference in recovery rates for the assumed field-emission arcs and thermionic arcs may be because of the higher anode temperature in the latter case at the higher currents, and the consequent ease with which this electrode, which becomes the cathode in the following half-cycle, can supply electrons for the reignition discharge. This suggestion is supported by the lower reignition voltages for the higher arc currents (see Fig. 10.28), although it would appear that more confirmatory experimental work is required.

Dielectric recovery of A C. arc channels has been studied by Browne [158] and others [159, 160], for the case where the recovery process was aided by the use of a gas blast. This is, of course, an accepted technical practice [161, 162]. Browne concludes that recombination is a relatively unimportant process in the de-ionization mechanism, and that the diffusion of ions is of more consequence. Recombination is considered to be negligible at temperatures of 1,000° K. or more and is of importance only near the boundary of the channel. It appears that the cooling of the gas is not given sufficient prominence in this argument [156].

Similar work to the above has also been carried out by Stekolnikov [163], who has published details of a circuit for investigating the recovery of dielectric strength of a spark gap, for currents of up to 160 kA.

Some experiments on the recovery of breakdown strength of short spark gaps in air were carried out in the Metropolitan-Vickers laboratories, first by Garfitt on a three-electrode triggered spark gap [58] and later on two-electrode gaps [164]. The unpublished work of C. J. Braudo and G. H. Webster [164] is described here by permission of the Metropolitan-Vickers authorities. Braudo worked with annular gaps

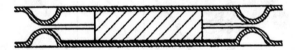

Fig. 10.29. The electrode arrangement of Braudo's spark gaps.

of the type shown in Fig. 10 29 and these were broken down with a 5,000-A pulse af 10–15 μsec. duration, the D.C. breakdown voltage of a single gap was 5–6 kV At a variable time t after the application of this initial pulse, the breakdown voltage V of the gap was determined by applying a second pulse with a slowly rising front (see Fig. 10.30) and measuring the voltage at which the gap then broke down. Recovery curves for air at 1 atm. are given in Fig. 10.31. Braudo found that the recovery voltage curve could be closely represented by

$$V = V_0(1 - e^{-at}), \tag{10.30}$$

where V_0 is the D.C. breakdown voltage of the gap. For a single gap $a = 1,515$ with t in seconds.

Since the initial discharge in the above experiments was of very short duration and could be described as a spark it was clearly important to determine the effect of a longer and more arc-like initial discharge. In the latter case, as longer times are involved, it is reasonable to suppose that relatively more energy would be used in heating the gas in the discharge channel, instead of causing excitation and ionization of high states. This comparison between discharges of short and long duration is, however, over-simplified, as the discussion in Chapter I on accommodation times shows that thermal equilibrium in a spark may be approached more quickly than in an arc It may be, for instance, that the 1 m sec. generally supposed to be necessary for attainment of thermal equilibrium in an arc [165] is much longer than the corresponding time for a spark [10, 166] since the electron temperature in a spark, especially during its early stages, may be much higher than in an arc.

McCann and Clark [156], whose work is described on p. 421, found
that recovery depended on the strength of current flowing during the

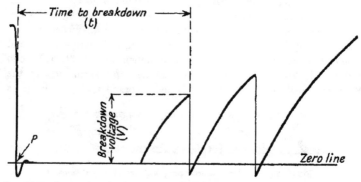

Fig. 10.30. A typical oscillogram from which the restriking voltage and time to breakdown are measured   P is the initial pulse

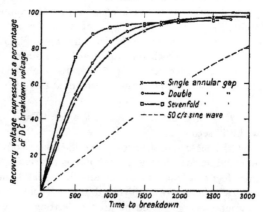

Fig. 10.31 Dielectric recovery characteristics of annular gaps,
with a 15 μsec., 5,000 A initial pulse.

initial breakdown Further, as Braudo pointed out, his experiments on
the recovery of a gap from spark discharges were of a preliminary nature
and had, for example, little relevance to the important technical case of
recovery of lightning arrestor gaps which, after breakdown by a surge,
carry a power frequency current for perhaps ¼ cycle or more. Webster,
therefore, carried out further experiments, using initial breakdown
impulses of about 3 kV, 68 A, which lasted for about 900 μsec. and

had again a rectangular shape (see Fig. 10.32)   Experiments were made
with spark gaps of various types, all in air at 1 atm.   For comparison
with the data of Fig. 10 31 it is sufficient to give the results of Fig. 10.33

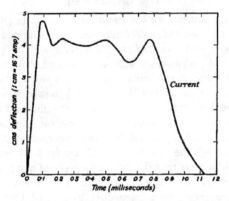

FIG. 10.32.   Current and voltage wave-forms from
the artificial line used in Braudo's measurements of
the dielectric recovery of spark gaps.

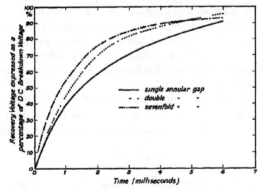

FIG 10 33   Dielectric recovery of single, double, and
sevenfold annular gaps using long-time low-current
pulses

as recorded with gaps of the type shown in Fig. 10.29.   It is apparent
that the rate of recovery is slower after the passage of the long (0·9
m.sec) discharges and that again the use of a multiple gap is beneficial,
although subdivision into more than seven gaps is unnecessary (the

total gap length in the multiple gap is equal to that in the single gap). Full recovery is obtained now in some 5 m.sec. or more. Webster showed that the use of the empirical relation

$$V = V_0(1 - e^{-0.85a_1t}) \qquad (10.31)$$

gave better agreement with the results than equation (10.30). Values for a_1 of 360, 511, and 560, with t in m sec , were obtained for the single gap, double gap, and sevenfold gap respectively.

The improved performance with a subdivided gap may be due to the increased possibility of heat transfer from the discharge channel, since Mackeown, Cobine, and Bowden [167] showed with 1 mm. gaps that the thermal conductivity of the electrodes influenced the reignition voltage. Also, it seems likely that even in non-attaching gases (see p. 19) recombination will be appreciable in times short compared with 1 m.sec. Therefore, although the data are not sufficient for final judgement to be made, it would appear that the rate of recovery of a gap might depend greatly on the rate of cooling of the hot channel, except for the earlier part of the recovery period. Slepian [159] showed that immediately on the extinction of a short arc the reignition voltage rises rapidly to 230 V and thereafter increases relatively slowly. If this voltage is equivalent to a cathode, and perhaps anode, fall of potential (see Chapter XI) then it is reasonable to suppose that in a multiple gap of n sections the total cathode fall would be increased approximately n times, and this total voltage would presumably be the minimum impulse required for breakdown This simple picture is clearly liable to complication by a non-linear distribution of voltage, which might occur in practice, or by other effects.

Bramhall [168] studied the de-ionization of a 5 A arc between copper electrodes in air with a platinum wire probe and found that the curve relating ion density with time could be accurately represented by the equation

$$N = \frac{N_0}{1 + N_0 \alpha t} \qquad (10.32)$$

which, of course, is the simple recombination equation given in Chapter I (pp. 19–20). The initial ion concentration N_0 was 7×10^{11} and was equal to the electron concentration. $\alpha = 5.7 \times 10^{-9}$ cm.^{3}/ion-sec and $N \sim 0.1 N_0$ at $t = 2$ 2 m.sec.

Later work on dynamic probe methods, which appears to throw doubt on the validity of the earlier experiments, is described by Koch [169] and Dürrwang [170]. The difficulties encountered when probes are used in high-pressure static discharges are described, for example, by Mason [171] and Boyd [172], but low-pressure decaying plasmas [173] may still be investigated this way. A novel method of studying de-ionization of thyratrons is described by Knight [174].

REFERENCES QUOTED IN CHAPTER X

1. L. B. LOEB and J. M. MEEK, *The Mechanism of the Electric Spark*, Stanford, 1941.
2. H. NORINDER and O. KARSTEN, *Art. Mat. Astr. Fys* **36** A, pt 4 (1949).
3. J. B. HIGHAM and J. M. MEEK, *Proc Phys Soc. Lond.* B, **63** (1950), 649
4 R D. CRAIG, in course of preparation.
5. R. D. CRAIG and J. D. CRAGGS, in course of preparation.
6. J. W. FLOWERS, *Phys Rev.* **64** (1943), 225
7. H. ALFVÉN, *Cosmical Electrodynamics*, Oxford, 1950.
8. P. L. BELLASCHI, *Elect. Engng. N.Y.* **56** (1937), 1253.
9. J E. ALLEN, *Proc. Phys. Soc. Lond.* A, **64** (1951), 587.
10. J. D. CRAGGS and J. M. MEEK, *Proc. Roy. Soc.* A, **186** (1946), 241.
11. E. O. LAWRENCE and F. G. DUNNINGTON, *Phys. Rev.* **35** (1930), 396.
12. E. FÜNFER, *Z. angew. Phys.* **1** (1949), 21.
13. J. J. THOMSON and G P. THOMSON, *Conduction of Electricity through Gases*, Cambridge, 1928.
14. A. L. FOLEY, *Phys. Rev.* **16** (1920), 449.
15 W. McFARLANE, *Phil. Mag.* **18** (1934), 24.
16 J. A ANDERSON and S SMITH, *Astrophys J.* **64** (1926), 295.
17. C. G. SUITS, *G.E. Rev.* **39** (1936), 194.
18. C G. SUITS, ibid. 430.
19. C. G. SUITS, *J. Appl. Phys.* **6** (1935), 190.
20. J. D CRAGGS, unpublished
21. D. R. DAVIES, *Proc Phys. Soc. Lond* **61** (1948), 105
22. G. I. TAYLOR, *Proc. Roy. Soc* A, **201** (1950), 159, 175.
23. F. KESSELRING, *Elektrotech Z.* **55** (1934), 92
24. V. KOMELKOV, *Bull. Acad Sci. U.S.S.R. (Tech. Sci.)*, No. 8 (1947), 955.
25. P. L. BELLASCHI, *Elec. Jour.* **32** (1935), 237.
26. N. M. SOLOMONOV, *J. Tech. Phys. U S.S R* **18** (1948), 395
27. J. DURNFORD and N. R. McCORMICK, *Proc. Instn. Elect. Engrs.* 99, Part II (1952), 27.
28. H. C. BIERMANNS, *A.E.G. Prog.* **6** (1930), 307.
29. J. B. HIGHAM and J. M. MEEK, *Proc. Phys. Soc. Lond.* B, **63** (1950), 633.
30. A. E J. HOLTHAM and H. A. PRIME, ibid. 561.
31. A. E. J. HOLTHAM, H. A. PRIME, and J. M. MEEK, in course of preparation.
32. R. D. CRAIG and J. D. CRAGGS, in course of preparation.
33. W. HOPWOOD, R. D. CRAIG, and J. D. CRAGGS, in course of preparation.
34. P. RAVENHILL and J. D. CRAGGS, in course of preparation.
35. T. B. GRIMLEY and R. F. SAXE, *E.R.A. Report*, L/T 211, 1948.
36. J. D. CRAGGS, W. HOPWOOD, and J. M. MEEK, *J. Appl. Phys.* **18** (1947), 919.
37. R. F. SAXE and J. B. HIGHAM, *Proc. Phys. Soc.* **63** (1950), 370.
38. S. D. GVOSDOVER, *Phys. Z. Sowjet.* **12** (1937), 164.
39. D. GABOR, *Z. Phys.* **84** (1933), 474.
40. H. EDELS and J. D. CRAGGS, *Proc. Phys. Soc. Lond.* A, **64** (1951), 574.
41. L. B. LOEB, *Fundamental Processes of Electrical Discharge in Gases*, Wiley, New York, 1939.

42. R. H. HEALEY and J. W. REED, *The Behaviour of Slow Electrons in Gases*, Amalgamated Wireless, Sydney, 1941
43. L. VON HÁMOS, *Ann. Phys. Lpz.* **7** (1930), 857.
44. A. C. LAPSLEY, L B SNODDY, and J. W. BEAMS, *J. Appl. Phys.* **19** (1949), 111.
45. A. M. ZAREM, F. R MARSHALL, and F L. POOLE, *Elect. Engng.* **68** (1949), 282.
46. H. BRINKMAN, *Dissertation*, Utrecht, 1937
47. C E. R BRUCE, *Nature*, Lond, **161** (1948), 521.
48. R. D. CRAIG, in course of preparation.
49. H. HORMANN, *Z. Phys.* **97** (1935), 539.
50. H. J. VAN DEN BOLD and J. A SMIT, *Physica*, **12** (1946), 475.
51. L. HULDT, *Ark. Mat. Astr Fys.* **36**, No 3, 1948.
52. R. ROMPE and W. WEIZEL, *Z. Phys.* **122** (1944), 636.
53. L BINDER, *Die Wanderwellen Vorgange*, Springer, Berlin, 1928.
54. T. R. MERTON, *Proc Roy. Soc* A, **92** (1915), 322.
55 E. O. HULBURT, *Phys. Rev.* **23** (1924), 106.
56. E O. HULBURT, *J. Franklin Inst.* **201** (1926), 777.
57. J. D. CRAGGS and W. HOPWOOD, *Proc. Phys. Soc.* **59** (1947), 755.
58. J. D. CRAGGS, M. E. HAINE, and J. M. MEEK, *J. Instn. Elect. Engrs.* **93** (IIa), (1946), 963.
59. S. VERWEIJ, *Publ Astr Inst Univ. Amsterdam*, No. 5, 1936.
60. A. UNSOLD, *Physik der Sternatmospharen*, Springer, Berlin, 1938.
61 R. F SAXE, Ph D. Thesis, Liverpool, 1948.
62 S. CHANDRASEKHAR, *Rev Mod. Phys.* **15** (1943), 1.
63. H. EDELS, *E.R.A. Report*, L/T 230, 1950.
64. E N. DA C. ANDRADE, *The Structure of the Atom*, Bell, London, 1927.
65. H S. TAYLOR and S. GLASSTONE, *A Treatise on Physical Chemistry*, vol. 1, *Atomistics and Thermodynamics*, van Nostrand, New York, 1942.
66 J. D. CRAGGS, G. C. WILLIAMS, and W. HOPWOOD, *Phil. Mag.* **39** (1948), 329.
67. A. VON ENGEL and M STEENBECK, *Elektrische Gasentladungen*, Springer, Berlin, 1934.
68. L. S ORNSTEIN and H. BRINKMAN, *Physica*, **1** (1934), 797.
69. M. KORNETZKI, V. FORMIN, and R STEINITZ, *Z. tech. Phys.* **14** (1933), 274.
70. O. LAPORTE, *C.R. Acad. Sci. Paris*, **201** (1935), 1108.
71 J. VAN CALKER and E. TACHE, *Z. Naturforsch.* **4a** (1949), 573.
72. W. G. STANDRING and J S. T. LOOMS, *Nature*, Lond, **165** (1950), 358.
73. J. W. BEAMS, A. R. KUHLTHAU, A. C. LAPSLEY, J. H. McQUEEN, L. B. SNODDY, and W. D. WHITEHEAD, *J. Opt Soc. Amer.* **37** (1947), 868.
74. S. M. RAISKIJ, *J. Tech. Phys. U.S.S.R.* **10** (1940), 529.
75. K. WOLFSON, *Bull. Acad Sci. U.R.S.S. Sér. Phys.* **9** (1945), 239.
76. G. C. WILLIAMS, J. D. CRAGGS, and W. HOPWOOD, *Proc. Phys. Soc.* B, **62** (1949), 49.
77. P. M. MURPHY and H. E. EDGERTON, *J. Appl. Phys.* **12** (1941), 848.
78 H. E. EDGERTON and K J. GERMESHAUSEN, *Electronics*, **45** (1932), 220.
79. J. N. ALDINGTON and A. J MEADOWCROFT, *J. Instn. Elect. Engrs.* **95**, Part II (1948), 671.
80. J. W. MITCHELL, *Trans Illum Engng. Soc. Lond.* **14** (1949), 91.
81. P. SCHULZ, *Ann. Phys. Lpz.* **6** (1949), 318.

82. W ELENBAAS, *The High Pressure Mercury Vapour Discharge*, North Holland, Amsterdam, 1951.

83. G GLASER, *Optik*, 7 (1950), 33, 61

84. H. E. EDGERTON, *J. Opt. Soc Amer* 36 (1946), 390.

85. W. D. CHESTERMAN, D. R CLEGG, G. T. PECK, and A. J. MEADOWCROFT, *Proc. Instn. Elect. Engrs.* 98, Part II (1951), 619.

86. S. DUSHMAN, *J. Opt. Soc Amer*. 27 (1937), 1.

87. F L. MOHLER, ibid 29 (1939), 152.

88. C. J. BRAUDO, J. D. CRAGGS, and C. G. WILLIAMS, *Spectrochim. Acta*, 3 (1949), 546.

89. H. KAISER and A. WALLRAFF, *Ann. Phys. Lpz.* 34 (1939), 826.

90 I. I. LEVINTOV, *J. Tech. Phys. U.S.S.R.* 17 (1947), 781.

91. F. LLEWELLYN JONES, *J Soc. Chem Ind.* 64 (1945), 317.

92 H. ISRAEL and K WURM, *Naturwiss.* 29 (1941), 778.

93. M. NICOLET, *Ciel et Terre*, 59, No. 3 (1943)

94. M. DUFAY, *C. R Acad. Sci. Paris*, 225 (1947), 1079.

95. P D. JOSE, *J. Geophys Res* 55 (1950), 39.

96. O. C DE C. ELLIS and W A. KIRKBY, *Flame*, Methuen, London, 1936

97 J. D. MORGAN, *Phil. Mag.* 11 (1931), 158.

98. E. JONES, *Proc. Roy. Soc.* A, 198 (1949), 523

99. B. W. BRADFORD and G. I. FINCH, *Chem. Rev.* 21 (1937), 221.

100 G. I. FINCH, *Proc. Roy Soc.* A, 111 (1926), 257.

101. E M. GUÉNAULT and R. V. WHEELER, *J Chem. Soc.*, Pt. II (1934), 1895.

102 G I FINCH, *Proc Roy. Soc* A, 133 (1931), 173.

103. G. I FINCH, ibid. 143 (1934), 482.

104. J. C. SLATER, *Introduction to Chemical Physics*, McGraw-Hill, New York, 1939.

105. W. JOST, *Explosion and Combustion Processes in Gases* (trans. by H. O. Croft), McGraw-Hill, New York, 1946.

106. A. KOWALSKY, *Z phys. Chem* B, 2 (1930), 56.

107. G. GORSCHAKOV and F. LAVROV, *Acta Physicochim. U.R.S.S.* 1 (1934), 139.

108 H. G. LANDAU, *Chem. Rev.* 21 (1937), 245

109. J. W. LINNETT, E. J. RAYNOR, and W. E. FROST, *Trans. Faraday Soc.* 41 (1945), 487.

110 W. E. FROST and J. W. LINNETT, ibid. 44 (1948), 416.

111. W. E. FROST and J. W. LINNETT, ibid. 421.

112. M. V. BLANC, P. G. GUEST, G. VON ELBE, and B. LEWIS, *J. Chem. Phys.* 15 (1947), 798.

113. B. LEWIS and G. VON ELBE, ibid. 803.

114. A. R. BOYLE and F. J. LLEWELLYN, *J. Soc. Chem. Ind.* 66 (1947), 99.

115. R. D. CRAIG, in course of preparation.

116. D. MULLER-HILLEBRAND, *Tekn. Tidskr.* 77 (1947), 761.

117. C. CIPRIANI and L. H. MIDDLETON, *Soc. Auto. Engrgs.* Preprint, 6–11 June 1948.

118. I. E. BALIGIN, *Elektrichestvo*, No. 4 (1949), 70.

119. V. S. KRAVCHENKO, ibid. No. 2 (1950), 70.

120. H. MACHE, *Öst. Ingen. Arch.* 1 (1947), 273.

121 R. VIALLARD, *J. Chem. Phys.* 16 (1948), 555.

122. M. LAPORTE and R. PIERREJEAN, *J. Phys. Radium*, 7 (1936), 248.

123. S. J. BOGDANOV and K. S. WOLFSON, *C R Acad. Sci U.R.S.S.* **30** (1941), 311.
124. J. D CRAGGS and W. HOPWOOD, *Proc. Phys. Soc.* **59** (1947), 771.
125. R B. BRODE, *Rev. Mod. Phys.* **5** (1933), 257
126. F. H MOHLER, *J. Res. Nat. Bur. Stand. Wash.* **2** (1929), 489.
127. F. L. MOHLER, ibid. **19** (1937), 559.
128. C. KENTY, *Phys. Rev.* **32** (1928), 624.
129. H. W. WEBB and D. SINCLAIR, ibid **37** (1931), 182.
130. H. ZANSTRA, *Proc. Roy. Soc.* A, **186** (1946), 236.
131. LORD RAYLEIGH, ibid, **183** (1944), 26.
132. R. B. HOLT, J M. RICHARDSON, B. HOWLAND, and B. T. McCLURE, *Phys. Rev.* **77** (1950), 239.
133 M. A BIONDI and S. C. BROWN, ibid. **75** (1949), 1700.
134 M. A BIONDI and S. C. BROWN, ibid. **76** (1949), 1697
135. R D. RAWCLIFFE, *Rev. Sci. Instrum.* **13** (1942), 413
136. J. M. MEEK and J D. CRAGGS, *Nature,* Lond , **152** (1943), 538.
137. G. H. DIEKE, H. Y. LOH, and H M. CROSSWHITE, *J. Opt. Soc. Amer.* **36** (1946), 185.
138. G H. DIEKE and H. M. CROSSWHITE, ibid **192**.
139. V. L. GRANOVSKY, *J. Phys. Acad. Sci. U.S.S.R.* **8** (1944), 76.
140. V. L. GRANOVSKY, *C. R. Acad. Sci. U S S R.* **28** (1940), 37.
141. H. MARGENAU, F. L McMILLAN, I H. DEARNLEY, C. H. PEARSALL, and C. G. MONTGOMERY, *Phys. Rev* **70** (1946), 349.
142. J. S TOWNSEND, *J Franklin Inst.* **200** (1925), 563
143. R. MANNKOPF, *Z. Phys.* **86** (1933), 161.
144. H. JAGODZINSKI, ibid **120** (1943), 318
145. R. MANNKOPF and H. PAETZ, unpublished (see *Rev. Sci. Instrum.* **18** (1947), 142).
146. G. O. LANGSTROTH and D. R McRAE, *Can. J. Res.* A, **16** (1938), 17.
147. G. O. LANGSTROTH and D R. McRAE, ibid. 61.
148. I. I. LEVINTOV, *J. Tech. Phys. U.S.S.R.* **17** (1947), 781.
149. J. SINGH and S. RAMULU, *Indian J. Phys.* **19** (1945), 235.
150. N. R. TAWDE, *Proc. Phys. Soc.* **46** (1934), 324.
151. R. LADENBURG, *Rev. Mod. Phys.* **5** (1933), 243.
152. H. EDELS and J D. CRAGGS, *Proc. Phys. Soc Lond.* A, **64** (1951), 562.
153. B. T. BARNES and E. Q. ADAMS, *Phys Rev.* **53** (1938), 545.
154 B. T. BARNES and E Q. ADAMS, ibid. 556.
155. C. J. BRAUDO, *Metro.-Vick. Research Report, No. 9461,* 1946.
156. G. D. McCANN and J. J. CLARK, *Trans. Amer. Inst. Elect. Engrs.* **62** (1943), 45.
157. J. D. COBINE, R. B. POWER, and L. P. WINSOR, *J. Appl. Phys.* **10** (1939), 420.
158. T. E. BROWNE, ibid. **5** (1934), 103.
159. J. SLEPIAN, *Amer. Inst. Elect. Engrs. J.* **47** (1928), 706.
160. F. C. TODD and T. E. BROWNE, *Phys. Rev.* **36** (1930), 732.
161. J. SLEPIAN and C. L. DENAULT, *Amer. Inst. Elect. Engrs. J.* **51** (1932), 157.
162. J. SLEPIAN, *J. Franklin Inst.* **214** (1932), 413.
163. I. STEKOLNIKOV, *Elektrichestvo,* No. 3, p. 67, March 1947.
164. C. J. BRAUDO, *Metro.-Vick. Research Report, No 4799,* 1948.
165. H. WITTE, *Z. Phys.* **88** (1934), 415.
166. S. LEVY, *J. Appl. Phys.* **11** (1940), 480.

167. S. S. MACKEOWN, J. D. COBINE, and F. W. BOWDEN, *Elect. Engng. N.Y.* **53** (1934), 1081
168. E. H. BRAMHALL, *Proc Camb Phil Soc.* **27** (1930–1), 421.
169. W KOCH, *Z. tech. Phys.* **16** (1935), 461.
170. J. DURRWANG, *Helv. Phys. Acta.* **8** (1935), 333.
171. R. C. MASON, *Phys. Rev.* **51** (1937), 28.
172. R. L. F. BOYD, *Proc. Roy Soc.* **201** (1950), 329.
173. V. L. GRANOVSKY, *C. R. Acad. Sci. U S.S.R.* **26** (1940), 873.
174. H. de B. KNIGHT, *J. Inst. Elect. Engrs.* **96**, Part II (1949), 257.
175. F. FRUNGEL, *Optik*, **3** (1948), 128
176 G. D HOYT and W. W. McCORMICK, *J. Opt. Soc. Amer.* **40** (1950), 658.
177. J. A. FITZPATRICK, J. C. HUBBARD, and W. J. THALER, *J. Appl. Phys.* **21** (1950), 1269.
178. R. A. JOHNSON, B. T. McCLURE, and R. B. HOLT, *Phys. Rev* **80** (1950), 376.
179. S M RAISKIJ and E. PUMPER, *J. Tech. Phys. U S S.R.* **20** (1950), 822.
180. M. GARBUNY and L. H. MATTHIAS, *Phys. Rev* **58** (1940), 182.
181. J M. RICHARDSON and R. B. HOLT, ibid **81** (1951), 153.
182. W. FINKELNBURG, *Z. Phys.* **70** (1931), 375.

PLATE 9

Fig. 10.1. Streak photographs of hydrogen spark channels. (200 A)

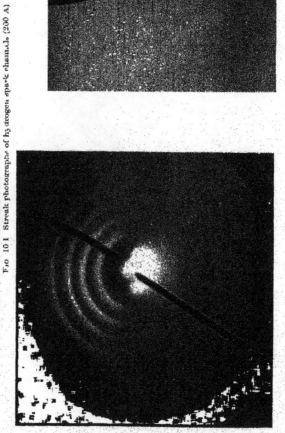

Fig. 10.3. Sound waves from spark discharges. (G. K. Adams, E.R.D.E. Technical Memorandum No. 3/M/39.)

PLATE 10

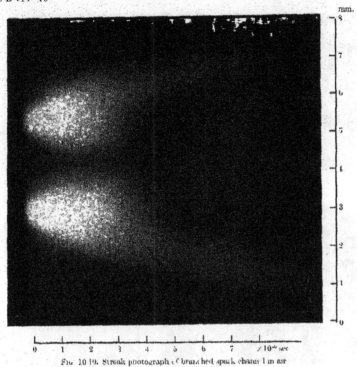

mm.

Fig. 10.10. Streak photograph of branched spark channel in air

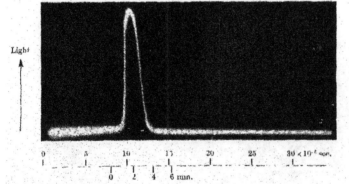

Light

Fig. 10.11. Typical record obtained with multiplier tube technique for study of
spark channel structure

PLATE 11

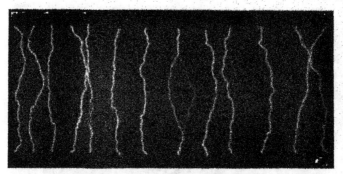

Fig. 10·12

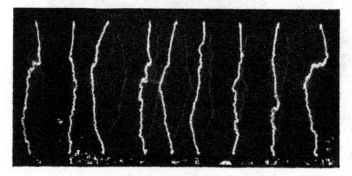

Fig 10·13

Figs 10·12 and 10·13 Photographs of 250 A spark discharges in air and hydrogen respectively to show the tortuous nature of the channels

(a)   (b)   (c)

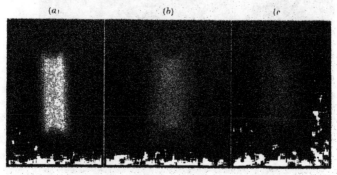

Fig 10·14 Kerr cell photographs of spark channels in air at different times following breakdown

PLATE 12

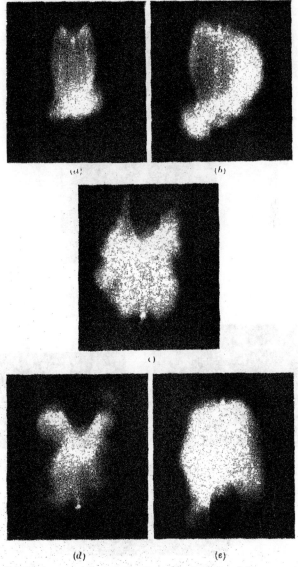

FIG. 10 15. Kerr cell photographs of spark channels in hydrogen at different times after breakdown.

PLATE 13

Fig. 10.20. Oscillogram of 200 A, 4 μsec current pulse used in spark studies at Liverpool

Fig. 10.26a. Afterglow in 110 A, 4 μsec spark discharge in hydrogen. The record shows the current pulse and the longer light emission pulse

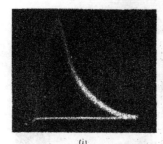 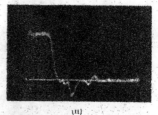

(i)   (ii)

Fig. 10.26b Afterglow in 110 A, 4 μsec spark discharge in argon. (i) and (ii) show, respectively, the light emission time pulse and the current pulse

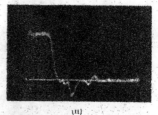

Fig. 10.27 Afterglows obtained with 110 A, 10 μsec neon sparks, using the streak camera technique

PLATE 14

Fig. 11.2. Cathode spots in 100 A sparks in hydrogen with carbon electrode and a spark duration of 10 μsec. Magnification × 30.

Fig. 11.3. Rotating mirror photograph of Mg cathode spots in 100 A, 10 μsec. spark in hydrogen. Time scale 5 μsec./cm, × 35 magnification.

Fig. 11.4. Five individual sparks showing cathode mercury vapour jets. The mercury pool (cathode) is at the bottom. (Retouched)

Fig. 11.5. Rotating mirror photograph of 100 A, 10 μsec. spark in hydrogen with tungsten cathode. The record was taken 2 mm. from the cathode (12 mm. gap length.)

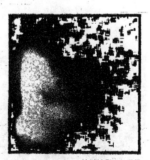

Fig. 11.6. Rotating mirror photograph of 100 A, 9 μsec. spark between Ba electrodes in air. Time scale 40 μsec./cm.

PLATE 15

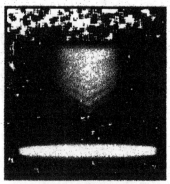

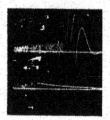

Fig. 12.18 The effect of a suitable non linear resistance (Metrosil) connected across a contactor switch gap

Fig. 12.3 Glow discharge in hydrogen at 1 atm pressure 3 mm. gap 0.3 A D.C, and between copper electrodes The upper electrode is the anode

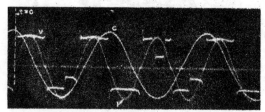

Fig. 12.8 Current and voltage relations during transition from glow to arc

(a)   *(b)*   *(c)*

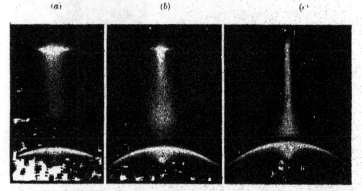

Fig. 12.14. Transitional discharges in hydrogen at 1 atm. pressure, 1.5 A r.m.s. (50 c.s.), and with a gap length of 0.5 cm. (a) shows the diffuse striated discharge before transition, (b) shows the discharge during transition, and (c) shows the discharge after transition

ELECTRODE EFFECTS

In Chapter X it is pointed out that the establishment of a spark channel, following the transition from the streamer discharge, is accompanied by a sudden rise in discharge current (for given circuit conditions) and the establishment of a cathode spot The high current density at the latter ($\sim 10^5$ A/cm.^{2}), and at the anode spot, during spark discharges necessitates the establishment of violent local conditions at the electrode spots and copious evaporation of metal from the electrodes, often in the form of luminous clouds of excited vapour. A study of these clouds and of the electrode spots is difficult, but is of considerable fundamental and technical interest. The evaporation process sets in after current has flowed for only a very short time Knorr [1] studied the time of appearance of the arc and spark lines of Cd and Zn in condensed sparks in air with a Kerr cell technique, and found that the air lines appeared $\sim 10^{-10}$ to 10^{-11} sec before the spark lines, which in turn preceded the arc lines (appearing in 10^{-8} sec) for each element. The Cd arc lines preceded the Zn arc lines. These results are of interest in connexion with the formation of cathode spots but it does not seem possible to interpret them fully in this respect at the present.

The present chapter will deal with certain aspects of electrode phenomena but it should be made clear that no final theory of cathode spots has yet been formulated. The small size of the spots, and their high current densities, make direct measurement difficult and therefore the above quantities can only be roughly estimated.

The theories of Finkelnburg [2] and Llewellyn Jones [3], who consider that the cathode currents are largely due to the incident positive ions, are in contrast with those of other authors [4, 5, 6], who consider that electron emission at the cathode is important. A major difficulty is that if the positive-ion current-density at the cathode is about 10^6 A/cm.^{2}, which requires $N_+ \sim 10^{19}$ ions/cm.^{3}, then an application of the space-charge equation [7] gives the cathode-fall thickness $d_c \sim 10^{-6}$ cm. so that field emission of electrons from the cathode should be appreciable. Further, the ion-current theories hardly explain the localized nature of the spots, but as the following discussion shows, any theory yet advanced of cathode spot mechanisms hardly explains all the known experimental data.

Finkelnburg's theory of the production of these high-speed vapour

TABLE 11.2

	Hg spark	C arc
Positive jet velocity (calculated) cm./sec.	$1\ 5\ \times 10^5$	$1\cdot 4 \times 10^5$
,, ,, (observed) ,,	$1\ 55 \times 10^5$	$1\ 5\text{--}4 \times 10^5$
Negative jet velocity (calculated) ,,	$2\ 1\ \times 10^5$	
,, ,, (observed) ,,	$1\cdot 9\ \times 10^5$	

Considering the assumptions made, the agreement shown in Table 11 2 between the calculated and observed jet velocities is excellent but perhaps partly fortuitous. Finkelnburg also points out that, whilst the jet velocity depends on current densities, the latter depend in practice to some extent on the nature of the electrodes. High current densities cannot be produced unless space charges are sensibly neutralized and this is possible in front of evaporable electrodes by virtue of ionization of the metal vapour. Hence vapour jet production should, on this argument, be relatively difficult with such refractory metals as tungsten or molybdenum. Recent experiments with sparks in hydrogen [9, 10] show copious jets of vapour even with tungsten electrodes Other data on current densities at cathode spots in arcs have been given by Cobine and Gallagher [11] for low-current arcs and yield current densities of 14,000–20,000 A/cm.^{2} for Hg or Cu with currents of 10–2·6 A. For Cu and W, and the same arc currents, j_c was found to be 50,000–120,000 A/cm.^{2} The agreement with Table 11 1 is not good, but discharge conditions were not strictly comparable in both cases. Froome [12], working with 1,400 A (peak) transient discharges, found $j_c \sim 10^5\text{--}10^6$ A/cm.^{2} for Hg, Cu, and Na cathodes, and similar data for sparks in air at 1 atm. have been obtained by Craig [13].

It should perhaps be emphasized that cathode spots are measured in terms of the dimensions of the luminous area of the discharge at or very near the cathode. The above papers (particularly those of Froome and Craig) discuss the implications of this limitation and other aspects of the work.

Somerville and Blevin [14] have also found apparent current densities of, at most, $1\ 6 \times 10^6$ A/cm.^{2} at the cathode spots of discharges carrying up to 200 A in square pulses varying in length from 1 to 200 μsec. In this work the crater markings were measured, and assumed equal in size to the cathode spots; usually [7, 12] the luminous area of the spots, during the discharge, is measured. Photographs of the luminous spots [13] are given in Figs. 11.2 and 11.3, Pl. 14. Fig 11 2 shows the cathode spots in a single 100 A, 10 μsec.-pulse, hydrogen spark discharge, with a carbon cathode. Fig. 11.3 shows a rotating-mirror photograph of a similar array of cathode spots for a magnesium cathode in hydrogen with the same current pulse.

Cobine and Gallagher [11] stress that their results for current densities can only be considered as lower limits and Froome's data may be more accurate even though, in the latter case, the effect of spot movements may have led to estimates of lower limits. Froome's later work [15] on liquid cathodes gives $j_c \sim 10^6$–5×10^6 A/cm.^{2} in general agreement with Craig's [13] data on spark discharges in hydrogen at 1 atm. with Mg, Cu, W, and C cathodes.

The following comments on Finkelnburg's work have been made by Craig:

(i) The expression $W_c = j_c V_c$ implies that the rate of heating of the cathode is greater than the total rate of generation of energy in the cathode region, whatever cathode mechanism is involved, for there is an energy drain $j_c \phi$ required to provide the plasma electron drift current.

(ii) Further discussion of ϕ_+, the heat of condensation of the positive ions, should probably have been given, with reference to the work of Compton and Van Voorhis[16], and Moon and Oliphant[17, 18].

(iii) The assumption that all the available energy goes to form vapour jets is not consistent with the work of Doan and others [19] on welding arcs, where it has been shown that most of the energy, dissipated at the electrodes, is used to cause melting. Craig [10] suggests that many of the properties of the vapour jets [23, 24] can be explained if the latter are composed initially of a liquid spray and compares them with the jets observed by Anderson and Smith in their work on exploded wires [20].

(iv) If it is assumed, as several workers [21] have shown for welding arcs, that the electrode metal leaves the parent surface in a liquid jet, then Q would be much smaller than the Finkelnburg theory suggests

Haynes [22] has reported experiments on vapour jets in hydrogen at 90 cm Hg for sparks between molybdenum electrodes and a mercury pool. Rectangular current pulses (0·25 to 5 μsec long) were used, with currents $\sim$ 100–300 A peak. The vapour jets, identified spectroscopically as being composed of mercury atoms, could be made to proceed from anode or cathode or from both together (since mercury drops could be made to hang from the molybdenum rod). They are not [22] caused by the propagation of high energy pulses along the spark column since they occasionally (Fig. 11.4, Pl. 14) project outside the spark channel and

appear to be directed approximately perpendicular to the Hg surface
When the jets meet, they scatter sideways Haynes measured the speed
of propagation of the jets across the discharge gap with the method used
by Williams, Craggs, and Hopwood [24] This method involves a photomultiplier, which observes the spark through a transverse slit and which
therefore responds when the vapour cloud reaches the slit from either

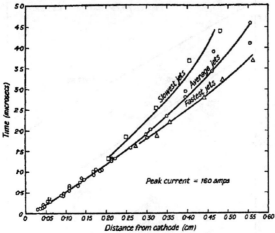

FIG 11 7   Time of arrival of cathode jet front as a function of distance from mercury cathode

electrode. Froome [7, 12, 15, 25] used a Kerr cell technique which
could be applied to the study of vapour jets Craig [10, 13] used various
rotating-mirror techniques for his work on vapour clouds and electrode
spots, in which the photograph shown in Fig. 11.5, Pl. 14, was
obtained (see p. 436) A rotating-mirror photograph of similar clouds
is shown in Fig. 11.6, Pl. 14 [24]. The jet speeds measured by Haynes
varied slightly (Fig. 11.7) and mean values are plotted in Fig. 11 8. The
initial velocities of anode and cathode jets are respectively $1·55 \times 10^5$
cm./sec. and $1·9 \times 10^5$ cm /sec. but these velocities decrease as the clouds
move away from the electrodes due presumably to collisions with
hydrogen atoms in the discharge channel. The initial velocities were
also found to be independent of discharge current and gas pressure.
Haynes remarks that the jets could hardly obtain their energy thermally,
in the regions near the electrodes, because, if they did so, the jets would
not be so highly directional. It is also difficult to account for them as
jets of positive ions of high velocity, as independently pointed out by

Williams *et al.* [24], since the jet speeds are too high (see Chapter I and Tyndall [26]). Cathode jets are also quite inexplicable in this fashion although it is possible (see below) that they are composed of negative ions. It is also possible [10, 21, 27] that the behaviour of these metal jets can be explained if it is assumed that they are initially composed of liquid

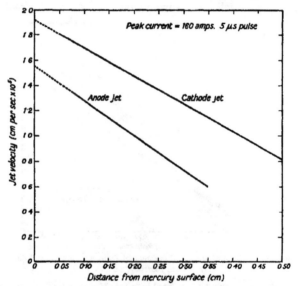

Fig. 11 8 Velocity of cathode and anode jet fronts as a function of distance from mercury surface

electrode metal (see p. 437) Haynes, further, found that oscillatory current pulses, in which the field direction changed in alternate half-cycles, had negligible effect on jet propagation.

The energy of the anode jet [22] can be accounted for by the passage of positive ions across the anode drop region. The velocity of the anode jet will then be

$$v_+ = \frac{M_+}{M} \frac{M}{M+M_h} \left(\frac{2eV_a}{m}\right)^{\frac{1}{2}} \text{ cm./sec.} \qquad (11.5)$$

where M_+ is the effective mass of the Hg atoms/sec. that cross the anode drop as positive ions, M is the total mass of the atoms in the jet/sec, M_h is the effective mass of hydrogen involved/sec., e is the electronic charge, and V_a is assumed to be less than 10·4 V (the ionization potential for Hg). M_+/M is likely to be much less than unity, as also will be the quantity $M/(M+M_h)$ due to the mass of the hydrogen. For $v_+ = 1·5 \times 10^5$

cm./sec. it is found that f, i.e $M_+/(M+M_h)$, must be < 0.5, which is not unreasonable, otherwise v_+ (calculated) becomes $> v_+$ (measured).

Compton had, in 1930, postulated in connexion with certain mechanical effects at arc cathodes that neutralized positive ions would rebound from the cathode with energy $(1-\alpha)$, where α is the accommodation coefficient of the positive ions. The physics of this secondary reflection process is further discussed by Oliphant and Moon [17, 18]. The velocity of the cathode jet v_- for such a mechanism to be operative is [22]

$$v_- = \frac{M_+}{M}\frac{M}{M+M_h}\left(\frac{2eV_c(1-\alpha)}{m}\right)^{\frac{1}{2}} \text{cm./sec.} \qquad (11.6)$$

where M_+ is the effective mass of mercury atoms which are positively ionized at the anode end of the cathode drop and escape into the jet by rebounding from the cathode while yet retaining $(1-\alpha)$ of their incident energy. V_c is the cathode drop. M_+/M and $M/(M+M_h)$ are again certainly < 1. Since $V_a+V_c \sim 40$ V, from early experiments on discharges of this type, $V_c \not> 30$ V if, as assumed, $V_a \not< 10$ V, since 10·4 eV is the ionization potential of Hg. If $\alpha = 0.9$ (which is not unreasonable, since $\alpha = 0.93$ for 50 V Hg+ ions on tungsten [22]) and $V_c = 30$ V, then $v_- = 1.7 \times 10^5$ cm./sec. Taking f_{max} from the anode jet as 0·5, $v_- = 0.85 \times 10^5$ cm./sec, which is about one-half of the observed jet velocity.

It appears, therefore, that other particles leave the cathode drop with higher energies. It is suggested that these might be negative ions, produced in the manner discovered by Arnot and Milligan [28, 29], although the low value for the probability of production of the ions found by those authors would need to be increased in order to account for the observed value of v_-. If this increase can reasonably be postulated on the grounds of the high field at the cathode of the sparks, then $v_- = 1.9 \times 10^5$ cm./sec. if $V_c = 22$ V, $\alpha = 0.93$, $\phi \sim 4$ V (the effective work function for a negative ion), $M_+/M = 0.4$, and $M_-/M = 0.36$ in the equation

$$v_- = \frac{M_+}{M}\frac{M}{M+M_h}\left(\frac{2eV_c(1-\alpha)}{m}\right)^{\frac{1}{2}} + \frac{M_-}{M}\frac{M}{M+M_h}\left(\frac{2e(V_c-\phi)}{m}\right)^{\frac{1}{2}}. \qquad (11.7)$$

Haynes finally states that the vapour jets from anode and cathode appear to be bent in a magnetic field in such a direction as to indicate respectively positive and negative charges on the particles (see the early work by Milner [30]).

Further measurements on vapour jets have been made by Langstroth and McRae [23] and Craig [10] (see also [24]) The former authors suggest after some discussion that neither diffusion nor the movement of charged particles from the electrodes can adequately account for the observed speeds but that the cloud movement may be due to some explosive effect at the electrodes.

A recent detailed theoretical treatment of the mechanisms involved in electrode erosion in sparks is given by Llewellyn Jones [32, 33], who also discusses the more technical consequences of the erosion processes, with special emphasis on the phenomena at electrical contacts [33]. Llewellyn Jones calculates the size of cathode spots by considering the propagation of a single electron avalanche across the spark gap from anode to cathode, followed by a positive ion cloud moving to the cathode and diffusing radially in transit. In typical short gaps the cathode-current cross-section is thus calculated to be $\sim 5 \times 10^{-5}$ cm.^{2} and typical data are given in Table 11 3 for air sparks, in parallel plate gaps. The last row of data refer to typical spark-plug conditions.

TABLE 11.3

Gap length in mm.	Air pressure in atm.	Sparking voltage in V	Cathode spot radius in cm
0 2	1	1,500	2 8 × 10^{-2}
0 5	1	2,700	4 0 × 10^{-3}
10	1	3,200	6 0 × 10^{-2}
0 3	3	4,800	1 5 × 10^{-2}

Llewellyn Jones [3, 32] has also calculated the rate of erosion from the hot spots formed in the manner briefly explained above. The result of a detailed treatment is that the volume in cm.^{3} of metal eroded per spark is

$$\text{volume} = \frac{cV^2 - b\theta^4 - gk(\theta - T)}{\rho[(\theta - T)S + 21\theta/M]}, \tag{11.8}$$

where V is the sparking potential of the gap in volts, θ is the boiling-point in °K of the electrode material, T is the steady electrode temperature at points remote from the hot spot, ρ is the electrode density, S is the mean specific heat in the metallic state of the electrode over the temperature range T to θ, and k is the thermal conductivity.

$$b = 2\ 4 \times 10^{-8} A\sigma t,$$
$$g = 2\sqrt{(\pi A)}.t \quad \text{and} \quad c = c_e/8\cdot4;$$

c_e is the effective gap capacity in farads discharged by the spark to the electrodes, and is considerably less than the total capacity in parallel with the gap. A is the hot spot area and, when the anode and cathode hot spot areas are A_1 and A_2, $A = A_1 + A_2$. σ is the value of the effective emissivity of the hot spot, and is less than Stefan's constant if the hot spot is not a black body. t is the positive ion transit time,

The constants c, b, and g were determined [3] from a knowledge of the required experimental data obtained on spark-plug electrodes for three widely differing metals, this can only be an approximation as, for example, the A values may show variations for different metals From these data the following are calculated·

$$c = 0{\cdot}95 \times 10^{-12} \text{ F},$$
$$b = 2 \times 10^{-20} \text{ cal.}/(°\text{K.})^4;$$
$$g = 0{\cdot}95 \times 10^{-8} \text{ cm./sec.}$$

Thus $c_e = 8\ \mu\mu\text{F}$ and, as the self-capacity of the spark plugs was $70\ \mu\mu\text{F}$, it appears, not unreasonably [8], that about 11 per cent. of the spark energy was released at the two electrodes.

This novel treatment constitutes one of the few attempts to explain, quantitatively, the processes of electrode erosion in spark gaps. It is not likely, in view of the complexity of the subject, to be correct in all details For example, the breakdown mechanism assumed as a basis of the theory may be incorrect as it tends to over-simplify the physics of spark channel growth and to describe it in terms of the movement of an electron avalanche and a retrograde positive ion cloud. Even for the short gaps described [3] it does not seem certain that this mechanism is finally established.

The theory seems to suggest that the cathode spot is determined by the breakdown conditions, which is in disagreement with the work of Froome and Craig (referred to above) carried out with discharges differing widely from those considered by Llewellyn Jones and in which cathode spots and current densities of the same order of size and magnitude were observed. Similar spots are also observed in arcs. Further, the theory does not seem to explain the multiple spots (Figs. 11 2 and 11.3 and the photographs in Froome's papers [7, 12, 15]) and their rapid movement on the electrode surface. However, conditions are different in the channels of sparks (see Chapter X) from those in the earlier stages of breakdown and, perhaps, in the channels of very weak discharges. It is probable that the work of Llewellyn Jones refers especially to the latter case.

Spark discharges are often used as sources in spectrographic analysis, and, in order to obtain greater accuracy than has hitherto been available, considerable efforts have been made during the last ten years to obtain reliable spark sources for analytical work, especially on steels and light metals.

Following very early work by Hartley in 1882 [34] and subsequently by many others on condensed spark circuits, Duffendack and his collaborators [35, 36] used a high-voltage A.C. arc and Pfeilsticker [37] in 1937 introduced the use of a controlled arc circuit in which the discharge gap was broken down by a high voltage impulse so that it could pass current from a relatively low-impedance, high-current circuit and so form an arc of short duration, say $\sim$ 1 m sec Much of the recent work has been carried out with sources of this type. Particular care is taken to ensure that spark breakdown of the gap occurs at constant voltage, either by means of external radiation (see Chapter VIII) or by the use of triggered gaps [38, 39] or with other means such as the synchronous interrupter of Hasler and Dietert [40]. The object of this work is to improve the accuracy of analysis as far as possible by ensuring constancy of electrical conditions in the discharge.

The results have been fairly satisfactory with certain sources for light metals and alloys (see, for example, the work of Braudo and Clayton [38]) but have been disappointing for steel [41], where the use of the older but simpler type of uncontrolled spark source continues to give comparable or better accuracy. Typical modern controlled sources are described by Braudo and Clayton [38], Walsh [43], Malpica and Berry [44], and Hasler and Dietert [40].

Besides errors due to temporal variations of the electrical characteristics of the discharge gap used in spectrographic spark sources, there are, of course, errors introduced by the photographic techniques involved. These errors have been reduced by the modern tendency to use multiplier phototubes for the measurement of spectral line intensity, and direct-reading spectrographs have been developed, notably by Dieke and Crosswhite [45]. Early work on the observation of radiation from discharges with multiplier tubes [46, 47, 48] lead to the observation of large fluctuations in electrode metal line intensities, even in cases where electrical variations were negligible [49]. Such fluctuations were also observed in the metal vapour clouds emitted at the electrodes in the much earlier work of Langstroth and McRae [23]. Further information on the discontinuous emission of metal vapour, especially for impulsive discharges, has been given by Braudo, Craggs, and Williams [50], Haynes [22], Craggs, Williams, and Hopwood [24], and others and is discussed elsewhere in this chapter (pp. 434 to 441). It seems unlikely that very high reproducibility of analytical results will be obtained in cases where such fluctuations are serious and this is particularly the case with easily vaporized metals.

For general accounts of the detailed theory of excitation processes in discharges, reference should be made to the papers of Llewellyn Jones [54], Mason [55], and Kaiser and Wallraff [51]. Levy [52, 53] has discussed the stability of condensed spark discharges, for use as spectroscopic sources, in terms of excitation processes (thermal excitation) but the application of this work to practical cases is rendered difficult by the uncertainties attached to the electrode processes.

REFERENCES QUOTED IN CHAPTER XI

1. H. V. Knorr, *Phys. Rev.* 37 (1931), 1611.
2. W. Finkelnburg, ibid 74 (1948), 1475.
3. F. Llewellyn Jones, *Brit. J. Appl. Phys.* 1 (1950), 60.
4 J. D. Cobine, *Gaseous Conductors*, McGraw-Hill, New York, 1941
5. L B. Loeb, *Fundamental Processes of Electrical Discharge in Gases*, Wiley, New York, 1939.
6 A. H Compton, *Phys. Rev.* 21 (1923), 266
7. K D. Froome, *Proc. Phys. Soc.* 60 (1948), 424.
8. R D. Craig, in course of preparation
9. A. E. J. Holtham, H. A Prime, and J. M Meek, in course of preparation.
10. R. D Craig, in course of preparation.
11 J D Cobine and C J. Gallagher, *Phys. Rev* 74 (1948), 1524.
12. K D. Froome, *Proc. Phys Soc.* B, 62 (1949), 805.
13. R D Craig, *E.R.A. Tech. Rep. L/T* 260 (1951).
14. J. M. Somerville and W. R. Blevin, *Phys. Rev.* 76 (1949), 982.
15. K D Froome, *Proc Phys. Soc.* B, 63 (1950), 377.
16. K. T Compton and C. C. van Voorhis, *Phys. Rev.* 27 (1926), 724.
17. M L. E Oliphant and P B. Moon, *Proc. Roy Soc.* A, 127 (1930), 388.
18 M L. E. Oliphant, ibid. 124 (1930), 228.
19. G. E. Doan and J L. Myer, *Phys Rev.* 40 (1932), 36.
20 J. A. Anderson and S. Smith, *Astrophys J.* 64 (1926), 295.
21. G E. Doan and A. M. Thorne, *Phys. Rev.* 46 (1934), 49.
22. J. R. Haynes, ibid 73 (1948), 891
23. G O. Langstroth and D. R. McRae, *Can. J. Res* A, 16 (1938), 17, 61.
24. G C. Williams, J. D. Craggs, and W. Hopwood, *Proc. Phys. Soc.* B, 62 (1949), 49.
25. K. D. Froome, *J. Sci Instrum.* 25 (1948), 37.
26. A. M. Tyndall, *The Mobility of Positive Ions in Gases*, Cambridge, 1938.
27. J. von Issendorf, *Phys Z* 29 (1928), 857.
28. F. L. Arnot and J. C. Milligan, *Proc. Roy. Soc.* A, 156 (1936), 538.
29. H. S. W. Massey, *Negative Ions*, Cambridge, 1950.
30. S. R. Milner, *Philos. Trans* A, 209 (1908), 71
31. S. M. Raiskij, *J. Tech. Phys. U S.S.R.* 10 (1940), 529.
32. F. Llewellyn Jones. *Nature*, Lond., 157 (1946), 480.
33. F. Llewellyn Jones, *J. Instn. Elect. Engrs.* 96 (1949), Part I, 305.
34. W. N. Hartley, *J. Chem. Soc.* 41 (1882), 90.
35 O S Duffendack and K. B. Thomson, *J. Opt. Soc. Amer.* 23 (1933), 101.
36. O. S Duffendack and R. A. Wolfe, *Ind. Eng. Chem. (Anal.)*, 10 (1938), 161.

37. K. PFEILSTICKER, *Z. Elektrochem.* **43** (1937), 719.

38. C. J. BRAUDO and H. R. CLAYTON, *J. Soc Chem Ind* **66** (1947), 259.

39. J. D. CRAGGS, M E HAINE, and J. M. MEEK, *J. Instn. Elect. Engrs.* **93** (IIIa), (1946), 963.

40. M. F. HASLER and H. W. DIETERT, *J. Opt. Soc. Amer.* **33** (1943), 218.

41. H. T. SHIRLEY and E. ELLIOT, *J. Iron and Steel Inst.* **147** (1943), 299, 323; Discussion 324–36

42. A. WALSH, 'Light Sources' (ch. 7 of *Metal Spectroscopy*, edited by Twyman, Griffin, London, 1951).

43 A. WALSH, *Metal Ind* **68** (1946), 243, 263, 293

44. J. T. N. MALPICA and T. M BERRY, *Gen. Elect. Rev* **44** (1941), 563.

45 G H DIEKE and H M CROSSWHITE, *J. Opt. Soc. Amer.* **35** (1945), 471.

46. D. H. RANK, R. J. PFISTER, and P. D. COLEMAN, ibid. **32** (1942), 390.

47 R. D. RAWCLIFFE, *Rev. Sci. Instrum.* **13** (1942), 413.

48. J. D. CRAGGS and J. M. MEEK, *Proc Roy. Soc.* A, **186** (1946), 241.

49. J. D. CRAGGS and W. HOPWOOD, *Nature*, Lond, **158** (1946), 618

50. C. J. BRAUDO, J. D. CRAGGS, and G. C. WILLIAMS, *Spectrochim. Acta*, **3** (1949), 546

51. H KAISER and A WALLRAFF, *Ann. Phys Lpz.* **34** (1939), 297.

52. S. LEVY. *J. Opt. Soc Amer.* **35** (1945), 221

53. S. LEVY, *J. Appl Phys.* **11** (1940), 480.

54 F. LLEWELLYN JONES, *J. Soc. Chem Ind* **64** (1945), 317.

55. R. C. MASON, *Symposium on Spectroscopic Light Sources*, Amer Soc for Testing Materials, Special Publ. No. 76 (1946), p. 25.

XII

GLOW-TO-ARC TRANSITIONS

GLOW–ARC transitions sometimes form an important part of the breakdown characteristics of high-pressure gases, as in the transient discharges occurring between switch contacts. Both glow and arc discharges may be defined as self-sustaining discharges, i.e. they are independent, once initiated, of external sources of electrons, and occupy, respectively, the region *EH* and the high current region beyond *K* of Fig. 2.42 (the complete discharge characteristic explained below).

A glow discharge maintains itself by virtue of the secondary electrons which are liberated at the cathode, usually, but not always, by the impact on it of positive ions. If, therefore, a single electron leaving the cathode of the discharge and passing to the anode liberates n ion pairs in the gas, then the n positive ions must between them liberate at least one electron from the cathode for a self-sustaining discharge to be established. Usually, but not always (see below), the glow discharge is encountered at low pressures (< 20 cm. Hg) and low currents (10^{-3}–1 A), but high voltages of 100 to 400 V are necessary since the positive ion mechanism is otherwise too inefficient. Glows are characterized by the non-uniformity of light emission from them, which is linked with the variations along the discharge of the local space charge conditions and excitation processes

A typical glow discharge is represented in Fig 12.1. Here the region between the cathode and a is the cathode-fall region or cathode dark space (a region of high positive space charge), ab is the negative glow which is due to collisions made by electrons accelerated in the cathode fall; bc is the Faraday dark space where the electrons, having lost much of their energy in ab, are too slow to cause appreciable excitation, cd is the positive column (a plasma, where the concentrations of electrons and positive ions are nearly equal); the region d to the anode is the anode glow Glow discharges are treated in detail by Thomson and Thomson [1], Darrow [2], Loeb [3], and Druyvesteyn and Penning [4] among others.

The cathode current in arcs, on the contrary, is largely due to mechanisms other than positive ion bombardment, e.g. to thermionic emission from the cathode [5, 6]. The voltage drop near the cathode need not be as large as in a glow discharge and the discharge tends to be more uniform from cathode to anode. Unless the cathode in an arc discharge is heated by some external means to ensure adequate thermionic emission, there will in general be a transition from a glow to an arc a short time after the voltage is applied to a discharge tube It should be stressed that cathode mechanisms in arcs do not appear to be understood in all cases, but it is generally accepted that thermionic emission from the cathode

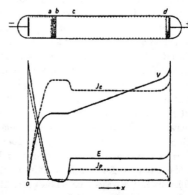

Fɪɢ 12 1 Typical longitudinal variations of luminosity, potential (V), electric field (E), electron current (J_e), and positive ion current (J_p) in a glow discharge.

is a dominant characteristic of arcs with high melting-point cathodes such as carbon or tungsten.

The general discharge characteristic

The subject of glow–arc transitions may be introduced by consideration of the general gaseous discharge characteristic [4, 7]. This is shown in Fig 2.42, as given in the review paper by Druyvesteyn and Penning [4] Their description of the various phases through which a discharge passes, as the current through it is gradually increased by a change in the applied voltage, will be followed in the next few paragraphs.

The discharge occurs in a tube with flat, parallel, electrodes and the value of pd, where p = pressure and d = electrode separation, is set so that no positive column or anode glow occurs, i.e. $2 < pd < 20$ mm. Hg cm. The voltage on the tube from E to K is almost equal to the cathode fall. For an applied voltage $V < 10$ volts the current, without photoelectric or field emission, is very small and is due to external causes such as cosmic rays, in air at 1 atm the current is $\sim 10^{-18}$ A/cm.^{2} Further, the current will naturally be subject to fluctuations (AB in Fig. 2.42).

If photo-emission is present, for which external illumination is necessary, the curve $A'B'$ is obtained. Then a region BC or $B'C$ is reached in which $\partial V/\partial \iota \to 0$, i.e. the current increases very rapidly for a small change in V, thus leading to a sharply defined breakdown potential V_B. If the discharge from B or B' onwards is non-intermittent it is termed a Townsend discharge Next, the discharge passes (perhaps through a region with a positive characteristic such as CD') to CDE and has then a negative characteristic so that, as may be expected, it often becomes intermittent. At these currents space charges begin to be appreciable and a cathode-fall region develops. From F to H a glow discharge is obtained followed by another transition region HK which is finally succeeded by a stable arc from K onwards. At higher pressures the discharge forms take apparently different forms for certain parts of the characteristic and corona and spark discharges are often obtained. Corona discharges are equivalent in certain of their aspects to the Townsend discharges and the latter replace for certain (but not all) circuit conditions the region HK. The arc is still the ultimate form of discharge provided that the external circuit is capable of sustaining it, but the means by which this state is reached in a gas at ~ 1 atm. is not always clear, in that the relative importance of high-pressure glow discharges, glow-arc transitions (region HK of Fig. 2.42), and sparks as preliminary stages of the discharge has not yet been fully assessed.

Other factors relating to high-pressure glow-to-arc transitions will be discussed below, and summaries of the main papers on low- and high-pressure work will be given. At the present time, there is considerable doubt in many cases as to the precise cathode mechanism involved and this situation unfortunately precludes any systematic account of glow-to-arc transitions.

High-pressure glow discharges

Von Engel, Seeliger, and Steenbeck [8] and Seeliger [47] have given a valuable account of some of the properties of high-pressure glow discharges, with a treatment also of glow–arc transitions. They worked with a spherical discharge vessel in which the cathode was composed of a water-cooled copper sheet of 1 mm. thickness. The anode was spaced 1 cm. from the cathode and was a spherically-ended brass tube, also water-cooled. The cathode area covered by the glow was magnified some 10–20 times on a ground-glass screen and measured visually. The authors give characteristics of glows produced in such a system and refer also to the curves of current density j A/cm.^{2} plotted against pressure in mm. Hg, as shown in Fig. 12.2 for air and hydrogen. Only at low pressures does j vary as p^2 as predicted from the Similarity Principle [5]. At higher pressures $j \propto p^{\frac{4}{3}}$ in accordance with theoretical calculations based on the following assumptions [8]. (a) the discharge is of such dimensions that variables need to be studied only in one dimension, (b) the gas heating in the cathode-fall region is due to collisions between positive ions and neutral gas atoms, and (c) the electric field characteristic falls linearly with distance from the cathode [9].

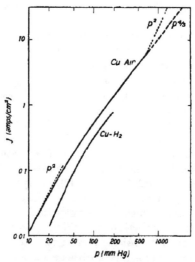

FIG 12 2. Variation of current density with pressure in high-pressure glow discharges.

It is thus possible to show that

$$p = \frac{T_k}{T_1} \sqrt{\left(\frac{j_0}{j_1}\right)} + \frac{2}{3T_1} \sqrt{\left(\frac{V_k j_1 d_1}{3\alpha}\right)} \sqrt[4]{\left(\frac{j_0}{j_1}\right)^3}, \qquad (12.1)$$

where p is the gas pressure, T_k is the cathode surface temperature, and V_k is the cathode fall in potential [5]. α is given by $\lambda(T) = \alpha T$, where λ is the thermal conductivity of the gas. The other terms, j_1 and d_1, are given on the Similarity Principle by the following relations:

$$j_0 = j_1(pT_1/\overline{T})^2, \qquad (12.2)$$

$$d = d_1\left(\frac{\overline{T}}{pT_1}\right). \qquad (12.3)$$

Here j_0 is the current density in the gas at the cathode surface. j_1 and d_1 are the values of current density and cathode-fall space thickness for a normal discharge for $p = 1$ mm. Hg and temperature T_1. $\bar{T}$ is the mean temperature for the whole cathode region.

For small values of j_0 equation (12.1) shows that p varies as $\sqrt{j_0}$, but for larger values of j_0 the second term predominates and p varies as $j_0^{\frac{1}{2}}$. This is in quantitative agreement with the characteristics given in Fig. 12.2.

For pressures above about 300 mm. Hg the calculated values begin to depart from the experimental data and the calculated current densities are too high. This is partly because the relation $\lambda(T) = \alpha T$ cannot be expected to hold at high temperatures. In air at 1 atm. the thickness of the cathode glow region is ~ 1–2×10^{-2} cm.

von Engel et al. [8] discuss glow–arc transitions, firstly in the light of thermal effects at the cathode surface. They stress the experimental difficulties that are encountered if tests are made by varying the cathode temperature and observing the changes in discharge characteristics. Some of these difficulties appear to arise because of the likely interaction of two or more mechanisms such as thermal and field current effects.

If the gas pressure in a glow discharge is increased the cathode current density increases as p^2, whilst the thickness of the cathode fall region decreases as $1/p$. The energy transferred per unit volume of the fall region thus increases as p^3. The cathode may therefore be so strongly heated at high pressures that the glow may change to an arc with hot cathode spots and an arc positive column. This transition may be prevented by adequate cooling, at least at 1 atm. pressure and with currents of 1 A. Fig. 12 3, Pl. 15, shows a high-pressure glow discharge in hydrogen photographed by W. A Gambling (unpublished). At the cathode a normal negative glow discharge is seen, and is accompanied by a Faraday dark space, a positive column and the anode fall region. The cathode current density and the fall space thickness obey the similarity principle [5, 6] with reference to the mean excess temperature of the gas. Assuming that the heat conductivity λ of the gas is proportional to the temperature T then we have $\lambda(T) = \alpha T$. The outflow of heat from the cathode region by gaseous conduction to the cathode with the given temperature T_K is governed by the temperature $\bar{T}$ in the fall region. It may then be deduced that

$$\sqrt{\bar{T}}\,(\bar{T}-T_K) = \frac{2}{3}\sqrt{\left\{\frac{1}{3\alpha}V_n\left(\frac{j_n}{p^2}\right)_K (d_n p)_K\right\}}\,\sqrt{(pT_K)}, \qquad (12.4)$$

where V_n is the cathode fall, $(j_n/p^2)_K$ and $(d_n p)_K$ are values of relative current density and dark space thickness at T_K, and the magnitudes are in volts, amps , cm., and mm. Hg. Thus one obtains, for normal discharges in air at 1 atm., $\bar{T} \simeq 1{,}100°$ K. In hydrogen $\bar{T} \simeq 700°$ K. for $T_K = 285°$ K. (with cooling water). Equation (12.4) shows that at higher gas pressures the excess temperatures cause such a gas rarefaction that the current density increases, not as p^2, but as $p^{\frac{1}{2}}$. Fig. 12.4 (see also Fig 12.2) shows a comparison between theory and experiment.

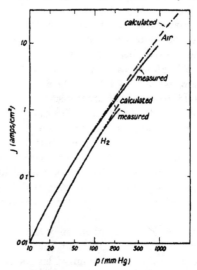

FIG 12.4. Variation of current density with pressure in high-pressure glow discharges.

Thermionic emission mechanism of glow–arc transitions

Von Engel and Steenbeck [5] developed a quantitative theory of the glow–arc transition process by introducing a quantity giving the number of electrons emitted by the cathode (either by thermionic or secondary emission) per positive ion incident on it. If i, i_-, and i_+ are the total electron and positive ion currents respectively at the cathode surface then

$$i_- = \gamma i_+ + i_t. \tag{12.5}$$

i_t is the thermionic emission current and is given by $i_t = f A T^2 e^{-B/T}$, where f is the cathode area which is emitting electrons, T is the cathode temperature, and A and B are the Richardson constants [10]. Also

$$\gamma' = \frac{i_-}{i_+} = \frac{i - i_+}{i_+}.$$

The total current is $\quad i = i_- + i_+ = i_+(1+\gamma) + i_t$

so that $\qquad\qquad \gamma' = \gamma + \dfrac{1+\gamma}{i/i_t - 1}.$ $\qquad\qquad$ (12.6)

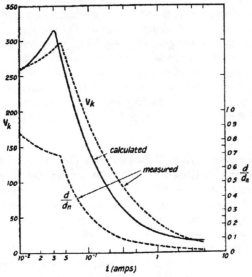

Fig. 12 5 Variation of cathode fall (V_k) and dark space thickness with current in glow-arc transitions of thermionic type. d/d_n is the ratio of dark space thicknesses in the transition region and in the normal glow

T may be found as a function of i by equating the energy transfer to the cathode and the radiated heat loss, and consequently

$$iV_c = f\sigma T^4,$$

where V_c = cathode fall and σ = Stefan's constant. V_c may be plotted as a function of i if γ' is found from γ and a knowledge of A and B. The full treatment for glow discharges is given by von Engel and Steenbeck ([5], pp 72–74).

The above theory was used successfully to interpret Wehrli's [11] data, which are shown in Fig. 12.5. The curve for V_k-i is given in Fig. 12.5 with the curve calculated by von Engel and Steenbeck. It only remains to be added that arcs between less refractory metals than tungsten (e.g. copper) may not be maintained by thermionic emission at the cathode but possibly by field emission or some other mechanism [12, 24, 25].

General studies of glow–arc transitions

The effect observed by Guntherschulze and Fricke [13], and described also later for different conditions by Malter [14] and Paetow [15, 19], is of some interest and importance [4]. It was found possible to operate glow discharges with cathodes of carbon, covered with a layer of insulating powder (Al_2O_3), or of oxidized aluminium at voltages < 40 V. The cathode structure of the glow [13] was found to be abnormal, in that the Crookes dark space was absent and the current densities were abnormally high. The phenomenon is termed a 'spray discharge'.

It appears that cold emission takes place from the cathode at the regions covered by the insulating grains. The current densities are given, as a function of the voltage V, by a relation of the form

$$i = AV^2 e^{-B/V},$$

which is the characteristic of field currents [16]. Often the insulating layer has to be quite thick ($\sim 10\,\mu$) and the average field strength at the cathode might then be $\sim 10^5$ V/cm. This low value implies that local higher fields possibly exist. By diminishing the thickness of the insulating layer a continuous transition from the 'spray discharge' to a normal glow could be obtained. In the vacuum discharge studied by Malter [14] the positive charge on the insulating particles is brought about by secondary emission from them and the effect seems closely akin to that observed by Güntherschulze and Fricke [13]. The extent of the Malter effect may be judged from Paetow's [15] results, the latter found that if $1\,\mu A$ is passed between nickel plates for 1 sec. the rate of emission of electrons was still about 300/sec. some 10 sec after the initial current was switched off. Recent work on this effect has been carried out by Haworth [48].

Druyvesteyn [17] carried out experiments over the pressure range 1–13 cm. Hg on glow–arc transitions in helium, neon, and argon. The electrodes were 1·8 mm tungsten spheres, spaced 0·5–1 mm. apart and the gas pressure was low so that no positive column was observed. The cathode temperatures were measured with a pyrometer and were found to be sensibly uniform over the spherical surfaces. The current-voltage characteristics showed a maximum voltage (Fig. 12.6) and the cathode temperatures for all gases were 2,000° K.$\pm$50° K. The discharge voltage was found to be almost equal to the cathode fall. If the cathode temperature was < 2,000° K. the thermionic emission from the cathode was about 0·1 mA and was then, according to Druyvesteyn, an appreciable part of the electron current from the cathode. As the current became greater,

the thermionic emission increased and the voltage fell, since the positive
ions had to liberate fewer electrons from the cathode in order to maintain the discharge. The voltage fell eventually to a value roughly equal
to the ionization potential of the gas and the thermionic current was then
greater than one-half of the arc current.

Recent experiments on low-pressure transitions, including an inves-

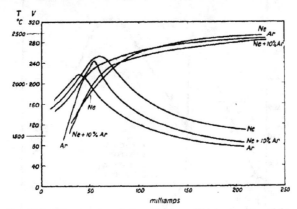

FIG. 12 6. Characteristics of glows at 50 mm. Hg as a function of discharge current. T is the cathode temperature and V the voltage.

tigation of the effect of a barium layer on the cathode have been
made by Pirani [18]. The work was designed particularly to investigate
starting conditions in low-pressure discharge lamps filled with the rare
gases. In neon, some experiments with tungsten wire cathodes (100, 250,
and 500 μ, and 1 mm. diameter) have been made [17]. For 100 μ wires,
no steady characteristics could be obtained (Fig. 12.7) but the characteristics were always steady for the thicker wires.

We shall proceed now to summarize various experimental researches
on glow–arc transitions in chronological order, and to draw very brief
general conclusions.

Mackeown [20] studied conditions in the region of the abnormal cathode
fall in glow discharges (FH of Fig. 2.42) and describes glow–arc transitions. The latter were measured oscillographically [21] with A.C. discharges (r.m s. current of 4 A). It is seen from a study of the voltage
wave, shown in Fig. 12.8, Pl. 15, that even when conditions are propi-
tious a transition to the low voltage discharge, described as an arc, does
not take place in every half-cycle of the wave. Mackeown inferred from
this fact that the transition mechanism takes place at localized regions

of the cathode from which electron emission is relatively copious   In
the case of non-refractory materials (e.g. Cu but not W) the mechanism
of emission might be due to field currents initiated by a localized positive
space charge near the active spots. This implies that these regions are
relatively active in the glow discharge phase, so that the positive ion

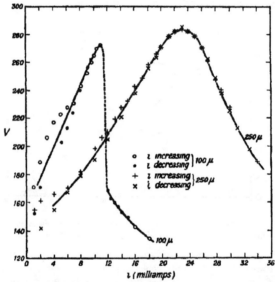

FIG. 12 7   Druyvesteyn's results on current-voltage characteristics
of wire cathode glow–arc discharges Wire diameters (in μ) shown
on curves

clouds are localized *ab initio* This is not unreasonable since it is known
from corona studies (Chapter III) that the γ mechanism [3] necessary
for the maintenance of a glow shows variations over the cathode surface.
Local impurities or oxide spots, etc., may be present, particularly in
metals which are cleaned with difficulty.

Mackeown, Bowden, and Cobine [22] studied glow–arc transitions
produced by 60 c./s. A.C. voltages in air at low pressures for gaps of
1 mm. between cylindrical electrodes 0·64 cm. in diameter. The method
used was as follows: the reignition potentials, i e. the voltages for each
half-cycle at which a discharge struck, were measured oscillographically
for a current of 1 5 A r.m.s. Copper and graphite electrodes were used,
and it was found (see, for example, Fig. 12.9) that the reignition potentials always fell below the sparking potentials for the particular set of

electrodes in use The discharge usually restruck as a glow and changed, if the circuit conditions allowed, to an arc later in the half-cycle The above authors discuss the conditions which could lead to this reduction in reignition voltage below the sparking value, and suggest that in no case was the cathode temperature high enough to give the requisite current density at the cathode for cathode spot formation ($\sim$ 1,000 A/cm.^{2})   Table 12.1 shows some reignition data [22].

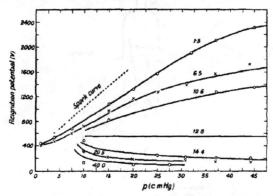

Fig 12 9 Arc reignition potentials for pure graphite electrodes
The arc currents (amps) are shown on the curves

TABLE 12.1

Material	Reignition potentials (volts) for various arc currents and pressures					
	1 5 A		8 1 A		12 3 A	
	25 cm. Hg	50 cm Hg	25 cm. Hg	50 cm Hg	25 cm Hg	50 cm Hg
Carbon .	1,400	2,330	450	400	200	150
Graphite	1,575	2,425	1,210	1,550	730	900
Copper	1,656	2,850	1,400	1,800	1,000	1,600

The effect of impurities, of low work function, present on the surface of the electrodes is brought out in a striking fashion by the results. Carbon cathodes, if treated with sodium hydroxide, gave reignition potentials at 30 cm. Hg of about 450 V whilst the value for pure graphite electrodes was 1,800 V A carbon-cathode and copper-anode system gave a reignition voltage of 1,000 V. It should be pointed out that material from the anode can easily be volatilized on to the cathode.

Plesse [23] investigated the spatial stability of glow and arc discharges between inclined rod-shaped electrodes, as shown diagrammatically in

Fig 12.10, using 50 c./s. A.C. with a maximum voltage of 10 kV and currents up to 2 A. He found that various forms of discharge could be maintained over various current ranges and tested a wide range of electrode materials. The results of Table 12 2 were obtained in nitrogen. Experiments in oxygen and hydrogen were also made.

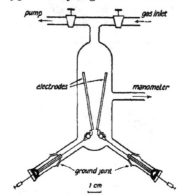

FIG 12 10   Plesse's experimental tube

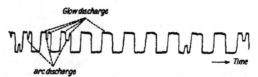

FIG. 12.11. Typical voltage oscillogram (Plesse) of discharge between Ni electrodes in nitrogen at 600 mm. Hg pressure, for a current of 0 65 A.

Plesse also made experiments with stationary discharges and found that if metal vapour was formed at the cathode in sufficient quantities a glow would change to an arc. A similar result was obtained if a foreign vapour was generated at the cathode   Plesse concluded that since the point of transition did not seem to depend directly on the cathode temperature, and also that since the arc could often pass at low currents between cold electrodes, the vapour generation was not in all cases caused by purely thermal effects. The occurrence of high-frequency oscillations in the glow–arc region, especially for a mercury cathode, was established. The oscillation frequency was 10^5 to 2×10^6 c./s., depending on the circuit parameters.

An example of the characteristic for a stationary arc between nickel electrodes in nitrogen at 600 mm. Hg, for a current of 0·65 A, is given

TABLE 12.2

Electrodes	Ca	Mg	C	Cd	Zn	Pb	Sn	Al	Fe	Cu	Ag	Ni	Pt	W
Glow discharge with stationary cathode spot	.	.		$i <$ 0·03	$i <$ 0·03 to 0·05	$i <$ 0·05 to 0·06	$i <$ 0·05 to 0·06	$i <$ 0·03 to 0·05	$i <$ 0·05 to 0·08	$i <$ 0·05 to 0·08	$i <$ 0·05 to 0·08	$i <$ 0·05 to 0·08	$i <$ 0·05 to 0·08	$i <$ 0·05 to 0·08
Glow discharge with moving cathode spot	.		..	→	→	→	0·05 to 0·06	→	0·05 to 0·20	0·05 to 0·24	0·05 to 0·28	0·05 to 0·3	0·05 to 0·6	0·05 to 0·75
Arc with moving cathode spot	.	.	..			→	→	0·03 to 0·12	$i >$ 0·15 to 0·20	$i >$ 0·18 to 0·24	$i >$ 0·20 to 0·26	$i >$ 0·22 to 0·30	$i >$ 0·4 to 0·6	$i >$ 0·6 to 0·75
Arc with stationary cathode spot	$i >$ 0·02	$i >$ 0·02	$i >$ 0·02	$i >$ 0·03	$i >$ 0·03 to 0·05	$i >$ 0·05 to 0·06	$i >$ 0·06	$i >$ 0·10 to 0·12					.	.

Current i is given in A.

in Fig. 12.11. The spasmodic formation of arc discharges is apparent. Further results are summarized in Table 12.3.

TABLE 12 3

Transition current in A for glow–arc transitions in N_2 at 500 mm Hg		Transition current in A for glow–arc transitions in H_2 at 500 mm Hg	
Ca	< 0 02	Ca	< 0 02
Mg	< 0 02	Mg	< 0 02
C	< 0 02	C	< 0 02
Hg	0 03	Hg	0 04
Al	0 05	Al	> 1 0
Cd	0 09	Cd	0 23
Zn	0 23	Zn	0 38
Pb	0 25	Sn	0 6
Pt	0 45	Pb	0 3
Fe	0 65		
W	0 9		
Ni	0 95		
Ag	1 15		
Cu	1 5		

Plesse does not discuss directly the mechanism later invoked by Suits, Cobine, and others, namely cold cathode emission caused by the high local electrostatic fields round small insulating particles on the cathode. He does, however, describe experiments with sulphur deposited on one of a pair of tungsten electrodes; in these experiments it was easy to produce an arc with the coated electrode as cathode, but not with the clean tungsten cathode. Plesse stresses the importance of the emission of metal vapour at the cathode, since a similar vapour emission would probably be found with the sulphur coating Unfortunately the experiment does not seem decisive now in the light of the later work [4, 6, 26] since the 'small particle mechanism' could have been operative.

Plesse's paper is perhaps the most detailed to date on glow–arc transitions since it covers a wide range of materials for several gases. His oscillographic records, of which Fig. 12.12 is a further example, are of exceptional clarity and interest.

Suits and Hocker [26] describe experiments with copper electrodes in air and other gases at 1 atm. pressure, using an experimental arrangement, similar to that of Stolt [27], in which the arc is moved over the cathode by means of a magnetic field. With pure hydrogen it was found that, after a preliminary period during which arcs could be struck, the only form of discharge which could be sustained for the current range 0–10 A was a striated glow. The arc would not strike to the cleaned copper, although at higher currents (10–100 A) an incipient discharge over a long path to uncleaned parts of the electrode was obtained. With silver electrodes in hydrogen it was impossible to produce anything but the glow form of discharge. Arcs in pure gases with C or W cathodes were stable Suits and Hocker believe that these results indicate the importance of oxide films in the case of cold cathode arcs, particularly

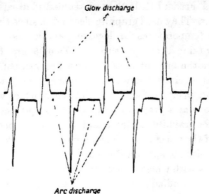

Glow discharge

Arc discharge

Fig 12 12. Enlargement of part of a record similar to Fig
12 11 showing glow–arc transitions (580 mm Hg pressure)

with cathodes of low melting-point (Ag or Cu) which are unable for low currents, of 1–10 A, to provide sufficient emission without an oxide mechanism, possibly identical with that of Guntherschulze and Fricke [13]. Suits and Hocker also point out that it is difficult to strike welding arcs in air between clean bright iron welding rods and clean steel cathode plates until a visible film of oxide has formed on the cathode surface.

Cobine [28] has also reported experiments, which substantiate the work of Suits and Hocker, on glow–arc transitions for 0·25 in. copper electrodes spaced 1 mm. apart in air at pressures < 50 cm. Hg. The power supply was from a 2·2-kV power transformer which could supply current up to 10 A. Freshly filed copper electrodes gave glow discharges which changed after some 30 cycles to an arc. Repeated runs with the initially clean electrodes eventually resulted only in the establishment of stable arcs, and oxidation by heating in a gas flame gave the same result. A new result was that cleaning of the copper in potassium dichromate, but not in chemically pure nitric acid, with a subsequent washing in distilled water, resulted in readily formed arcs. The nitric acid treatment gave glow discharges throughout the entire cycle for all pressures up to 50 cm. Hg.

Cobine pointed out that, even with careful washing, the arcs between dichromate-treated electrodes showed, spectroscopically, the presence of potassium. He suggested that, whilst the oxide mechanism (see above and Druyvesteyn [29]) probably holds for oxidized copper, the additional presence of some low work function material such as potassium might modify the mechanism in detail.

Maxfield and Fredenhall [30] have studied glow–arc transitions in mercury vapour. They used graphite electrodes, since the latter may be heated to high temperatures for degassing and are not affected by the vapour. By working at low pressures (0·005–0·05 mm. Hg) the authors considered that the current densities of a few mA/cm.^{2} would be too small to give appreciable thermionic emission. The tube used in these experiments had, in effect, four electrodes, one of which was the mercury pool. Transitions at a graphite probe electrode projecting into an arc between another graphite electrode and the mercury pool were studied, as a function of the arc current. The fourth graphite electrode was used to maintain a cathode spot on the mercury pool

The frequency with which an arc was struck to the probe electrode, as cathode, was then studied as a function of the various tube parameters, including the vapour pressure, current in the priming arc, and probe electrode temperature

Maxfield and Fredenhall give a detailed analysis of their results and finally conclude that they are best explained if the transitions are caused by discontinuous emission of bursts of gas from the cathode giving local high current densities. The electrodes used were carefully baked but it is suggested that much more cleaning of the cathodes would be necessary to remove virtually all the gas. This gas-burst mechanism had previously been suggested by von Engel and Steenbeck [5, vol. 2, p. 123].

It is difficult indeed to disprove this theory, which can only be completely untenable in the case of carefully outgassed refractory cathodes. Further, the exact consequences of the production of high local current densities, which Maxfield and Fredenhall term the transition, need to be carefully considered since the cathode spots so formed might then be described as giving a transition after producing local field currents or thermionic emission at the cathode. Thus, the gas-burst mechanism may perhaps be considered only as an initial step in the transition process, which would again invoke one of the mechanisms discussed elsewhere in this chapter. Further, although stable discharges are not considered in this book, the gas-burst mechanism can hardly account for the maintenance of arcs between non-refractory electrodes.

The above work was extended by Maxfield, Hegbar, and Eaton [31], with the conclusion again that gas bursts were important. The size of a gas burst may be equivalent to 10^8 positive ions and, on the assumption of 100 per cent. ionization, this is equivalent to 3×10^{-14} gm. Hg (4×10^{-12} cm.^{3} at 1 atm.).

Hsu Yun Fan [21] has described experiments on glow discharges and glow-arc transitions in air, oxygen, and nitrogen, at 1 atm., and in hydrogen from 1 to 13 atm., using various combinations of electrode materials. The experiments showed that glow discharges could be obtained in hydrogen with currents up to 14 A and, though not simultaneously, for pressures up to 13 atm. Glows were exceptionally stable in hydrogen. The D.C discharges were passed between water-cooled electrodes and were initiated by a high voltage impulse. The nitrogen and hydrogen were about 99·9 per cent. pure and the oxygen was of cylinder grade.

Hsu found that intermittent changes accompanied by sudden changes in discharge voltage, i.e. from glow to arc and back again from arc to glow, took place at rates depending upon the experimental conditions and these disturbances were particularly violent in air and oxygen. (Similar experiments have been described also by Wheatcroft and Barker [33].) In nitrogen the voltage disturbances were less violent and comparatively stable glows could be obtained for currents up to 2 A, in contrast with the behaviour of discharges in hydrogen where the glows were relatively very stable. It was possible to maintain a glow between copper electrodes in hydrogen at 2 A for several hours without the occurrence of disturbances, and even at 10 A glow discharges could last for 15 minutes. In order to obtain these high current glows it was necessary to increase the discharge current slowly, in this manner the current in one case was raised to 14 A before the transition to an arc took place. Cathodes of Pt, Mo, and Al gave more disturbances than Cu and it was found that the anodes (Cu in all cases) were not appreciably affected by the discharges. The characteristics of high-pressure glows in hydrogen are shown in Fig. 12.13. An interesting series of photographs is given by Hsu, depicting the hydrogen glows in 3 mm. gaps at various currents and pressures. At lower values of these parameters (e.g. 1 atm , 0·25-2 A) the positive column shows striations, at higher currents or pressures (e.g. 4 atm., 1 A) the column concentrates into a narrow line with a simultaneous drop in the discharge voltage of about 70 V. Hsu suggests that this change may be due to the onset of thermal ionization in the column although he still describes the discharge as a glow [21]. Even at

2 atm. pressure and 0 25 A the discharge shows a bright narrow and well-defined positive column separated from the cathode and the cathode glow by a dark space. Hsu's two photographs at 1 atm. with 0·25 and 2 A, both show striated positive columns and his discussion does not specifically state that the sharp column [34] was observed at 1 atm. It is clear that Bruce's later results [34] confirm this work of Hsu's and that there is some as yet unexplained transition possible in a positive column.

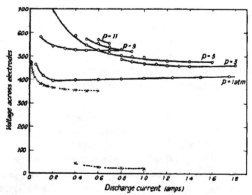

Fig. 12 13. Volt-ampere characteristics of a discharge in air at atmospheric pressure between water-cooled copper electrodes separated 1 0 mm. (dotted curve), and of a glow discharge in hydrogen between water-cooled copper electrodes separated 0 9 mm at several pressures

Bruce has recently published [34] some interesting photographs of 50 c./s. A.C. discharges in hydrogen at 1 atm. between water-cooled Cu electrodes 0·5 cm. apart, passing 1·5 A (r.m.s.). It is found that as the electrode temperature increases the discharge column near the anode begins to change from a diffuse glow to a narrow, light crimson column (Fig. 12.14, Pl. 15). At 1·5 A the narrow column reaches the vicinity of the cathode, which has throughout been covered by the widespread glow visible in Fig. 12.14. The cathode fall remains constant throughout at about 350 V, whilst the potential gradient along the column falls from about 650 V/cm. to 350 V/cm., the latter corresponding to the narrow column. Bruce interprets these results as corresponding to a high-pressure glow (diffuse column) changing to an arc (narrow bright column). A discussion of Bruce's theory of glow–arc transitions in long spark discharges is given on p. 229.

In earlier work by Suits [35], on high-pressure arcs and glows in hydrogen, it was found that with pure carbon electrodes, at 1,500 V D C., it was impossible to obtain satisfactory electrical data in the 1–10 A current range for pressures greater than 1 atm. since, besides other forms of instability, the arc column developed rapid movements due to turbulent convection currents. The effect was very pronounced at 20 atm However, the arcs with thermionic cathodes were stable at 1 atm. and the following information was obtained. There appear to be, for the conditions of the experiment, three forms of discharge depending on the current, viz. two kinds of glow and an arc discharge. The voltage drops for a gap length of 0·5 cm. between carbon electrodes, and the corresponding voltage gradients, are shown in Figs. 12.15 and 12.16 Suits interprets Type I as an arc, and suggests tentatively that Type II may be governed by the normal glow mechanism with positive ion bombardment of the cathode as the source of electrons. Type III appears to be difficult to interpret.

In a comprehensive paper by Hofert [36] the relation between thermionic transitions and field emission transitions for a tungsten cathode is clearly shown. As the cathode temperature was raised, to the point where thermionic emission was appreciable, the time of transition from glow to arc, as measured by a sudden fall in discharge tube voltage, increased from $\sim 10^{-8}$ sec. to $\sim 10^{-4}$ sec. Transitions with low melting-point cathodes, including Hg, were also studied, in various atmospheres. A special study was made of the effect of varying, with a Hg pool cathode, the Hg vapour pressure in various gases (N_2, H_2, A) While the results were too complicated to interpret, Hofert supports the suggestion that dense vapour clouds near the cathode are sometimes important in governing the transition, but it should be emphasized that experiments with Hg pools in various gases, and in which the Hg vapour pressure is altered by temperature control, do not strictly compare with conditions near the cathode during the transition process, when highly localized and dense vapour clouds may be formed [12].

A photograph by Suits [35] of the stable hydrogen discharge in the arc mode (Type I) shows a uniform and narrow discharge column extending across the gap. No information relating to the transition region (II–I of Figs. 12.15 and 12.16) is given. Bruce found the transition to take place in 50 c./s. A.C. discharges at 1·5 A, with a column gradient of 350 V/cm. in excellent agreement with Suits's results, although the difference in electrodes somewhat vitiates a comparison (see below). Suits determined his gradients with a vibrating electrode chamber and oscillograph, but Bruce does not state his technique or how the cathode fall was separated from the column gradient. Suits states also that preliminary results indicate four discharge modes for W electrodes in H_2, four modes for Cu in N_2, and two modes for Cu and Ag in H_2.

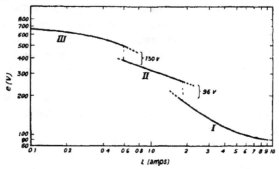

FIG 12 15 Voltage-current curves for 5 mm. gap length discharges in hydrogen at 1 atm pressure between C electrodes

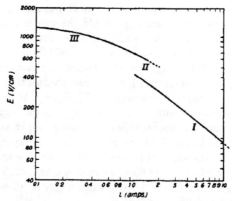

FIG. 12.16. As for Fig. 12 15, but showing electric field as a function of discharge current.

It is clear from the above discussion of cathode mechanisms in glow–arc transitions that there are, to summarize the position, two general types of arc, viz. thermionic and field emission arcs (the latter including for the present discussion such mechanisms as the Malter, Paetow, and Guntherschulze–Fricke effects). It should also be made clear that new mechanisms are not to be excluded [12], as indicated by the recent work on cathode spots by Froome and others (discussed in Chapter XI).

Rothstein's new mechanism for arcs with non-refractory cathodes [12] involves the suggestion that a region, perhaps 10^{-5} cm. thick, of very dense metallic vapour is formed near the cathode spot. The inter-atomic fields in this region are so high that the electron energy levels become bands and metallic conduction takes place from the cathode into the vapour cloud, maintained by the ions impinging on the cathode. The consequences of this suggestion are developed briefly by Rothstein, with reference to the work of Smith [37, 38], Mierdel [39], and Birch [40, 41], and it may prove to be of importance in the study of glow–arc transitions. It has been shown by von Engel et al. [8] (see the discussion by Druyvesteyn and Penning [4, pp. 152–5]) in studies of the movement of the cathode spot, either through the influence of an external magnetic field or by striking an arc between the rims of two spinning disks moving relatively to each other, that some arcs will not readily move (if between pure W, pure C, or impure Cu electrodes), whilst others will do so (if between pure Cu or Fe electrodes). It would be expected that thermal arcs with a hot tungsten cathode would be extinguished on moving surfaces, since the heating process should take an appreciable time, and that field-current arcs would not be extinguished. However, this simple picture is confused by an experiment of von Engel et al. [8], who found that in a particular case of discharge between copper electrodes, spaced 1 mm. apart in air, only a glow discharge was found to pass between the stationary disks but an arc was produced at 0·4 A with a disk velocity of $\sim$ 0·5 m./sec.

Glow–arc transitions in switches

Glow–arc transitions are encountered in the studies of the interruption of D.C. circuits by switching. Various authors, notably Holm [42] and more recently Hamilton and Sillars [43] and Garfitt [44], have published detailed accounts of the phenomena associated with the opening of switches in inductive circuits (see also Betz and Karrer [45]). The work involved a study of the voltage pulses developed across the switch contacts after the latter opened, with and without a condenser connected across the switch or across a load in series with the switch. The condenser reduces the peak transient voltage V, which is given by

$$V = I\sqrt{(L/C)}, \qquad (12.7)$$

where I is the steady current flowing before the break and L and C are the circuit inductance and capacitance. A discharge, passing between the switch contacts, may absorb a proportion of the energy $\frac{1}{2}LI^2$; and then the rate of opening of the switch contacts, since it will clearly affect the nature and duration of the discharge, will also affect the transient voltage developed across the switch. The use of a large quench condenser will tend, other things being equal, to reduce the transient voltage and so shorten the length of the discharge since the length of

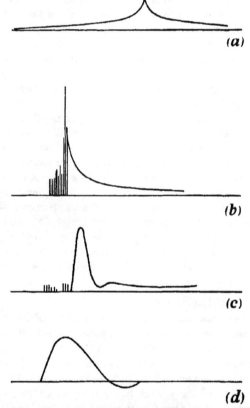

Fig. 12 17. Typical transient voltage-time curves for the breaking of inductive circuits.

discharge that can be maintained by the transient will naturally tend to increase with increasing voltage. It appears that no information is available on the voltage gradients in switching discharges, although many oscillograms have been taken by the various workers and some few photographs of discharges but without the necessary information to enable instantaneous gradients to be found. This information would be of interest for comparison with the work described elsewhere in this chapter on high-pressure glow discharges and glow–arc transitions.

Typical oscillograms [44] taken with a switch breaking an inductive circuit are shown in Fig. 12.17 (a)–(d). The peak voltage for Fig. 12.17 (c) may be $\sim$ 1·5–2 kV with a visible discharge perhaps $\sim$ 1 or 2 mm. long.

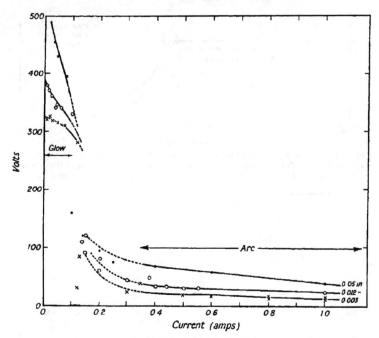

FIG. 12 19   Relation between voltage and current for discharges in air between silver alloy electrodes at three spacings.

The small, isolated peaks in Fig. 12.17 (b) and (c) show where the gap between the separating contacts breaks down, clears as the gap increases, and then breaks down again. The times elapsing between successive discharges is so short (perhaps $\sim$ 0·5 m.sec., and often much less) that the discharge probably assumes the form of a series of high-pressure glows (Figs. 12.17 (b) and (c)) whereas, with the lowest or no capacitance, an arc is formed. The time base for Fig. 12.17 was about 30 m.sec. Fig. 12 17 (a) was taken with a small shunt capacitance, Figs. 12.17 (b) and (c) with a larger capacitance, and finally Fig. 12.17 (d) was taken with a capacitance so large that the rate of rise of voltage across the contacts was less than the rate of increase of breakdown voltage between them. In Fig. 12.17 (d) the voltage shown is that found across the capacitance, in shunt with the contacts. The magnitude of the switching

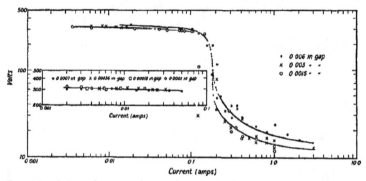

FIG. 12.20. Relation between voltage and current for discharges of various lengths in air between silver alloy electrodes. The dotted curve gives the range for which the validity of the results is doubtful because of the tendency for the arc to take a longer path.

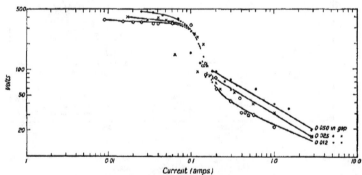

FIG 12.21. Relation between voltage and current for discharges of various lengths in air between silver alloy electrodes. The dotted curve gives the range for which the validity of the results is doubtful because of the tendency for the arc to take a longer path.

transients can be greatly reduced by the use of a non-linear resistance connected across the capacitance [43] as shown in Fig. 12.18, Pl. 15.

Glow–arc transitions have also been studied further by Sillars [46]. Typical volt-current curves of a steady discharge between a pair of silver contacts in air are shown in Fig. 12.19 for separations of 0·003, 0·012, and 0·050 in. Consistent readings could not be obtained in the transition region since the discharge was very unstable. Other data with silver alloy (95–99 per cent. Ag) contacts are given in Figs. 12.20 and

12.21 [46]. The glow voltage drop is sensibly independent of current with the exception of silver alloy contacts at the widest spacing of 0·050 in. The glows are stable up to 100–200 mA, and tungsten contacts appear to sustain a glow for rather larger currents than the other metals. It is also apparent that the glow voltage depends very little on gap length even for the very short gap lengths (down to 0·0001 in.; see Fig. 12.20). This implies that the high field region of the discharge is short compared with gap lengths of this order. From Fig. 12.19 it is seen that the mean voltage gradient for the shortest gap of 0·0015 in. is about 8 kV/cm , and it is likely that in the space charge regions of the discharge it will be much larger In the arc range the voltage drops increase with increasing gap length and, as usual for arcs [3, 5, 6], the voltage drops decrease with increasing current.

REFERENCES QUOTED IN CHAPTER XII

1. J J. THOMSON and G. P. THOMSON, *Conduction of Electricity through Gases*, Cambridge, 1928

2. K. K. DARROW, *Electrical Phenomena in Gases*, Williams and Wilkins, Baltimore, 1932.

3. L. B LOEB, *Fundamental Processes of Electrical Discharge in Gases*, Wiley, New York, 1939

4. M. J. DRUYVESTEYN and F. M PENNING, *Rev. Mod. Phys* **12** (1940), 87

5. A. VON ENGEL and M STEENBECK, *Elektrische Gasentladungen*, Springer, Berlin, 1934.

6. J. D. COBINE, *Gaseous Conductors*, McGraw-Hill, New York, 1941.

7. R. SEELIGER, *Einfuhrung in die Physik der Gasentladungen*, Barth, Leipzig, 1934.

8. A. VON ENGEL, R SEELIGER, and M. STEENBECK, *Z Phys.* **85** (1933), 144.

9. F. W. ASTON, *Proc. Roy. Soc.* A, **84** (1911), 526.

10. T. J. JONES, *Thermionic Emission*, Methuen, London, 1936.

11. M. WEHRLI, *Helv. Phys. Acta*, **1** (1928), 323.

12. J. ROTHSTEIN, *Phys. Rev.* **73** (1948), 1214.

13. A. GUNTHERSCHULZE and H. FRICKE, *Z. Phys.* **86** (1933), 451, 821.

14. L. MALTER, *Phys Rev* **50** (1936), 48

15. H. PAETOW, *Z. Phys* **111** (1939), 770

16. R O. JENKINS, *Rep. Phys. Soc. Progr. Phys.* **9** (1942–3), 177.

17. M. J. DRUYVESTEYN. *Z. Phys.* **73** (1932), 727.

18. M. PIRANI, *Proc. Phys. Soc.* **55** (1943), 24.

19. L. R. KOLLER and R. P. JOHNSON, *Phys Rev* **52** (1937), 519.

20. S. S. MACKEOWN, *Elect Engng. N.Y.* **51** (1932), 386.

21. HSU YUN FAN, *Phys. Rev* **55** (1939), 769.

22. S S. MACKEOWN, F. W. BOWDEN, and J. D. COBINE, *Elect Eng.* **53** (1934), 1081.

23. H. PLESSE, *Ann. Phys. Lpz.* **22** (1935), 473

24. J. NIKLIBORC, *C. R. des Sci. de la Soc. Polonaise de Phys.* **5** (1930), 425.

25. C. Reczynski, *Acta. Phys. Polon.* **5** (1931), 287.

26. C. G. Suits and J. P. Hocker, *Phys Rev.* **53** (1938), 670.

27. H Stolt, *Ann. Phys Lpz.* **74** (1924), 80

28. J D. Cobine, *Phys Rev* **53** (1938), 911.

29. M J Druyvesteyn, *Nature*, Lond , **137** (1936), 580

30. F. A. Maxfield and G. L. Fredenhall, *J. Appl. Phys.* **9** (1938), 600

31. F A Maxfield, H. R. Hegbar, and J. R. Eaton, *Trans. Amer. Inst Elect. Engrs.* **59** (1940), 816.

32 W. A. Gambling and H. Edels, in course of preparation.

33. E. L. E. Wheatcroft and H Barker, *Phil. Mag* **29** (1940), 1.

34 C E R. Bruce, *Nature*, Lond , **161** (1948), 521.

35 C G. Suits, *J. Appl Phys* **10** (1939), 648.

36. H J. Hofert, *Ann. Phys. Lpz.* **35** (1939), 547.

37 C. G. Smith, *Phys. Rev.* **62** (1942), 48.

38. C. G Smith, ibid. **69** (1946), 96

39. G Mierdel, *Z Tech Phys.* **17** (1936), 452.

40. F. Birch, *Phys Rev* **40** (1932), 1054.

41 F. Birch, ibid. **41** (1932), 641

42. R Holm, *Electric Contacts*, Gebers, Stockholm, 1946.

43. A. Hamilton and R W Sillars, *J. Instn. Elect. Engrs* **96** (I), (1949), 64.

44. D. F M. Garfitt, *M Sc Thesis*, Manchester, 1947.

45 P L Betz and S Karrer, *J Appl Phys.* **8** (1937), 845.

46. R. W Sillars, *Metro.-Vick Research Report*, No 9192, 1944.

47. R. Seeliger, *Phys Z.* **27** (1926), 730.

48. F. E. Haworth, *Phys Rev.* **80** (1950), 223

APPENDIX. RECENT DEVELOPMENTS

CHAPTER I

A COMPREHENSIVE work on collision processes has been published by Massey and Burhop [Oxford University Press, 1952] in which the experimental studies in this field are fully treated

Photoionization

Largely because of increased interest in the physics of the upper atmosphere, stimulated greatly by the Gassiot Committee of the Royal Society, theoretical and experimental work on this subject is increasing in volume.

The experimental techniques have been reviewed by Price [*Phys Soc. Rep. Prog. Phys* **14** (1951), 1] and several papers by Weissler and his colleagues on experimental determinations of absorption cross-sections have been published. Ditchburn and Jutsum [*Nature*, **165** (1950), 723] have made measurements on Na.

Weissler, Po Lee, and Mohr [*J. Opt. Soc Amer.* **42** (1952), 84] have measured absolute absorption coefficients in nitrogen for $\lambda = 300$ to 1,300 angstroms in a vacuum spectrometer, using a grazing incidence grating technique. For λ less than 796 angstroms continuous absorption is found with a first maximum value of the absolute coefficient (k) of 680 cm.$^{-1}$ near 760 angstroms (at N T.P.) Other regions of diffuse absorption between this wave-length and 661 angstroms show larger values of k, up to 2.76×10^3 cm.$^{-1}$ Exhaustive data are given over the above wave-length range. Similar detailed work in oxygen is reported by Weissler and Po Lee [*J. Opt Soc Amer* **42** (1952), 200] for the wave-length range 300–1,300 angstroms, continuous absorption below 740 angstroms is found, with $k \simeq 700$ cm^{-1} between 400 and 600 angstroms, and with $k = 530$ cm^{-1} at 303 angstroms. These large values of k thus occur over a wider range of wave-lengths than in nitrogen Po Lee and Weissler [ibid. 214] have also published a brief communication on neon. The absorption at the spectral head series limit rises very steeply to give $k = 155$ cm.$^{-1}$ ($\sigma = 5.7 \times 10^{-18}$ cm^2, for the cross-section per atom) at about 574 angstroms. This value is in excellent agreement with one of the values calculated by Bates [*Monthly Notices, Roy. Astron Soc.* A, **106** (1946), 432; **100** (1939), 25] and Seaton [*Proc Roy Soc* A, **208** (1951), 408], namely $\sigma = 5.8 \times 10^{-18}$ cm.^{2} The other value of Bates and Seaton, obtained with a different treatment, is $\sigma = 4.4 \times 10^{-18}$ cm.^{2}

Finally, Po Lee and Weissler [*Astrophys. Jour* **115** (1952), 570] have published a short note on the absolute absorption in the H_2 continuum, where the ionization limit is 803·7 angstroms ($H_2^+ \times {}^2\Sigma_g^+$). Large absorption coefficients are found near the dissociation limit (748 6 angstroms) where k is ~ 100–300 cm.$^{-1}$ at N.T P. Similar values of k are found at about, and on the short wave-length side of, 800 angstroms Dalgarno [*Proc. Phys. Soc.* A, **65** (1952), 663] has calculated the photoionization cross-section for methane, which is found to be 9.4×10^{-17} cm.^{2} at the spectral head (105,850 cm.$^{-1}$). In the same paper a brief mention is made of calculations for argon, which give a cross-section of 3×10^{-17} cm^2 at the spectral head (127,110 cm.$^{-1}$).

These results confirm general expectations from the early work of Vinti and Wheeler on helium (see pp 12–19) and the theoretical work of Bates and others, and show that the absorption coefficients for photoionizing radiations are very high.

Further theoretical work has been carried out by Bates and Seaton [*Monthly Notices, Roy. Astron Soc* **109** (1949), 698, for O, N, and C] and by Seaton [*Monthly Notices,* ibid. **110** (1950), 247, for K+, *Proc. Roy. Soc.* A, **208** (1951), 408, 418, for B to Ne, and the alkali metals].

Recombination, Diffusion Processes, and Mobilities for Positive Ions

Massey [Faraday Society Discussion, April 1952] has summarized the present position on gaseous ions and their reactions. The studies by Biondi and Brown and others (see p. 21) of helium afterglows show recombination coefficients for electrons to be $\sim 2 \times 10^{-8}$ cm ^{3}/sec. for pressures > 20 mm. Hg Bates [*Phys. Rev.* **77** (1950), 718, and **78** (1950), 492] has suggested that this large coefficient is due to dissociative recombination involving He_2^+, i.e.

$$He_2^+ + e \to He^* + He^{**},$$

and that the He_2^+ content increases with increasing pressure. Biondi [ibid. **83** (1951), 1078] has confirmed this in an ingenious manner suggested by Holstein. Helium was contaminated with argon and the afterglow was then found to give a low recombination coefficient and the ions therein to have a mobility characteristic of A^+. The latter ions are formed by the reaction

$$He^* + A \to He + A^+ + e$$

(see also Biondi and Holstein [ibid **82** (1951), 962]). Ne_2^+ and A_2^+ have also been studied [Biondi and Brown, ibid. **76** (1949), 1697; Redfield and Holt, ibid **82** (1951), 874; and Hornbeck, ibid. **84** (1951), 615]. It thus appears likely that the discrepancy between observed recombination coefficients and those calculated for radiative single atom-electron processes can be resolved in many, if not all, cases by consideration of the dissociative recombination process, involving molecules, suggested by Bates. The full implications of this effect in discharge physics have not yet been worked out, but they are likely to be very important

Data on mobilities and diffusion coefficients for positive ions are to be found in the above papers Hornbeck and Wannier [ibid. **82** (1951), 458] have also reported on measurements of drift velocities of He+, Ne+, and A+ in their respective parent gases, and further work is described by Hornbeck [ibid **84** (1951), 615] The ions He_2^+, Ne_2^+, A_2^+, Kr_2^+, and Xe_2^+ have been identified in a special mass spectrometer by Hornbeck and Molnar [ibid. **84** (1951), 621]. Phelps and Brown [ibid **86** (1952), 102] have studied He afterglows with a mass spectrometer technique.

Diffusion is an important factor in the study of the life of decaying meteor trails and investigations in this field have been made by Huxley [*Aust. J. of Sci. Res.* A, **5** (1952), 10], Herlofson [*Ark Fys.* **3** (1951), 15], Kaiser and Closs [*Phil. Mag.* **43** (1952), 1], and Feinstein [*J Geophys. Res* **56** (1951), 37].

Studies of afterglows and decaying plasmas, using the u h f. and optical techniques mentioned above and in Chapter I have also been made by Dandurand and Holt [*Phys. Rev.* **82** (1951), 278 (Cs); ibid. 868, (Hg)], by Redfield and Holt [ibid. 874] for A, and by Richardson [ibid **88** (1952), 895] for Kr. Earlier studies of He (cf. the work of Biondi and Brown, quoted above and in Chapter I) have been made by Johnson, McClure, and Holt [ibid. **80** (1950), 376] (see Chapter I) and of H_2 by Richardson and Holt [ibid. **81** (1951), 153].

An exhaustive review of recombination processes covering the u.h.f. work very fully has been given by Massey [*Advances in Physics,* **1** (1952), 395].

Attachment Processes

Geballe and Harrison [*Phys Rev* **85** (1952), 372] have measured Townsend's ionization coefficient α in oxygen and have shown, as Penning [*Ned. Tijd. voor Natuurkunde*, **5** (1938), 33] had done earlier, that at low E/p (50 V/cm /mm. Hg) α decreases more rapidly with decreasing E/p than is the case at higher E/p. This is attributed to attachment and, from the α/p–E/p plots, it is possible with this assumption to calculate attachment cross-sections. Geballe and Harrison, and Penning, have made such calculations, and have compared the results with the direct measurements of Bradbury [*Phys. Rev.* **44** (1933), 883]. Penning, using Masch's data for α in oxygen (see reference [161] of Chapter I), obtains better agreement than that found by Geballe and Harrison. This work is of interest in connexion with the study of gases of high dielectric strength, referred to in references [110] and [175] of Chapter I Further, the above critical value of E/p (45 V/cm /mm Hg) above which detachment commences is to be preferred to the value of 90 V/cm /mm. Hg quoted in Chapters I and III

An important paper on ionization by electron impact in CO, N_2, NO, and O_2 has been published by Hagstrum [*Rev Mod Phys* **23** (1951), 185] who used a modified mass spectrometer for the work. Critical appearance potentials for positive and negative ions have been accurately measured, together with the values of initial kinetic energy possessed by the products of dissociation at the critical potentials. Hasted [*Proc. Roy. Soc* A, **212** (1952), 235] has made further important contributions to the study of collision processes involving negative ions, with special reference to the determination of detachment cross-sections from the reaction

$$X^- + Y \rightarrow X + e + Y - \Delta E.$$

Studies were made, for example, with O^-, Cl^-, and F^- in the rare gases.

Mobilities of Electrons

Klema and Allen [*Phys Rev.* **77** (1950), 661] have measured drift velocities of electrons in A, N_2, and A/N_2 mixtures for E/p varying from 0 to 2 4 V/cm /mm. Hg and for pressures of 0·5–3 atmospheres Data are also given for electron drift velocities in A (containing 1 per cent O_2) at 3 atmospheres pressure. The method involves essentially the observation of pulse shapes in ionization chambers Addition of 1 per cent N_2 to A results, at $E/p = 0$ 2, in an increase in drift velocity from about 0 5 to 0 6 cm./μsec. This paper should be consulted for a discussion of the effects of impurities on electron drift velocities in various gases

Similar experimental work on A and A/N_2 mixtures has been carried out by Kirshner and Toffolo [*J. Appl Phys* **23** (1952), 594] whose results agree fairly well with those of Klema and Allen for A containing about 1 per cent N_2. Kirshner and Toffolo give reasons for suggesting that their uncontaminated argon was very pure.

Barbière [*Phys Rev.* **84** (1951), 653] has calculated, following Holstein's treatment, energy distributions, drift velocities, and electron temperatures in He and A for E/p varying from 1 to 4 V/cm./mm. Hg.

Townsend's Ionization Coefficients α and γ

Experiments have been made by Hornbeck [*Phys Rev* **83** (1951), 374] and by Molnar [ibid. 933, 940] to analyse the γ-processes of electron emission as caused by bombardment of the cathode by positive ions, metastable atoms, and photons.

Discharges in argon, with several cathode materials, have been studied. The method used is similar to that described by Engstrom and Huxford [ibid **58** (1940), 67] A Townsend discharge is stimulated by photo-electrons generated by a shuttered light beam which illuminates the cathode of a gas-filled tube The transient character of the resultant current between the electrodes is then observed oscillographically. The current is found to consist of a component in step with the stimulating light pulse and a component which lags by a time of about 1 m sec This second component is caused by metastable atoms. The fast component includes the primary electron current amplified by gas ionization and electron emission from the cathode caused by ion and photon effects, all of which reach a steady-state value in a time of about 10 μ sec From analyses of the oscillograms the fractions of the electron emission produced by metastables and by ions and photons is obtained. The relative emission efficiencies of photons and of metastable atoms is measured by an experiment in which metastable atoms are converted into radiating atoms by irradiation of the discharge space with light of the proper wave-length

For the purpose of the above analyses γ is defined as the number of electrons which are liberated at the cathode and enter the discharge stream per ion formed in the gas, rather than as the number of electrons liberated at the cathode per ion striking it It is then possible to write γ as

$$\gamma = f_{esc}\left\{\gamma_i + \frac{\alpha_r}{\alpha_i} f_{rk}\gamma_r + \frac{\alpha_m}{\alpha_i}[f_{mk}\gamma_m + f_{mr}f'_{rk}\gamma_r]\right\},$$

where $\gamma_i, \gamma_r, \gamma_m$ = number of electrons liberated at the cathode per ion, per photon, and per metastable, respectively, $\alpha_i, \alpha_r, \alpha_m$ - number of ions, photons, and metastables produced per ion per electron, f_{esc} = fraction of the electrons liberated at the cathode which escape the back diffusion effect and enter the discharge stream; f_{rk}, f_{mk} fraction of the photons and metastables generated in the gas which reach the cathode, f_{mr} - fraction of the metastables generated in the gas which are converted to radiating atoms, f'_{rk} = fraction of the photons from these radiating atoms which reach the cathode In the above expression for γ, f_{rk}, f_{mk}, f_{mr} and f'_{rk} are not independent of the field strength E and therefore γ is not strictly a constant when one measures i/i_0 as a function of E as in a normal Townsend measurement

In the experiments by Hornbeck and Molnar the fast component of current is assumed to be described by a Townsend equation with the γ coefficient given by γ_f, where

$$\gamma_f = f_{esc}\left[\gamma_i + \frac{\alpha_r}{\alpha_i} f_{rk}\gamma_r\right].$$

The total current is described by a Townsend equation with

$$\gamma = \gamma_f + \gamma_s,$$

where $$\gamma_s = f_{esc}\frac{\alpha_m}{\alpha_i}[f_{mk}\gamma_m + f_{mr}f'_{rk}\gamma_r]$$

The techniques used in their measurements, and the methods of analysing the experimental data, are described in detail by Hornbeck and Molnar. Some of their results are listed in Table 1 The values for the two higher values of E/p were obtained with $p = 1\,535$ mm. Hg and with the surfaces in the same condition. The values for the lowest value of E/p were obtained with $p = 4\,135$ mm Hg and with the surfaces having a reduced efficiency. The quantities α_m and α_r could not be measured, and the values given in Table 1 are as calculated and are equal.

The values of γ_i, γ_m, and γ_r are not considered to be accurate to better than 10 or 20 per cent Subject to this limitation γ_r is much smaller than γ_m and γ_i which are closely equal.

TABLE I

Data Concerning the Various factors Constituting γ

Gas-dependent factors

E/p V/cm /mm. Hg	f_{esc}	α_s cm $^{-1}$	$\alpha_r = \alpha_m$ cm $^{-1}$
195 4	0 97	6 63	2 6
117·2	0 91	3 72	2 4
72 6	0·86	5 10	7 0

Cathode-dependent factors

E/p V/cm /mm Hg	Cathode material	γ_f	γ_s (near break-down)	$\alpha_m \gamma_m$	$\dfrac{\alpha_r}{\alpha_s} f_{rk} \gamma_r$	γ_s	γ_m	γ_r
195 4	Ta	0·026	0 0025	0 062	0 0014	0 026	0 023	0 009
195 4	Mo	0 076	0 0074	0 160	0 0008	0 071	0 060	0 005
195 4	BaO–Ta	0 23	0 050	0 83	0 005	0 23	0 31	0 031
117·2	Ta	0·022	0 0026	0 056	0 002	0 022	0 023	0·009
117 2	Mo	0 065	0 011	0 158	0 001	0 071	0 065	0 005
117 2	BaO–Ta	0 21	0 065	0 65	0 007	0 22	0 27	0 027
72 6	Ta	0 0060	0 0009	0 025	0 0017	0 0053	0 0035	0 0021
72 6	Mo	0 030	0 0025	0 14	0 001	0 034	0 020	<0 001
72 6	BaO–Ta	0 073	0 013	0 54	.	0 085	0·078	..

Measurements have been made by Llewellyn Jones and Parker [*Proc. Roy Soc.* A, **213** (1952), 185] of α and γ $(= \omega/\alpha)$ in air for relatively low values of E/p. Nickel electrodes shaped to a Rogowski profile, with an overall diameter of 15 cm. are used Nineteen holes, each of 1 mm. diameter, are drilled in the centre of the anode, inside a circle of 8 mm. diameter, to admit ultraviolet light to the centre of the cathode. The initial photocurrent I_0 is between 10^{-15} and 10^{-14} A and the gas-amplified current I is between 10^{-12} and 10^{-9} A. All the measurements are made at a pressure of 200 mm. Hg, with spacings between the electrodes ranging from 1·0 to 3·8 cm. and with potential differences up to 30 kV.

Curves showing the variation of $\log_e I$ with the gap length d, for several values of E/p, are given in Fig 1 The results show a clearly defined departure from linearity in the relation between $\log_e I$ and d when the voltage is within 2 per cent. of the sparking value, and consequently a secondary ionization effect must then be operative. The authors point out that the failure by Sanders and others to observe a secondary effect at these values of E/p is due to the fact that no measurements of the current I were made at voltages near to the spark breakdown voltage

Analysis of the curves gives values for α and γ as listed in Table 2, which includes also values of the calculated sparking distance d_s at a gas pressure of 200 mm. Hg Two values of I_0, namely 6×10^{-15} and 20×10^{-15} A, are obtained at $E/p = 40$ for a constant incident light intensity. The corresponding values of γ are $2 3 \times 10^{-5}$

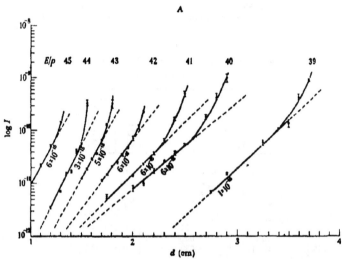

Fig 1 log I-d relations in air for values E/p from 39 to 45 V/cm /mm. Hg. Values for I_0 are indicated on the curves

and 15×10^{-5}. The reason for the difference in values observed is attributed to a change in the nature of the cathode surface, probably by the formation of an oxide layer. The significance is that when the number of photoelectrons produced per second is increased the value of the secondary coefficient γ also increases. This result is evidence for the view that the secondary process is in part due to secondary emission from the cathode, by photoelectric effect and/or by positive ion bombardment.

TABLE 2

Values of α/p and γ in Air as a Function of E/p when $p = 200$ mm. Hg

E/p V/cm /mm. Hg	$10^{15} I_0$ A	α/p (cm mm Hg)$^{-1}$	$10^6 \gamma$	d_s cm
39	1	0 0161	8	3 73
40	6	0 0181	23	2 94
40	20	0 0181	$\geqslant 150$	.
41	6	0 0196	40	2·53
42	6	0 0224	46	2 22
43	5	0·0252	113	1·83
44	3	0 0295	105	1 57
45	6	0 0345	84	1 37

The influence of an oxide layer on the cathode is discussed by Llewellyn Jones and Parker. Positive ions reaching the cathode fall on the outer surface of the partly insulating oxide film and set up an intense local electric field. This field may extract electrons from the metal, or provide electrons at the oxide surface from which they could easily be removed by bombardment by other positive ions.

The formation of an oxide layer on the cathode then produces an increase in I at gaps near the sparking distance because of enhanced secondary electron emission from the cathode, this enhanced emission being due to the positive charge acquired by the oxide surface However, the oxide layer can also produce a reduction in I_0 by lowering the photoelectric efficiency of the cathode surface

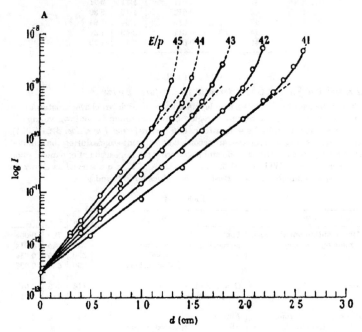

Fig 2. log I–d curves in nitrogen for values of E/p from 41 to 45 V/cm /mm. Hg at a pressure of 300 mm.

Investigations of α/p and γ in nitrogen have been made by Dutton, Haydon, and Llewellyn Jones [*Proc Roy Soc.* A, **213** (1952), 203], using the same equipment as that developed by Llewellyn Jones and Parker but with certain modifications giving improved stability and accuracy of measurement at lower currents. Curves showing the results obtained for log$_e$ I as a function of d in nitrogen at a pressure of 300 mm Hg are given in Fig. 2. The results show that the relation between log$_e$ I and d is not linear right up to the sparking distance, as had been concluded by earlier workers. Derived values of α/p and γ, and of the calculated sparking distance d_s, are given in Table 3. The values are the same, within experimental error, throughout the whole range of currents measured, namely 6×10^{-15} to 10^{-7} A, and the authors therefore state that this proves conclusively that space charges did not effectively distort the field under the conditions of their experiments

TABLE 3

Values of α/p and γ in Nitrogen as a Function of E/p when p = 300 mm Hg

$$I_0 - 3 \times 10^{-13} A$$

E/p V/cm /mm Hg	α/p (cm mm Hg)$^{-1}$	10^{4}γ	d$_s$ cm.
41	0 011	1 31	2 71
42	0 013	1 16	2 30
43	0 015	1 54	1 94
44	0 017	3 49	1 59
45	0 019	3 72	1 39

CHAPTER II

Breakdown Voltage Characteristics in Uniform Fields

THE influence of cathode surface layers on the breakdown characteristics of air and hydrogen, particularly in the region of the minimum breakdown voltage V_m have been measured by Llewellyn Jones and Davies [*Proc Phys. Soc.* B, 64 (1951), 397] Values of V_m in hydrogen are recorded at different stages during the removal of oxide layers from the cathode and also during the deposition of one metal upon another as base. Wide variations in the values of V_m are observed as shown in Table 4, which includes the corresponding values for E/p and γ

TABLE 4

Gas	Cathode	V_m	E/p	γ
Air (contaminated with mercury vapour)	Copper amalgam	460	720	0·004
	Mercury film on aluminium	390	885	0 014
	,, ,, ,, nickel	390	885	0 014
	,, ,, ,, Staybrite steel	390	585	0 006
Air	Oxidized aluminium	416	905	0 01
	,, nickel	421	957	0 01
Hydrogen (electrode surfaces treated by glow discharge)	Aluminium	243	200	0·1
	Aluminium deposited on nickel	212	200	0 15
	Nickel	289	180	0 075
	Nickel deposited on aluminium	390	245	0 015
	Commercial aluminium	225	200	0 125
	Aluminium on Staybrite steel	205	210	0 15
	Staybrite steel	274	190	0 075
	Steel deposited on aluminium	282	190	0·075

In further experiments to determine the mechanism of secondary ionization in low-pressure breakdown in hydrogen Llewellyn Jones and Davies [ibid. 519] have examined the dependence of the shape of the curves relating γ with E/p on the deposition of electro-positive or electro-negative atoms on cathodes of nickel, aluminium, silver, copper, and molybdenum Deposition of electropositive atoms on the cathode produces photoelectric peaks in the curves in the region $E/p \sim 150$ V/cm./mm Hg, but γ is not greatly affected at higher values of E/p. The results lead to the conclusion that the high photoelectric emission from cathodes of low effective work function is due to low-energy photons produced during the growth of the pre-breakdown ionization currents in the gap, but that for clean metals the secondary emission is due to impact of positive ions. Photo-emission due to high-energy photons is negligible.

Breakdown in Vacuum

Electrical breakdown over insulating surfaces in vacuum has been studied by Gleichauf [*J Appl. Phys.* **22** (1951), 535, 766]. In the investigated range of gas pressures, from 5×10^{-8} to 10^{-7} mm. Hg, the breakdown voltage is independent of pressure   Some indication is found that the breakdown voltage increases with increasing surface resistivity of the insulator, but no apparent correlation is found between breakdown voltage and dielectric constant. A roughening of the surface of the insulator in the region of the cathode electrode increases the break-down voltage   For electrode spacings of > 1 mm. the breakdown voltage does not increase linearly with the length of the insulator   The results of experiments in which one of the electrodes is separated from the insulator show that the critical gradient at breakdown in the space between the cathode and the insulator is not as large as would be required in gaps without insulators   These gradients are about the same for copper as for stainless-steel electrodes. The breakdown voltage over an insulator is raised when the edge of the insulator close to either electrode is rounded   When a layer of glass, thin compared with the gap between the electrodes, is fused to the cathode, breakdowns occur at lower voltages than for an identical vacuum gap.

A hypothesis is put forward by Cranberg [*J. Appl. Phys* **23** (1952), 518] that electrical breakdown in vacuum is caused by the detachment by electrostatic repulsion of a clump of material loosely adhering to one electrode, but in electrical contact with it, and the subsequent traversal of the gap by this clump. From considerations of the energy delivered to the target electrode Cranberg deduces that the breakdown voltage should be proportional to the square root of the gap length   An analysis of the experimental evidence available, for voltages from 20 kV to 7,000 kV and for gaps from 0 02 cm. to 600 cm., is considered to support the theoretical proposals

Geiger Counters

Further work on the discharge propagation mechanism in counters containing elementary gases, notably argon, has been carried out by Colli, Facchini, and Gatti [*Phys. Rev.* **80** (1950), 92]. These authors have studied the structure of discharge pulses and show that they give separate subsidiary pulses (fine structure) that can be correlated with electron transit times, taken radially across the counter. It follows that, in counters containing such simple gases, the discharge propagates along the counter by successive avalanche processes that are initiated by photo-electrons from the cathode, and not from the gas.

New measurements on ion mobilities in Geiger counters have been made by Den Hartog and Muller [*Physica,* **15** (1949), 789] and by Den Hartog, Muller, and Van Rooden [ibid. 581] on electron mobilities, using alcohol-argon mixtures.

CHAPTER III

THERE appears to have been little of interest published on point-plane coronas during the past two years. The short light pulses measured by English (reference [44] of Chapter III) have also been found in Liverpool by Murphy and Craggs who have observed that the duration of the light pulses increases as the voltage is raised towards breakdown. The longer light pulses (see reference [42] of Chapter III) occur nearer to breakdown. The physical significance of the short light pulses does not yet seem to be fully understood.

The great importance of small amounts of impurities found by Weissler (reference [2] of Chapter III) has been amply confirmed for positive wire-cylinder geometries by Miller and Loeb [*J. Appl. Phys.* 22 (1951), 494] who worked with N_2, O_2, and N_2-O_2 mixtures. A 0·006-in. diameter Pt wire, inside a 28·5 mm. internal diameter Ni cylinder, was used, and pressures ranged from about 27 to over 700 mm. Hg. Current–voltage curves show again (see p. 165) that a self-maintained discharge is established at currents < about 10^{-10} A, at which a sudden large increase in current to about 10^{-5} A takes place. Miller and Loeb state that breakdown streamers, leading to spark breakdown, are found at higher currents ($\sim 10^{-4}$ A) and voltages, even at pressures as low as 27 mm. Hg, although it is necessary, they suggest, to produce sufficient dissociation in pure N_2 to enable photoionization to become effective. The implication is, then, that photoionization in pure molecular N_2 is insufficient to allow streamer formation. The distinction drawn between a self-maintained Townsend discharge and a spark breakdown in Chapter III (pp. 157, 158) should again be emphasized, although these remarks refer only to inhomogeneous field breakdown processes. Miller and Loeb (loc. cit.) state, citing much experimental evidence, that in mixed gases (where the impurity may be ≪ 1 per cent.) streamer formation is readily observed and attribute this to greater photoionization probabilities, since photons produced from excited atoms in one gas may have energies greater than the ionization potential of the other gas. It is, of course, possible to produce photoionizing photons in a pure gas by recombination but the efficiency appears to be so low as to be negligible in these corona discharges and Miller and Loeb ignore this mechanism. The above paper lists the various corona thresholds as follows:

1. The glow discharge threshold

$$\gamma \exp \int_a^r \alpha \, dx = 1 \, ;$$

r is the cathode radius, a is the distance from the wire beyond which ionization by collision is negligible.

2. The Townsend discharge threshold.

$$\frac{\eta \theta g}{\alpha} \exp \int_a^r (\alpha - \mu) \, dx = 1 \, ;$$

μ is the absorption coefficient for photons in the gas; η is the number of photoelectrically active photons, that can act on the cathode, produced per electron in the avalanche, g is the fraction of the photons reaching the cathode and θ is the efficiency of these photons in liberating electrons from the cathode.

3. The burst pulse threshold (see p. 160).

$$\beta f \int_a \alpha \, dx = 1 \, ;$$

f is the fraction of the photons created by the avalanche electrons which are capable of liberating a photoelectron in the gas; β is the fraction of the photons liberated in a region where they can contribute to discharge propagation.

4. Steady burst corona onset:

$$\beta' \nu f' \exp \int_{a}^{r} \alpha \, dx = 1;$$

f' is the fraction of the electrons in the last avalanche of a decaying burst pulse that liberate a photon capable of giving photo-emission at the cylinder. β' is the chance that this photon reaches the cathode, where it has a chance ν of liberating an electron.

5. The streamer threshold

This is given as an extremely complicated expression by Loeb and Wijsman [*J. Appl Phys* **19** (1948), 797]

Similar work with the same gap geometry has been carried out by Miller and Loeb [ibid. **22** (1951), 614, 740] on negative corona in N_2, O_2, and N_2–O_2 mixtures. The effect of impurities is again found to be extremely important. In pure N_2, virtually no corona is found and the first visible discharge form is an arc The problems associated with negative corona appear to be more complicated than those found with positive corona, because of the importance in the former of the cathode (wire or point) surface Miller and Loeb describe in detail several such complicating features There is general agreement with the main results of Weissler's paper (reference [2], Chapter III) the importance of which should again be emphasized.

Lauer [*J. Appl. Phys.* **23** (1952), 300] has published a paper on wire-cylinder corona in H_2, A, and various mixed gases, at pressures ranging from 25 to 650 mm. Hg The work is of interest in connexion with Geiger counter mechanisms and with the work of Miller and Loeb (cited above) on N_2, with which it is in general agreement Mobility data, derived from pulse shapes, show the predominating ions in A to be A_2^+ with a mobility of 1 94 $\pm$ 0 08 cm ^{2}/V sec. In pure H_2 and A self-maintained discharges are established, with Miller's gap geometry (see above), at currents of about 0 1–1 μA. In argon sparks occur at $\sim$ 5–10 μA, but sparking is not observed in hydrogen even at 100 μA. Lauer cites the view of Fisher and Kachickas that at higher pressures (lower E/p), where γ is too small to give breakdown to a glow discharge, space charges lead to a spark by streamer formation once a self-sustaining discharge is established, and suggests that this interpretation is consistent with his experimental results.

The formative time lags for corona in point-plane gaps in air have recently been measured by Menes and Fisher [*Phys Rev* **86** (1952), 134] The gaps studied range from 0 5 to 1 5 cm. in length, gas pressures from atmospheric down to a few cm. Hg were used At atmospheric pressure the lags are too short ($<$ 0 1 μsec.) to be resolved from the statistical scatter, but with reduced pressures they become resolvable. With a 1·0-cm. gap in dry air at 422 mm. Hg the formative time falls from about 0·11 μ sec to 0 07 μ sec. as the percentage overvoltage is increased from zero to 1 5 This result is typical of the various gap geometries and gas pressures used. The intensity of ultraviolet illumination of the gap has no appreciable effect on the time lag

Visual observation of the gap shows that, under pulsed conditions, a filamentary streamer appears Oscillographic records of the light issuing from the streamer, as detected with a photomultiplier tube, confirm that the light both near and far from the point is coincident (within 0 1 μsec.) with the current pulse.

Menes and Fisher conclude that the formative lags observed are to be associated with the filamentary streamer and also that there is insufficient time for the cathode to play a role in the mechanism of formation of the discharge.

Low-pressure corona discharges of a special type occur in ozonizer tubes, generally in the form of discharges occurring recurrently at certain parts of an A.C voltage wave. Harries and von Engel [*Proc. Phys Soc.* B, **64** (1951), 916] have studied this form of discharge in a simplified tube system, and their paper should be consulted for details of an investigation into the effects of external irradiation, with light in the visible spectrum, on the discharges (Joshi effect).

CHAPTER IV

AN extensive investigation has been made by Norinder and Salka [*Arkiv for Fysik*, **3** (1950), 347] of the growth of spark discharges in sphere-plane and point-plane gaps in air with positive impulse voltages applied to the sphere or point. Gap lengths from 5 to 155 cm. are used   The technique adopted is similar to that of Torok and Holzer (see pp. 177–9) in that the discharges are arrested in the mid-gap region at different stages in their growth by the operation of a chopping-gap in parallel with the gap under observation. Some excellent photographs are given in Norinder and Salka's paper and the results show that the discharges grow in three phases The first phase appears suddenly at the corona onset voltage, the time of development being of the order of 10^{-6} sec. For the sphere-plane gap the first phase consists of a short, thick stem headed by branches of long, faintly luminous filamentary discharges or streamers During the second phase these streamers extend in steps and are accompanied by a diffuse discharge of luminous rays extending in the direction of the plate and generally following the electric field lines. This diffuse discharge corresponds to that also recorded by Saxe and Meek and by Komelkov using different techniques (see pp. 199–205) When a streamer branch reaches the cathode plate the third phase occurs, namely the main stroke.

In further work by Komelkov [*Izv Akad. Nauk, SSSR , Otdel Tekhn. Nauk*, No. 6 (June 1950), p. 851] the dimensions of leader channels and the potential gradients and current densities in leader strokes have been determined from photographs and oscillograms It is concluded that the leader channel is surrounded by an ionized region consisting of streamers of low luminosity and that the charge conveyed into the gap is mainly concentrated in the space surrounding the leader. Suggestions are made concerning the mechanism of leader growth. Other experimental studies of leader strokes are described by Akopian and Larionov [*Elektrichestvo*, No 7 (1952), 31, 46] and by Stekolnikov [*Dokl. Akad. Nauk*, **85** (1952), 1013]

To examine the influence on lightning discharges of geological discontinuities in the earth's surface experiments have been made by Norinder and Salka [*Tellus*, **1** (1949), 1], with sparks to a dried sand surface containing a vein of iron ore and other materials Spark lengths up to 170 cm. were used. Sparks from a positive electrode were markedly influenced by the presence of the ore vein, but those from a negative electrode hit the sand surface at random, independent by of discontinuities

CHAPTER V

IN a paper by Malan and Schonland [*Proc. Roy Soc.* A, **209** (1951), 158] the heights of the charges involved in the separate strokes of lightning discharges to ground have been estimated by five different methods. The results show that successive strokes tap progressively higher and higher regions of the cloud; the average height for first strokes is 3 6 km. and rises to a maximum of about 9·0 km. for later strokes, the average increase being about 0 7 km. per stroke.

Malan and Schonland conclude that the negative charge taking part in a discharge to ground is contained in a nearly vertical column, which is up to 6 km. long and whose base is about 1 km above the base of the cloud The temperature at the bottom of the column is about 0° C. and that at the top reaches to the −40° C. level. It is suggested that the charge is generated by freezing or sublimation processes at the top of the column, and that the lower negative pole is subsequently carried downwards by large hydrometeors falling under gravity.

CHAPTER VI

FOLLOWING the experimental results obtained by Ganger (reference [63] of Chapter VI), Fisher and Bederson (references [62, 87] of Chapter VI), and Kachickas and Fisher [*Phys Rev* 79 (1950), 232], Loeb [ibid 81 (1951), 287] has put forward suggestions for the modification of the streamer theory. He points out that above the threshold, $\gamma e^{ad} = 1$, for a Townsend discharge, positive ions accumulate in the gap and the resultant space-charge distortion enhances the development of a spark Near the threshold the space-charge build-up takes long time-intervals as shown by Schade (reference [1] of Chapter VI), Bartholomeyczyk [*Z. Phys.* 116 (1940), 235] and by von Gugelberg [*Helv Phys. Acta*, 20 (1947), 250, 307]. Calculations by Kachickas and Fisher, using Schade's theoretical approach, qualitatively predict the course of the observed curves relating time lag with overvoltage remarkably well, in both air and argon.

Loeb concludes that with low values of γ as in air, and especially with low values of α and high values of γ as in argon, the Townsend threshold for a space-charge build-up sets in at values of E/p below those which would initiate streamer formation. He suggests that sparks in uniform fields occur at thresholds for low-order Townsend discharges by creating field distortions. These distortions increase to the point where avalanches reach streamer-forming proportions in mid-gap and yield sparks in the streamer-type breakdown manner. As the overvoltage applied to the gap is increased the temporal rate of space-charge production is increased and the streamer-forming avalanches are produced in progressively shorter time intervals In air or oxygen the low values of γ require high voltages to give the correspondingly large values of e^{ad} for Townsend thresholds. Hence only slight overvoltages are needed in air so that the streamer criterion nearly holds In argon appreciable values of α occur at such low values of E/p, and α is so small while γ is large, that the Townsend thresholds fall far below streamer thresholds and the streamer theory fails to apply.

In the highly distorted fields near positive points and wires Loeb considers that streamers form and lead to breakdown, the threshold being set directly by the streamer theory and being entirely gas-dependent at high pressures.

A later paper by Kachickas and Fisher [*Phys. Rev.* 88 (1952), 878] describes measurements of the formative time lags for spark breakdown of uniform fields in nitrogen as a function of overvoltage, gas pressure (150 to 700 mm. Hg) and gap length (0 3 to 1·4 cm.). The curves relating time lag with overvoltage are almost identical with those for air (reference [87] of Chapter VI); for a 1-cm. gap in tank nitrogen at 725 mm. Hg the time lag is about 10 μ sec for an overvoltage of 0·2 per cent. and about 1 μsec. for an overvoltage of 1·5 per cent. The time lags are unaffected by changes in the primary cathode photocurrent i_0 ranging from 20 to 200 electrons/μ sec.

Kachickas and Fisher draw attention to the differing influence of i_0 on the breakdown characteristics of tank nitrogen and 'pure' nitrogen, the latter having been passed over hot copper and through a liquid nitrogen trap. The breakdown voltages for both tank nitrogen and pure nitrogen without ultraviolet illumination of the cathode have about the same values in close agreement with those given by Ehrenkranz (reference [18] of Chapter II). Some lowering of the breakdown voltage V_s occurs in tank nitrogen as i_0 is increased; for a gap of 1 4 cm. at 730 mm. Hg V_s is reduced by about 1 per cent. when i_0 is increased to 200 electrons/μ sec. Measurements of the lowering of V_s could not be made for the product pd less than about 400 mm. Hg × cm. as then ultraviolet illumination initiates a partial breakdown giving rise to a glow on the anode, and only when the voltage is raised further does a spark occur. The influence of i_0 is much more marked in pure nitrogen, the decrease in V_s being about 23 per cent for an i_0 of 200 electrons/μ sec.

The average value of the amplification factor $e^{\alpha d}$ at the threshold breakdown voltage V_s is about 1,000, which is too low to give rise to a streamer as the result of a single avalanche. Kachickas and Fisher conclude that their results can be explained by the assumption that secondary electron emission occurs either by photoionization in the gas near the cathode or most probably by photo-emission from the cathode, and that there is therefore a breakdown threshold set by a Townsend secondary process The secondary mechanisms permit space charges to build up to a critical value by successive electron avalanches, each avalanche starting at or near the cathode at about the same time as the preceding one reaches the anode. For the longest time lags observed perhaps 1,000 or more successive avalanches cross the gap before breakdown occurs. It is assumed that space charge distortion may be neglected until a critical space charge is built up in the gap, at this time breakdown may be completed by a streamer-like process in an interval short compared with the build-up time.

Kachickas and Fisher have derived an expression for the formative time lag on the assumption that a spark occurs when the number of electrons liberated from the cathode as the result of a single electron being emitted at an earlier time reaches a value N. It is assumed also that all photons are created near to the anode and that the time for a photon to cross the gap is negligible. The formative time lag T is given by

$$T = \frac{\log_e N}{v(\alpha - \alpha_s)},$$

where v is the electron velocity across the gap, α is the value of the Townsend ionization coefficient in the applied electric field, and α_s is the value of α in the threshold breakdown field. Good agreement is obtained between the experimental and theoretical curves for time lags as a function of overvoltage when the value of N is chosen to be about 10^8.

The cathode photoelectric effect has some advantage over the gas photoionization mechanism as the former requires photons only one-quarter as energetic as those required in the latter process. For positive ion bombardment of the cathode the calculated time lags are some 200 times greater than those observed.

In measurements of the ionization coefficients α and γ in air Llewellyn Jones and Parker [*Proc. Roy. Soc* A, **213** (1952), 185] have established that values of γ are measurable for E/p between 45 and 39 V/cm./mm. Hg. This range corresponds to breakdown voltages from 12 to 30 kV and values of pd up to 760 mm. Hg × cm. The results show that the Townsend mechanism of the spark can explain the breakdown of gaps up to 1 cm. in air at atmospheric pressure, for the nickel cathode used in the experiments. The agreement obtained between the calculated and measured breakdown voltages is consistent with the view that the static spark breakdown of the gaps studied is brought about by the same α- and γ-mechanisms as amplify the pre-breakdown current. The experiments also support the view that in uniform fields cathode emission plays an important part in the secondary ionization process, and that in air it is probable that emission is produced by the incidence of both photons and positive ions

Allied experiments for discharges in nitrogen have been made by Dutton, Haydon, and Llewellyn Jones [*Proc. Roy Soc.* A, **213** (1952), 203] with similar results to those for air. The presence of a secondary mechanism is confirmed for values of E/p ranging from 45 to 41 V/cm./mm. Hg and pd up to 810 mm. Hg × cm. with breakdown voltages up to 38 kV. The spark breakdown mechanism is again considered to be a natural consequence of the development of the pre-breakdown currents, according to the Townsend mechanism, and the experimental results 'in no way support the theory of static breakdown based on the sudden introduction of a so-called streamer'

It is clear from this work that the threshold for the breakdown process, over the range of variables studied, is set by a Townsend γ-mechanism, though the exact manner in which the discharge finally grows across the gap to form the spark channel is not described. The results are in agreement with those of Fisher *et al.* whose most recent conclusion is that cathode photo-emission is the most probable secondary mechanism in the breakdown of gaps subjected to low overvoltages. It is not yet clear whether this mechanism provides an explanation of the breakdown of gaps subjected to high overvoltages, when the formative time may be less than the electron transit time across the gap. Difficulties also arise in the explanation of the breakdown of longer gaps in uniform fields, and of the breakdown in non-uniform fields, for which further experimental evidence is required to determine the influence of the cathode, particularly in the case of breakdown with short impulse voltages applied to the gaps.

Dickey [*J. Appl. Phys.* **23**, 1336, 1952] has suggested that when formative times, as indicated by oscillograph measurements of the kind described by Fletcher (see pp. 275–9) are less than electron transit times, they may be explicable in terms of displacement currents, and a streamer mechanism need not then be invoked.

In the experiments by Llewellyn Jones and Parker the values of the product αd across the gap are such that $e^{\alpha d}$ ranges from about 10^4 to 10^5. Under these conditions the space charge field in an avalanche cannot attain a value of the order of the external field. Breakdown by streamer formation could occur in such gaps only if αd is about 18 or more, i.e for $e^{\alpha d} > 10^8$.

From considerations of other measurements of the breakdown voltage gradients of uniform fields in air at atmospheric pressure, and the corresponding values of α, it is found that the product αd increases with increasing gap length, as shown in Table 6.8. For gaps > 2 cm the value of αd across the gap is such that the space charge field in an avalanche, calculated from equation (6.7), reaches a value equal to the external field before the avalanche reaches the anode. Whether breakdown then occurs by a Townsend cathode mechanism, or by the transition from an avalanche to a streamer, remains to be proved. Experimental evidence in favour of the avalanche-streamer transition is given by the cloud-chamber experiments of Raether, as described on pp. 179–83. According to Raether's interpretation of his cloud-chamber photographs, for gaps in which $pd > 1,000$ mm. Hg $\times$ cm in air, the breakdown is initiated by an avalanche which develops across the gap at a speed of about 10^7 cm /sec and suddenly changes to a streamer developing at a speed of about 10^8 cm /sec.

The streamer theory has been examined in detail only for spark discharges in air. No results have been published concerning its possible application to other gases, with the exception of SF_6, for which Hochberg and Sandberg (reference [41] of Chapter II) show that there is reasonable agreement between theory and experiment In those cases where breakdown voltage data are available for high values of pd, the gases are usually of tank purity and it is doubtful whether the values of α obtained in pure gases should be applied (see pp. 54–9 and p. 83).

Some calculations have been made of the breakdown voltage of hydrogen on the basis of the streamer theory, using Hale's values for α (see p. 57). It is assumed that the space charge field in an avalanche is that given by equation (6.7) This assumption is not fully justified, but the cloud-chamber results given on p. 181 show that the avalanche diameters are about the same in hydrogen as in air at the same gas pressure The calculated breakdown voltages for a 1-cm. gap in hydrogen at pressures of 380, 250, and 100 mm. Hg are 8,900, 6,250, and 4,600 V respectively. The corresponding experimentally determined values given by Ehrenkranz (reference [18] of Chapter II) are 8,800, 6,200, and 3,100 V. It is not possible to calculate the breakdown voltages for higher gas pressures as the necessary data for α/p are not available. However, the above calculations over a limited pressure range show that the space charge field in an avalanche may attain a value about equal to the externally applied field

Calculations for nitrogen, based on the experimental data given by Llewellyn Jones et al , show that the space charge field attained in an avalanche over the range of pd studied, up to 800 mm. Hg $\times$ cm , is several orders of magnitude less than the external field Similar results are obtained in calculations for argon, for pd up to 250 mm. Hg $\times$ cm It follows, therefore, that streamer formation by a transition from an avalanche cannot occur in these gases for the stated ranges of values for pd.

The effect of recombination between electrons and positive ions during the growth of an electron avalanche has been calculated by Petropoulos and Ampariotis [*Phys. Rev.* 83 (1951), 658], who show that the number of ion pairs formed in an avalanche may be reduced appreciably below the value given by the relation $n = e^{\alpha d}$ in the case of gaps greater than about 1 cm. in air. For a 5-cm gap, under breakdown conditions, the reduction in the number of ion pairs is estimated to be about 40 times; for a 10-cm. gap the reduction is about 5×10^4 times.

The secondary processes active in the electrical breakdown of gases have recently been reviewed by Loeb [*Brit. J. Appl. Phys.* 3 (1952), 341].

CHAPTER VII

MEASUREMENTS have been made by Boulloud [*C. R. Acad. Sci.* (Paris), 231 (1950), 514, 232 (1951), 958] of the breakdown between parallel plates in compressed air, nitrogen, hydrogen, and carbon dioxide The influence of the cathode material on the breakdown voltage and on the time lags to breakdown is discussed

by Boulloud, who attributes the observed effects to the field emission of electrons
Allied studies of the breakdown between coaxial cylinders in compressed air are
described by Bright [ibid. **232** (1951), 714]. The effect of an oxide layer on the
cathode surface in enhancing field emission is referred to by Llewellyn Jones and
Morgan [*Phys. Rev.* **82** (1951), 970]

Breakdown voltages between spheres of 2 cm. diameter, in air and in hydrogen
have been measured by Neubert [*Arch. Elektrotech.* **40** (1952), 370] using D C.
voltages up to about 28 kV. The influence of different concentrations of other
gases, including CCl_4, $CHCl_3$, CH_2Cl_2, CH_3Cl, and CCl_2F_2 on the breakdown is
shown.

CHAPTER VIII

LLEWELLYN JONES and de la Perrelle [*Nature*, **168** (1951), 160, *Proc. Roy. Soc.*
A, **216** (1953), 267] have measured electron emission of 10^4 to 10^5 electrons/sec.
from cathodes of oxidized nickel and tungsten with electric fields of about 10^5
V/cm By relating this emission to the electric field by the Fowler–Nordheim
equation, estimates of the work function and of the emitting area of the source of
the electrons have been made. The effective work function for the emitting source
on oxidized nickel and tungsten is roughly 0·5 eV, and the emitting areas are about
10^{-13} cm ^{2} This is consistent with the view that the electrons are obtained from
the oxide layer and not from the underlying metal. The presence of tarnish films
and oxides on electrodes enhances the cold electron emission greatly and plays an
important part in the mechanism of electron production. The presence of metallic
dust enhances the emission by factors up to 10^3.

The influence of ultraviolet illumination on the D C. breakdown of gaps between
spheres has been studied by Jørgensen [*Elektrische Funkenspannungen*, E.
Munksgaard, Copenhagen, 1943] and Claussnitzer [*Phys. Z.* **34** (1933), 791]
Results obtained with a positive D C. voltage applied to one sphere, with the
other sphere earthed, show that at larger spacings, when the field between the
spheres is non-uniform, the breakdown voltage of the gap is increased rather than
decreased by the presence of illumination. In Jørgensen's experiments, with
spheres of 3 5 cm. diameter, a lowering up to 1·5 per cent. is observed for spacings
< 1 5 cm. when the gap is illuminated by a mercury-arc lamp. For spacings > 1 5
cm. illumination of the gap causes an increase in the breakdown voltage, the
increase being 8 per cent. at 3 cm and 10 per cent. at 4·5 cm. spacing

CHAPTER IX

THE relative breakdown stresses, statistical time lags, and formative time lags
in several gases have been studied by Prowse and Jasinski [*Proc I.E.E.*, Part IV,
vol. **99** (1952), 194] for a gap of length 1 4 cm. subjected to single pulses. The
gap is in a nosed-in-cavity type of resonator, the frequency being 2,800 Mc./s. Gas
pressures ranging from atmospheric down to about 200 mm. Hg are used. The
gas in the resonator is irradiated by light from an auxiliary spark fired just before
the beginning of the microwave pulse. The results show that the formative time
lag for air, nitrogen, oxygen, and hydrogen is $< 5 \times 10^{-8}$ sec., the limit of the
recording system used For neon, argon, and helium the formative time lag is
appreciable; in neon at atmospheric pressure an applied stress of 4·7 kV/cm.
produced breakdown after about 5×10^{-8} sec., but with a decreased stress of 2·2
kV/cm. breakdown occurred after 1·4 μ sec. A statistical time lag is not observed
in neon but is noticeable in helium and is appreciable in argon. Statistical time lags

are recorded in the polyatomic gases studied and are much greater in air and nitrogen than in oxygen and hydrogen.

The results lead Prowse and Jasinski to suggest that in the polyatomic gases studied, electrons are present in the gap throughout the pulse and that breakdown occurs as a consequence of a single event in the life of an electron rather than by a progressive growth of ionization, whereas, in the monatomic gases, it is thought that there is a steady growth of electron population, as by collision ionization, to the critical value necessary to cause instability and breakdown.

The breakdown of neon at frequencies of the order of 3,000 Mc./s. has also been investigated by MacDonald [*Phys. Rev.* 88 (1952), 421] over a pressure range from 0·5 to 300 mm Hg.

A theoretical treatment by Allis and Brown [ibid 87 (1952), 419] of the high-frequency breakdown of gases provides a simpler solution than that previously developed by Brown and his colleagues, and enables the breakdown characteristics of any gas to be calculated over a wide pressure range. Good agreement is obtained between the calculated breakdown voltages and the measured values for high-frequency discharges in hydrogen

The two principal effects of a magnetic field on high-frequency breakdown of gases, the energy resonance with transverse fields and the reduction of diffusion, have been investigated by Lax, Allis, and Brown [*J. Appl. Phys.* 21 (1950), 1297] The diffusion effect is shown to exist by itself when the electric and magnetic fields are parallel. The resonance effect could not be separated because of the presence of diffusion at all times, but is made prominent by reducing the diffusion loss in a large cavity. Experimental results are given for the breakdown field in helium, at pressures ranging from 1 to 30 mm. Hg, as a function of the applied magnetic field (up to 3,000 gauss)

The breakdown between wires and coaxial cylinders in air and in hydrogen at pressures < 20 mm Hg has been studied by Llewellyn Jones and Morgan [*Proc Phys. Soc* B, 64 (1951), 560, 574] for frequencies f ranging from 3 5 to 70 Mc./s. They find that the breakdown voltage for geometrically similar systems is the same provided that the parameters ap and f/p are invariant, where a is the wire diameter The results are in agreement with the view that the discharge is determined by primary ionization in the gas and loss of electrons to the walls by diffusion only, provided that the electron mean free path is small compared with the linear dimensions of the discharge tube and that the collision frequency is much greater than the oscillation frequency.

CHAPTER X

THE theory of the shock-wave régime at the beginning of the spark channel expansion process, illustrated for example in Fig. 10 4 of Chapter X, has been exhaustively studied by Drabkina [*J. Exp. Theor Phys. USSR*, 21 (1951), 473] following earlier work by Abramson, Gegechkori, Drabkina, and Mandelshtam [ibid 17 (1947), 862] Firstly the problem is simplified in the manner used also by Taylor (reference [22] of Chapter X), i e. by assuming instantaneous liberation of energy, E_0. The radius R of the shock wave, which will be the radius of the luminous channel for the early part of the expansion and for a subsequent time depending on the initial shock wave conditions, is given by

$$R = \frac{(\alpha E_0)^{\frac{1}{5}}}{\rho_0} t^{\frac{2}{5}},$$

where t is the time, ρ_0 is the initial gas density, and α is a complicated function of γ, the ratio of the specific heats at constant pressure and constant volume for the gas involved   From the above formula $R^2 \propto t$, for given initial conditions   Making further use of the above assumption of instantaneous energy liberation, Drabkina has calculated the radial variations of pressure and density and shows that the central part of the channel behind the shock wave has a density small compared with that in the front or even with that outside the expanding shock front.   This shows up as the dark central part of the record of Fig. 10 1 which was taken with a hydrogen spark (reference [5] of Chapter X).   The Russian workers do not appear to have recorded this phenomenon since it seems to occur only in hydrogen, and certainly the rare gases have not so far shown the effect [Tsui-Fang, Craig, and Craggs, unpublished]   A reason for this might be that if the central part of the channel is almost completely ionized, as it must be from the ion density data mentioned in Chapter X and from the low central density resulting from the shock process, then excitation in dissociated hydrogen (it has been shown spectroscopically that molecular hydrogen is not abundant in our spark channels) will be less, but in the rare gases the excitation of ions is possible.   Further differences, of course, between the special cases of atomic hydrogen and the rare gases may also be important.

Drabkina extends his treatment to take account of the special equation of state of the plasma, i.e. the dissociation of molecules and resulting effects, and of the finite time required for the liberation of energy in the spark channel   From consideration of the rate of energy input it is found that

$$R(t) = \left(\frac{\alpha}{\rho_0}\right)^{\frac{1}{4}}\left[\int_0^t E_0(t)^{\frac{1}{2}}\,dt\right]^{\frac{1}{2}}$$

to be compared with the more approximate expression given above

Drabkina thus finds

$$R(t) = K\left[\int_0^t E_0(t)^{\frac{1}{2}}\,dt\right]^{\frac{1}{2}}$$

for the channel radius,

and

$$D(t) = \frac{K}{2}E_0(t)^{\frac{1}{2}}\left[\int_0^t E_0(t)^{\frac{1}{2}}\,dt\right]^{-\frac{1}{2}}$$

for the shock-wave velocity.   Drabkina also defines for convenience the spark channel as that zone which has a temperature of 10^4 °K. at its surface (radius r) and shows that r is not sensitive to variations in this arbitrary boundary temperature.   Then it can be shown that

$$r(t) = LE_0(t)^M\left[\int_0^t E_0(t)^{\frac{1}{2}}\,dt\right]^N.$$

The various data, in c g s units, are given in Table 5 for air, argon, and hydrogen at the pressures given   Drabkina concludes this interesting and important paper by discussing in detail the density discontinuity, at the surface of the luminous zone of expanding gas, which decreases gradually with time.

Following Drabkina's paper, data on energy dissipation in spark channels, as a function of time, during the discharge have been given by Abramson and Gegechkori [*J. Exp Theor Phys. USSR.* **21** (1951), 484] and a detailed study of the optical phenomena associated with expanding spark channels has been published

TABLE 5

Constant	Air			Argon 1 atm.	Hydrogen 1 atm.
	760 mm.	200 mm.	3 atm		
ρ_0	$1 \cdot 29 \times 10^{-3}$	$3 \cdot 39 \times 10^{-4}$	$3 \cdot 87 \times 10^{-3}$	$1 \cdot 78 \times 10^{-3}$	$8 \cdot 99 \times 10^{-5}$
α	$5 \cdot 5 \times 10^{-1}$	$5 \cdot 5 \times 10^{-1}$	$5 \cdot 5 \times 10^{-1}$	$3 \cdot 32 \times 10^{-1}$	$5 \cdot 4$
K	$4 \cdot 55$	$6 \cdot 35$	$3 \cdot 46$	$3 \cdot 70$	$8 \cdot 85$
L	$1 \cdot 9 \times 10^{-1}$	$2 \cdot 92 \times 10^{-1}$	$1 \cdot 33 \times 10^{-1}$	$1 \cdot 1$	$4 \cdot 4 \times 10^{-1}$
M	$1 \cdot 25 \times 10^{-1}$	$1 \cdot 25 \times 10^{-1}$	$1 \cdot 25 \times 10^{-1}$	$4 \cdot 3 \times 10^{-2}$	$1 \cdot 38 \times 10^{-1}$
N	$3 \cdot 76 \times 10^{-1}$	$3 \cdot 76 \times 10^{-1}$	$3 \cdot 76 \times 10^{-1}$	$4 \cdot 6 \times 10^{-1}$	$3 \cdot 63 \times 10^{-1}$

by Gegechkori [ibid. 493] The latter paper gives measurements of $D(t)$ (see the above discussion), $r(t)$ and current densities as a function of time, a novel optical system is used which enables the luminous excited gas channel to be observed with a streak camera (as described in Chapter X) and at the same time enables a Schlieren record of the shock front to be obtained.

Radial expansion rates for rare gas channels have been measured by Wolfson and Libin [ibid. 510].

Olsen, Edmonson, and Gayhart [J. Appl. Phys. 23 (1952), 1157] have studied spark channel expansion processes, with reference to thermal theories of spark ignition of inflammable gases [see Fenn, Ind Eng. Chem. 43 (1951), 2865]. Olsen et al. used the schlieren method to observe the propagation of the shock wave from the initial constricted channel and its eventual separation from the hot gas core From measurements of the latter, the authors show that the expansion of the hot gas core (as shown, for example, in Fig 10 1) is adiabatic after the time of shock-wave separation. This is deduced by showing that the pressure in the channel at the instant of shock-wave separation is 2 56 atmospheres, from the Mach number for the shock wave at that time, which agrees well with the figure 2·52 atmospheres deduced from

$$\left(\frac{V_f}{V_s}\right) = \frac{\text{final core volume at pressure equilibrium}}{\text{core volume at instant of shock wave separation}}$$

From this, the above authors show that for argon a considerable part of the spark energy is used in expanding the channel, leaving only the remaining energy to appear as heat in the channel. The ignition processes in propane-air mixtures are briefly discussed in the above paper.

In Fenn's paper, cited above, the thermal theory of spark ignition is summarized as follows. It is assumed that almost all the spark energy becomes thermal in a very short time. The basic mechanism (due to Lewis and von Elbe) is then that there is a critically small volume of combustible gas in which the propagation criterion, that the rate of heat loss equals the rate of heat generation, obtains. The ignition energy appears as the excess enthalpy of a plane combustion wave. Fenn extends and simplifies this treatment but, since the physics of spark channels is not directly invoked, the discussion will not be taken further here.

The optical characteristics of spark channels have been further investigated by various authors. Thus Glaser [Optik, 7 (1950), 33, 61, and Zeits. f. Naturforsch 6a (1951), 706] has given detailed results for oscillatory discharges in various gases and has measured light intensities, channel radii, and emission spectra as a function of time. Tsui Fang [Brit. J. Appl Phys. 3 (1952), 139] and others have also described techniques for time resolved spectroscopy. Some techniques used in spark channel studies have been reviewed by Schardin and Funfer [Z angew. Phys. 4 (1952), 224], and Bayet [Rev. Sci. (Paris), 89 (1951), 351] has also discussed photomultiplier methods of studying light emission from spark channels and from afterglows. (See also Huyt and McCormick [J. Opt. Soc. Amer 40 (1950), 658].)

Measurements of electron densities in hydrogen discharges from the Stark effect have been made by Ware in the course of his work on high current transient discharges in toroidal tubes at low pressure [Phil. Trans. A, 243 (1951), 197]. The latter have also been further studied by Cousins and Ware [Proc. Phys. Soc. B, 64 (1951), 159], who report interesting results on shock-wave propagation in ionized gases Fuchs [Z. Phys 130 (1951), 69] has made a study of temporal variations of hydrogen Balmer line widths in spark channels

Olsen and Huxford [Phys. Rev. 87 (1952), 922] have studied electron densities in neon and argon sparks and afterglows at low pressures by measuring the Balmer line widths obtained by adding about 1 per cent of hydrogen to the carrier gas. This paper includes a further discussion of the inhomogeneous field first-order Stark effect (see Chapter X) as used in measurement of electron densities, and details of the strong continuous spectra observed, which are attributed to free-free transitions.

Fowler, Atkinson, and Marks [ibid. 966] have studied ion densities and recombination continua in expanding hydrogen sparks, at pressures ranging from about 0·3 to 8 mm Hg. They suggest that processes additional to radiative capture might contribute to the continuous radiation observed near the head of the Balmer and Paschen series, and cite a bremsstrahlung process as a possibility. They find that the Balmer continuum is the same in water vapour or hydrogen discharges, thus, they suggest, eliminating the possibility of its being caused by molecular dissociation in H_2.

Excitation temperatures of electrode vapour clouds in spark discharges have been studied by Blitzer and Cady [J. Opt. Soc. Amer. 41 (1951), 440] using a photographic technique for recording time-resolved spectra, previously described by Gordon and Cady [ibid 40 (1950), 852]. Crosswhite, Steinhaus, and Dieke [ibid. 41 (1951), 299] have used photomultiplier tubes, gated to respond only at certain times during the discharge, and show that this technique has certain advantages in that the sensitivity of spark methods of spectroscopic analysis can thereby be increased, due to the elimination of air lines and continuous background in the spectra.

The work of Blitzer and Cady, cited above, shows that for their discharges, which have a much higher peak current than those of Williams, Craggs, and Hopwood (reference [76] of Chapter X), a Boltzmann distribution of intensities exists, consistent with a single excitation temperature for certain iron lines Values for the excitation temperatures vary for various conditions, from about 5,500 to 8,000° K.

CHAPTER XI

Somerville, Blevin, and Fletcher [Proc Phys. Soc. B, 65 (1952), 963] have published a further study of electrode phenomena in transient arcs (or spark channels) in air with a detailed investigation of both anode and cathode spots. The cathode spot temperatures do not exceed about 1,200° C even with refractory metals; interesting results are described with tin cathodes at various temperatures ranging from − 50 to 220° C., the cathode marks increasing in size with increase in initial temperature. Calculations of the average cathode surface temperature give results agreeing satisfactorily with those inferred from experiments.

A theoretical study of plasma conditions in the region near the cathode, where channel contraction occurs, has been made by Ecker [Z. Phys. 132 (1952), 248].

INDEX OF AUTHORS

INDEX OF SUBJECTS

PRINTED IN
GREAT BRITAIN
AT THE
UNIVERSITY PRESS
OXFORD
BY
CHARLES BATEY
PRINTER
TO THE
UNIVERSITY

Printed in the USA
CPSIA information can be obtained
at www.ICGtesting.com
LVHW011308211123
764544LV00005B/31